Principles of

SEED SCIENCE AND TECHNOLOGY

Third edition

Principles of
SEED SCIENCE
AND TECHNOLOGY

Third edition

LARRY O. COPELAND
PROFESSOR OF CROP AND SOIL SCIENCES
AT MICHIGAN STATE UNIVERSITY

MILLER B. MCDONALD
PROFESSOR OF HORICULTURE AND CROP SCIENCE
AT OHIO STATE UNIVERSITY

CHAPMAN & HALL

New York • Albany • Bonn • Boston • Cincinnati • Detriot • London • Madrid • Melbourne
Mexico City • Pacific Grove • Paris • San Francisco • Singapore • Tokyo • Toronto • Washington

Cover design: Trudi Gershenov

Copyright © 1995
By Chapman & Hall

Printed in the United States of America

For more information, contact:

Chapman & Hall
115 Fifth Avenue
New York, NY 10003

Chapman & Hall
2-6 Boundary Row
London SE1 8HN
England

Thomas Nelson Australia
102 Dodds Street
South Melbourne, 3205
Victoria, Austrailia

Chapman & Hall GmbH
Postfach 100 263
D-69442 Weinheim
Germany

Nelson Canada
1120 Birchmount Road
Scarborough, Ontario
Canada M1K 5G4

International Thomson Publishing Asia
221 Henderson Road #05-10
Henderson Building
Singapore 0315

International Thomson Editores
Campos Eliseos 385, Piso 7
Col. Polanco
11560 Mexico D. F. Mexico

International Thomson Publishing - Japan
Hirakawacho-cho Kyowa Building, 3F
1-2-1 Hirakawacho-cho
Chiyoda-ku, 102 Tokyo
Japan

1 2 3 4 5 6 7 8 9 10 XXX 01 00 99 98 97 96 95

Library of Congress Cataloging-in-Publication Data

Copland, L. O. (Lawrence O.), 1936-
    Principles of seed science and technology / Lawrence O. Copeland
and Miller B. McDonald. -- 3rd ed.
        p.   cm.
    Includes bibliographical references and index.
    ISBN 0-412-06301-8
    1. Seeds.   2. Seed industry and trade   I. McDonald, M. B.
    II. Title.
    SB117.C73 1995   94-32630
    631.5'21--dc20   CIP

British Library Cataloguing in Publication Data available

Please send your order for this or any other Chapman & Hall book to
Chapman & Hall, 29 West 35th Street, New York, NY 10001, Attn: Customer Service Department.
You many also call our Order Department at 1-212-244-3336 or fax you purchase order to 1-800-248-4724.

For a complete listing of Chapman & Hall's titles, send your request to
Chapman & Hall, Dept. BC, 115 Fifth Avenue, New York, NY 10003.

Contents

Preface

This Third Edition of Principles of Seed Science and Technology, like the first two editions, is written for the advanced undergraduate student or lay person who desires an introduction to the science and technology of seeds. The first eight chapters present the seed as a biological system and cover its origin, development, composition, function (and sometimes nonfunction), performance and ultimate deterioration. The last seven chapters present the fundamentals of how seeds are produced, conditioned, evaluated and distributed in our modern agricultural society. A new chapter on seed enhancement has been added to reflect the significant advancements made in the last 10 years on new physiological and molecular biology techniques to further enhance seed performance.

Because of the fundamental importance of seeds to both agriculture and to all of society, we have taken great care to present the science and technology of seeds with the respect and feeling this study deserves. We hope that this feeling will be communicated to our readers. Furthermore, we have attempted to present information in a straight-forward, easy-to-read manner that will be easily understood by students and lay persons alike. Special care has been taken to address both current state-of-the-art as well as future trends in seed technology.

We believe this Third Edition represents a new level in presenting information that appeals to advanced undergraduate students as well as to those desiring more fundamental information on seed form and function. At the same time, it continues to have the strengths of the first two editions, in its readability as well as its comprehensive coverage of the broader area of seed science and technology.

Acknowledgements

This Third Edition is dedicated to all of the students, lay persons, and practitioners throughout the World who have made "Principles of Seed Science and Technology" successful since publication of the first edition in 1976.

Introduction

One for the buzzard,
One for the crow,
One to rot, and
One to grow!

The farmer hopes for better seed germination rates than the gardener in this old poem by Fay Yauger. But the fact is that most seed plants compensate beautifully by producing seeds in great abundance to assure survival of their species despite the formidable odds. And each seed that does survive is capable of producing a plant, more seeds, and still more plants and seeds to come. Deep within the seed are its own development forces, nutritive elements, and time and place mechanisms that signal the next growth stage. The seed itself is designed to disperse and scatter, using other forces of nature—wind, water, insects, birds, and other animals.

Seed husbandry formed the basis for early agriculture and eventual civilization. As people learned to plant, harvest, and preserve the seeds of certain grasses for winter, they abandoned the nomadic life to build permanent settlements. All the major civilizations throughout history have been founded on the culture of cereal gains, because these staples have high food value and are easily stored. The Mesopotamians planted wheat along the banks of the Tigris and Euphrates. The Chinese grew rice along the banks of the Hwang Ho and Yangtze. And the Mayans cultivated corn along the dry flat plains of the Yucatan.

Seeds for Survival and Subsistence

Seeds have been, and still are, the mainstay of the world's diet. The *Poaceae,* or large-seeded grasses, collectively known as cereals, contribute more food seeds than any other plant family. Cereal grains comprise approximately 90% of all cultivated seeds. They provide people with their most important source of carbohydrates, as well as some protein and other vital substances. As in ancient times, rice, wheat, and corn are the three major grains. Oat, barley, sorghum, millet, and rye are other important food and feed grains.

The second most critical food family, the *Fabaceae,* provides crops such as peanut, soybean, lentil, bean, pea, and chickpea. Legume seeds generally contain more protein

The Mayans cultivated corn along the dry, flat plains of the Yucatan. Improved varieties of corn are now the major cereal grain of the United States, with production exceeding 6 billion bushels annually.

than cereal seeds, and the protein has a better balance of amino acids for human nutrition than that in cereals.

In addition to being used directly as foods, seeds play other roles in human diets. Many seeds are used whole or ground as spices. The popular beverages coffee, cola, and cocoa are derived from seeds. Beers and ales are brewed from barley, and whiskey and gin are distilled from fermented mashes of cereal grains. Edible oils are obtained from seeds of corn, soybean, canola/rapeseed, cotton, peanut, coconut, palm, sunflower, and safflower. And seeds are used in the manufacture of some drugs and medicines.

In addition to being the most important source of food for human beings, seeds serve many commercial functions as well. Cotton, a major fiber, is spun from the cellulose hairs that surround the seeds of cotton plants. Seeds are also used in the manufacture of soaps, paints, varnishes, linoleum, jewelry, buttons, and many other products.

The beginnings of seed life lie deep within the burgeoning flower, where tiny structures begin to grow and develop to form the integument—the seed coat. This outer covering provides protection for the mature seed; it may even contribute nutritive support for the embryonic seed. This package also furnishes the ideal environment for the development of the embryo and the endosperm of the young seed.

From Orchids to Coconuts

Seed size is every bit as varied as the end products of the seed. An orchid species boasts the smallest known seed—a dustlike particle hardly visible to the naked eye. Seeds of tobacco and Kingston velvet bent grass are so minute that one-half million of them weigh only an ounce. Perennial plants (usually woody) claim the largest seeds, including acorns, walnuts, and coconuts.

Seed shape runs the gamut from round, oval, triangular, elliptic, and elongated to irregular. Predominant colors are black and brown, but hues of red, yellow, purple, green, and white are also found. Some seeds are multicolored. Seed surfaces may be smooth, rough, or textured, with silky hairs, cottony masses, hooks, bristles, or winglike structures. The intricacies of seed surfaces are often functional, serving as the mechanisms that ensure seed dispersal and survival.

The legendary tumbleweed breaks off at the ground surface and tumbles across the fields with the wind, dispersing seeds as it goes. Some of the weeds have lodged against barbed wire fence in this field in southwestern Nebraska.

Blowin' in the Wind

Not all seeds are distributed in brightly colored packets from Burpee Seed Company. The wind is probably the most effective agent in seed dissemination. It blows and scatters seeds in various ways—the cottony masses attached to cottonwood seeds, the hairy tufts of the dandelion and milkweed, the winglike appendages of the ash and maple seed, and the dustlike seed of the orchid. When the Russian thistle, the legendary tumbleweed, breaks off at the soil surface, the wind catches at the plant, blowing it across the western plains; it disperses seeds as it tumbles.

Since seeds of almost all species will float on water, many that land in streams will sail from their home site. Farmers and gardeners extend the territory of useful plant seeds. Animals and insects may also help seeds along their travels. Squirrels and packrats are notorious seed gatherers, while ants seem to be the most active seed collectors in the insect world. Innumerable plant species have fruits that are eaten by birds and animals. If the seeds pass through the digestive tracts, they may be transported great distances before being dropped to the ground to give rise to new plants.

Birds relish the purplish-black fruit of the mulberry tree, and they drop the seeds in fencerows and hedges, giving rise to new trees that grow as weeds.

When the seed pods of the milkweed dry and split open in the fall, the white hairs attached to the brown seeds spread out into rounded tufts and are wafted away in the breeze, carrying some of the seeds a long distance.

Some tenacious seeds manage to adhere to people's clothing or to animal's furry coats. Those who have tried to disengage the seeds of the beggar-tick from their clothes, cheatgrass from their socks, or cockleburs from the coat of their Irish setter know just how effective this dispersal method can be. Other seeds come equipped with spines or thorns that enable them to attach themselves to animals' feet and spread that way. The fruit of the mistletoe scatters its seed in a remarkable fashion. It is equipped with a propulsion mechanism that ejects the seed into the air. The seed is covered with a sticky substance that enables it to adhere to branches of trees and to the feet of birds. Birds carry the seed from one tree to another, where it germinates and begins to grow.

Just as seeds feature external structures that aid in dissemination, so they are equipped to germinate and resume their growth at a time and place that ensures the survival of their kind. Some seeds possess the ability to germinate and grow the instant they are dispersed. Others display almost uncanny mechanisms that prevent their germination until the time and place are right for continued growth. Some, especially weed seeds, may lie dormant in the soil for many years before they germinate.

In many ways, the seed is a microcosm of life itself. The seed is a neatly wrapped package containing a living organism capable of exhibiting almost all of the processes found in the mature plant. By studying the seed or the germinating seedling, we have gained much of our knowledge about growth regulators, respiration, cell division, morphogenesis, photosynthesis, and other processes.

But most of the seed research has been done in just the past century. While we no longer pray to the goddesses of grain (Demeter the Greek, and Ceres the Roman), we have a long way to go to unravel all the mysteries of the seed.

The winged seeds of the Ural maple may blow many yards in a strong wind, but most reach the ground near the tree.

Principles of

SEED SCIENCE AND TECHNOLOGY

Third edition

1

Reproductive Processes in Plants

Plant growth originates within the buds in regions known as *meristems*. In the meristems, cell division and elongation occur, and these processes produce tissues that soon develop into specific plant parts. *Vegetative meristems* give rise to parts such as stems, leaves, and roots, while *reproductive meristems* give rise to floral organs that ultimately produce fruits and seeds.

Within every meristem are minute *primordia* that resemble knobby outgrowths or ribbed inverted cones. Although hardly distinguishable to the naked eye, the configurations of the primordia become visible when the bud scales are removed and examined under magnification. As growth proceeds, the configurations enlarge and differentiate into recognizable plant organs.

FLORAL INDUCTION

The ability to support reproductive processes requires tremendous energy. Often, many crops do not begin to form flowers, and eventually seeds, until substantial vegetative growth has been accomplished. In some cases, as with most annuals, this is at the end of the life cycle. In other cases, the plant may not become reproductive for several growing seasons as with many fruit trees. During this phase, in which the plant is unable to form flowers because it does not possess sufficient vegetative structure, it is said to be *juvenile*. However, at some point there is enough vegetative growth present and the plants are ready to flower.

At that stage, certain external stimuli trigger *floral induction,* a physiological change that permits the development of reproductive primordia. This change may precede actual flowering by several days, weeks, or even months.

Temperature Stimuli

For floral induction to occur, many plants require exposure to low temperatures. This process has been called *vernalization*. In its narrowest sense, vernalization means

1

the promotion of flowering in some winter cereals by cold treatment of the moistened or germinating seeds. In its broader sense, vernalization means the induction of flowering in any winter annual, biennial, or even perennial species through exposure to low temperature. For example, rye (*Secale cereale*), a winter annual, and perennial ryegrass (*Lolium perenne*) both must undergo prolonged exposure to low temperatures before they can produce flowers. Sugar beets and carrots are examples of biennial species that grow vetetatively the first year, after which they are vernalized by exposure to winter temperatures. The optimum temperature for vernalization lies between 1°C and 7°C (Figure 1.1). These temperatures must be experienced by the vegetative meristems for periods between 10 and 100 days before a reproductive meristem is initiated when the crop is returned to warm temperatures.

In the chrysanthemum and tomato, floral induction is accomplished by repeated exposure to low night temperatures, separated by periods of higher temperature. This phenomenon occurs in many plants and has been called *thermoperiodism.*

Day-Length Stimuli

In many species, floral induction occurs in response to day length, or *photoperiod.* Thus, plant species have been categorized according to their day-length requirements as *short-day, long-day, intermediate-day,* or *day-neutral;* however, it is really the length of the night, or dark period, that is the critical factor that influences flowering. Table 1.1 provides examples of crops which require photoperiod and vernalization to induce flowering.

The photoperiod requirements for flowering may be qualitative or quantitative. Some short-day plants such as the Biloxi variety of soybean and cocklebur are unable to flower except under short-day treatments; in other short-day species, such as sunflower,

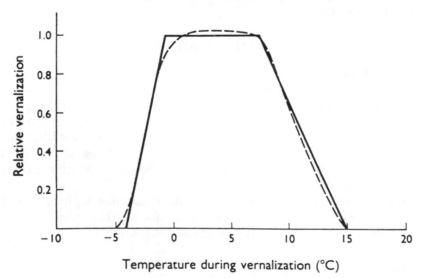

Figure 1.1. *Vernalization response of flowering in winter cereals (based on data for "Petkus" rye from Salisbury (1963)).*

Table 1.1. Photoperiodic and Vernalization Responses of Some Agricultural Species

	Short-Day Plants	Day-Neutral Plants	Long-Day Plants
Obligate photoperiodic response	soybean rice dry bean maize coffee	soybean cotton potato rice sunflower tobacco	oat annual ryegrass canary grass red clover timothy grass spinach radish
Facultative photoperiodic response	soybean cotton sugarcane rice potato sunflower		cabbage spring barley spring wheat spring rye potato sunflower red clover
Positive vernalization requirement	onion	onion carrot broadbean	winter oat winter barley perennial ryegrass winterwheat sugarbeet

flowering is hastened by the appropriate short-day conditions, although it eventually occurs without them.

Since the original discovery of photoperiod control of flowering by Garner and Allard in 1920 and the discovery of temperature, or thermal, induction by a Russian scientist, Lysenko (1932), there has been a widespread search for the existence of a universal flowering hormone, or *florigen,* in plants. It now appears that flowering is controlled not by one, but by several different hormone-like substances.

Phytochrome. Research with plant responses other than flowering—for example, seed germination, bud dormancy, stem elongation, and petiole development—have shown almost identical responses to light in different plant parts, suggesting that plant reactions are controlled by the same light-receptive substance. In 1959, this substance was finally isolated, identified, and named *phytochrome.*

Two photoreversible forms of phytochrome exist in plants. P_R phytochrome is receptive to red light (600–680 nanometers [nm]) and inhibits flowering while P_{F-R} phytochrome is receptive to far-red light (700–760 nm) and induces flowering. The conversion from P_{F-R} phytochrome to P_R phytochrome takes place in the dark, but at a much slower rate than that induced by far-red light. This is the basis for a "day-length," or photoperiodic light response, as well as a response to light quality (color), in the control of flowering. By successive exposures to red and far-red light, flowering of light-sensitive plants can be repeatedly induced or inhibited.

Chemical Stimuli

Certain natural and synthetic chemical substances can cause floral induction. Some are auxinlike compounds—for example, indoleacetic acid, naphthaleneacetic acid, or the common herbicide, 2,4-dichlorophenoxyacetic acid (2,4-D). At certain concentrations, gibberellic acid may also cause floral induction. It promotes flowering of long-day plants held in short-day conditions; however, it inhibits flowering of short-day plants under the same conditions. It has been demonstrated that the gibberellin content increased markedly during floral induction of *Hyoscyamus niger;* this is consistent with the effects of gibberellic acid in promoting floral induction.

Other substances known to cause flowering or to increase flower production are cytokinins, ethylene, acetylene, ethylene chlorohydrin, and 2,3,5-triiodobenzoic acid. In contrast, maleic hydrazide inhibits flowering.

With our growing knowledge about plant flowering responses and our increasing capability for producing synthetic hormones, it is becoming convenient and commercially profitable to manipulate flowering and fruit development in certain crops.

Nutritional Status

In floral induction, the nutritional status of a plant is also important, since construction of the flowering parts is dependent on food availability and translocation. The carbon-nitrogen ratio is particularly influential; in some species, such as holly, that bear male and female flowers on separate plants, a high nitrogen to carbon ratio favors pistillate rather than staminate flowers. In tomatoes, carbohydrate deficiencies cause microspore degeneration, leading to pollen sterility; however, a nitrogen deficiency has no such effect.

FLORAL INITIATION

Following floral induction, which may be triggered by external stimuli, *floral initiation* is the morphological expression of the induced state and usually occurs more or less deeply within the meristems of a plant. In monocotyledonous species, or flowering plants in which a single embryonic seed leaf appears in germination, floral initiation begins in specialized meristems called *dermatogens*, which also give rise to the epidermis. In dicotyledonous species, or flowering plants in which a pair of embryonic seed leaves appear at germination, floral initiation occurs in the lateral, terminal, or axillary buds.

Early in their development, reproductive meristems are similar to vegetative meristems, appearing as knobby or ribbed configurations on an inverted cone or pedestal. As development proceeds, the configurations develop into recognizable flower parts. The structure, development, and closure of the carpels to form the ovary can be traced in Figures 1.2 and 1.3.

FLORAL MORPHOLOGY

The typical flower of an angiosperm, or plant whose seeds are enclosed in an ovary, is composed of *petals, sepals, stamens,* and a *pistil.* The petals, often the most conspicuous, collectively are called the *corolla.* Sepals, usually (but not always) less conspicuous, are known collectively as the *calyx.* The stamens are the male pollen-bearing organs, and each consists of an *anther* and *filament.* The pistil, sometimes called the *gynoecium,* is the female part of the flower and consists of the *stigma,* which receives the pollen,

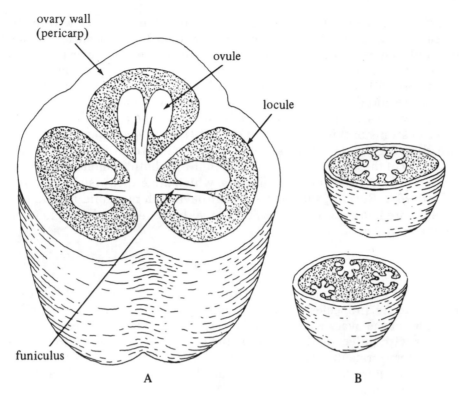

Figure 1.2. *Arrangement of fruit into locules: (A) a fruit arrangement with three locules, (B) other arrangements.*

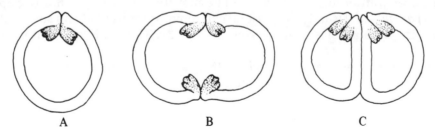

Figure 1.3. *(A) a simple carpel with one locule, or cavity, (B) a compound carpel with one locule, (C) a compound carpel with two locules.*

the *style,* and the *ovary.* The ovary may be composed of one or more *carpels,* which may be considered as highly modified leaflike structures. When only one carpel forms the ovary, it is termed simple and usually contains only one locule, or cavity. A compound ovary is made up of two or more united carpels and may contain one or more locules, depending on their arrangement (see Figure 1.3). The outermost wall of the ovary is called the *pericarp.*

The manner in which the seeds are attached to the placenta within the ovary locules is called *placentation*. Placentation occurs in one of three basic forms (see Figure 1.4). *Parietal* placentation occurs when the seeds are attached to the ovary wall, usually to both sides of the seam where the carpels fuse to form the ovary. *Axile* placentation occurs in flowers with ovaries divided by partitions, called *septa,* in which the placental attachment arises along the central axis of the ovary. When no septa are present in the ovary and the seeds are attached along the central axis, the placentation is termed *free central;* modifications of this occur in the case of *basal* or *apical* placentation.

Flowers having pistils, stamens, petals, and sepals are termed *complete. Incomplete* flowers lack any of these four parts. Flowers containing both stamens (male) and pistils (female) are termed *perfect;* unisexual flowers, which are either pistillate or staminate, are called *imperfect*. Species, such as corn, that have both male and female flowers on the same plant are known as *monoecious;* those that have unisexual flowers on different plants such as holly are *dioecious*.

MEGASPOROGENESIS

The seeds of angiosperms originate from meristematic tissue of the ovary wall called *ovule primordia*. In species with simple ovaries, these primordia are usually located near the suture of the ovary wall where the carpel is fused. In species with more than one carpel, or with polycarpellate ovaries, the seeds form at the fusion of the carpels or along the septa, or central carpel axes, depending on the type of placentation (see Figure 1.4). In some fruits (e.g., tomato), a well-developed placenta arises from which many ovule primordia develop.

Within the *nucellus,* or specialized tissue of the carpel, one cell, known as the *archesporial* cell, develops special characteristics that distinguish it from adjacent cells. As this cell increases in size, its nucleus becomes larger and its cytoplasm grows more dense in preparation for cell division. The first division results in a *megaspore mother cell* and a *parietal cell*. Usually the parietal cell remains undivided and soon deteriorates; however, in some species, it undergoes further division and contributes to seed formation.

The megaspore mother cell is *diploid* (2N), having the same number of chromosomes as the parent plant. However, it soon undergoes a two-step cell division known as

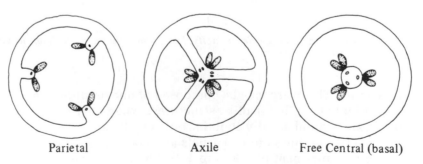

Parietal   Axile   Free Central (basal)

Figure 1.4. *Types of placentation.*

meiosis (see Figure 1.5). This process gives rise to four *megaspores,* each having one-half the chromosome complement of the mother plant; these are thus *haploid* (1N) cells. Normally, only one megaspore is functional, while the other three degenerate.

MEGAGAMETOGENESIS

The development of the female gametophyte, or embryo sac, from the functional megaspore is known as *megagametogenesis,* which is a process of successive nuclear divisions within an enlarging cell that becomes the embryo sac. Three successive free nuclear divisions (mitosis) occur (Figure 1.6), culminating in eight haploid (1N) nuclei. Soon these nuclei arrange themselves within the enlarging embryo sac and cell walls form, resulting in three antipodal cells at one end, two polar nuclei (without cell walls) near the center, and the egg apparatus (composed of the egg between two synergid cells) at the other end. After the two polar nuclei fuse to form a diploid (2N) nucleus, the resulting seven-celled structure is known as the mature female gametophyte (embryo sac), or *megagametophyte,* which is ready to receive the mature male gametophyte.

This describes the normal embryo sac development as it occurs in most species. Variations to this pattern occur in certain species, especially in the polar nuclei and antipodal development. With few exceptions, the egg apparatus development is as described.

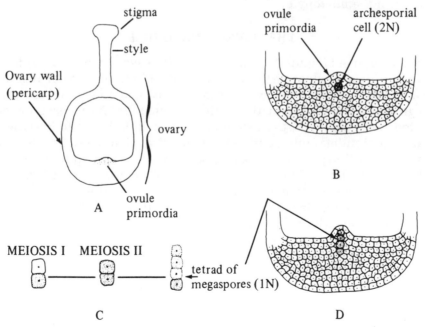

Figure 1.5. *Megasporogenesis: (A) location of ovule development, (B) cutaway section of the lower region of the ovary wall (pericarp), showing origin of the archesporial cell; note that it is larger than surrounding cells, having a larger nucleus and denser cytoplasm, (C) cell division during megasporogenesis, (D) cutaway section of lower part of the ovary, showing location of the four megaspores, three of which normally degenerate.*

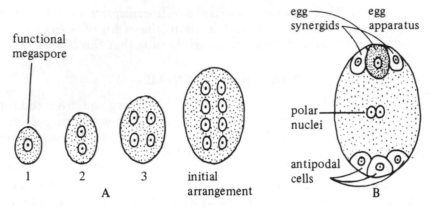

Figure 1.6. *Megagametogenesis: (A) three normal mitotic nuclear divisions leading to one large cell enclosing eight nuclei. Later, cell walls, enclose the nuclei and the entire structure becomes the female gametophyte, or embryo sac. (B) mature female gametophyte.*

The egg cell comprises most of the egg apparatus. It is a complete cell containing a haploid (1N) nucleus, with surrounding cytoplasm enclosed in a thin wall, or *fellicle*. The egg cell is positioned near the small opening (micropyle) of the ovule formed by the surrounding integuments. A small vacuole may be present near the point of attachment away from the micropyle.

THE DEVELOPING OVULE

Ovule development (Figure 1.7) occurs within the ovary, which provides a location for nurture and development of the female gametophyte, its sexual fusion with the male gametophyte, and embryo development, survival, and eventual regrowth. Ovule growth begins as a small outgrowth within the nucellus. As megasporogenesis and megagametogenesis continue, the region of the nucellus that is to become the ovule enlarges and differentiates into definite morphological characteristics. Secondary outgrowths, or collars (integuments), soon appear around the periphery of the nucellar outgrowths and envelop it. These usually consist of the inner and outer integuments and ultimately become the *testa* (seed coat) of the mature ovule.

The developing ovule is commonly attached to the placenta by the *funiculus*. The scar on the ovule made where the funiculus detaches at maturity is known as the *hilum*. The point where the integuments meet at the nucellar apex is the *micropyle,* and the region of integumentary origin and attachment, usually opposite the micropyle, is the *chalaza*. Between the chalaza and the hilum of many species is an area known as the *raphe*. The raphe may be visible on the seed coat of some species.

The Nucellus

The nucellus provides tissue for the origin and nurture of the female gametophyte, from archesporial cell to the mature megagametophyte. It originates from ovary tissue and provides the site of archesporial cell origin. Subsequently, part of it becomes

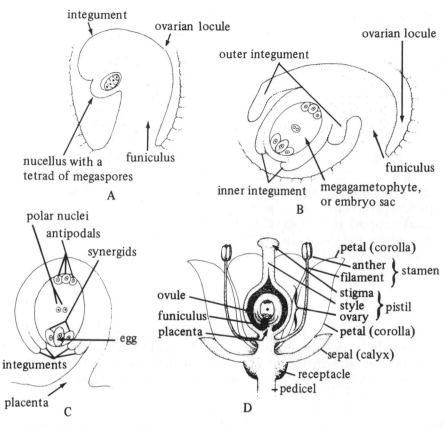

Figure 1.7. *Ovule development and its location in the flower: (A) longitudinal section through the ovary showing the developing ovule, (B) a later stage, (C) a still later stage showing the mature female gametophyte, (D) a generalized diagram of a complete flower showing the location of the ovule.*

trapped within the integuments as the ovule develops. Normally, no further growth occurs, and the nucellus is at least partially consumed, since it supplies nutritive support to the developing embryo sac. However, in some species it undergoes considerable development and contributes substantially to the storage tissue as the *perisperm*. Examples of species with well-developed perisperm are sugar beets (*Beta vulgaris*) and leafy spurge (*Euphorbia esula*).

Integuments

The nature and thickness of the integuments vary considerably among species, depending on their role in contributing to embryo sac and ovule development. In *Apiaceae,* the inner integument is completely absorbed and only two or three cellular layers of outer integument persist. In *Asteraceae,* most cells of both integuments are absorbed, leaving only a thin layer of crushed or disorganized integumentary tissue on the inner side of the pericarp. Practically no integumentary tissue remains in the fully developed corn caryopsis, and in *Symplocarpus* both integuments and endosperm are completely consumed by the developing embryo, leaving it naked inside the pericarp.

Integumentary Outgrowths.   Two types of integumentary outgrowths may occur in certain species, giving rise to special structures not found in most seeds. A third integument, or *aril,* may either arise from the base of the nucellus or split off from the outer integument. *Elymus,* for example, has a well-developed aril. Another type of integumentary outgrowth, a *caruncle,* arises as a proliferation of the outer integument in the region of the micropyle. Seeds of *Euphorbia esula* have a well-developed but fragile caruncle that extends back over the seed and appears to have no function. Still another type of appendage arises from the seed coat over the area of the raphe in some species (e.g., *Stylophorum* and *Trillium*) and is known as the *strophiole.*

Integumentary Tapetum.   In some species, the cells of the inner integument serve as nutritional support for the developing embryo sac and later harden and act as a protective layer for the ovule. In *Lobelia,* the cells of the inner integument take on a pronounced radial elongation and become binucleate before becoming hardened as the *integumentary tapetum.*

Micropyle

The micropyle is an integumentary pore or opening in the ovule through which the pollen tube grows to fertilize the egg cell of the female gametophyte. The micropyle may assume one of several configurations, depending on the closure of the inner and outer integuments (Figure 1.8).

Epistase

Epistase is the development of well-defined nucellar or integumentary tissue in the micropylar region of the seed of certain species. In *Castalia* (water lily) and *Costus* (spiral flag) species, epidermal cells of the nucellus proliferate and form a plug beneath the micropyle, which remains after the rest of the nucellar tissue is gone. Cells adjacent to the micropyle may show a marked radial elongation. Another type of epistase, an *operculum,* develops when cells of the integument proliferate and form a tightly compacted micropylar plug, as in species of *Lemna* and *Acorus.*

Figure 1.8. *Types of micropyle arrangements, showing different closure of the inner and outer integuments.*

Ovule Arrangement

Mature ovules are classified into five different types based on their arrangement within the ovary (Figure 1.9). The difference in arrangement begins at the time of archesporial cell development and becomes well defined by the time of embryo sac maturity. Ovule types have been determined for most well-known plant species and serve as a means of plant classification.

The effect of the ovule arrangement is often visible externally. For example, the relative position of the hilum (funicular detachment scar), chalaza, and micropyle of many legumes can be easily seen.

MICROSPOROGENESIS AND MICROGAMETOGENESIS

The period of flower development when the stigma is ready to receive the pollen is known as *anthesis*. Pollen is usually produced in four sacs, or *microsporangia* (Figure 1.10), of the anther, although occasionally fewer sporangia may occur. Within the sporangia, certain cells become the *microspore mother cells* and undergo a two-step reduction division (meiosis), or *microsporogenesis*, to yield four *microspores*, each of which is haploid (1N). Each of the four microspores is normally functional and undergoes two divisions, known as *microgametogenesis*, giving rise to a *microgametophyte*, or mature pollen grain.

FRUIT DEVELOPMENT

To understand seeds and seed formation one must have a basic knowledge of fruit development and morphology. The botanical definition of fruit is much broader than that conveyed by popular usage of the term. Actually, a fruit is a mature or ripened ovary that usually contains one or more *ovules* that develop into true seeds. Legume pods, peppers, and cereal grains are fruits, as are apples, oranges, and peaches.

The pericarp, or ovary wall of angiosperm fruits, is composed of three different

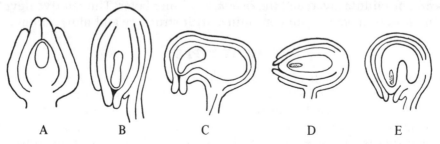

A   B   C   D   E

Figure 1.9. *Types of ovules as seen in vertical longitudinal section: (A) atropous (or orthotropous)—nucellar apex points away from the funiculus as in **Polygonaceae**, (B) anatropous-ovule completely inverted so that nucellar apex is turned toward the funiculus as in **Sympetalae**, (C) campylotropous—ovule is slightly curved, with the nucellar apex and funicular end pointed slightly downward as in **Fabaceae**, (D) hemianatropous— ovule is straight with axis lying perpendicular to the funiculus, as in **Ranunculaceae**, (E) amphitropous— ovule has a pronounced curve, with the nucellar apex near the funiculus, as in **Botomaceae**. (From P. Maheshwari (1950).)*

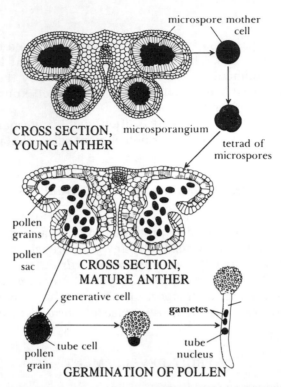

Figure 1.10. *The anther and the pollen grain. Each microspore mother cell within a microsporangium divides to form a tetrad of microspores that soon separate. The nucleus of each microspore then divides, and a tube cell and generative cell are formed within the wall of the microspore, which is thus converted into a pollen grain. Following pollination, the pollen grain generates, producing a pollen tube, and the generative cell gives rise to two male gametes. (From Wilson and Loomis (1952).)*

layers, which are more or less distinct in various species: the *exocarp,* or outer layer; the *mesocarp,* or middle layer; and the *endocarp* or inner layer. The relative development of each in various species often determines fruit structure and morphology.

FRUIT TYPES

Pseudocarpic fruit consists of one or more ripened ovaries attached or fused to modified bracts or other nonfloral structures. Examples: burdock, sandbur.

Multiple fruit is composed of the ovaries of more than one flower. Each unit of these fruits may be berries, drupes, or nutlets. Examples: fig, mulberry, pineapple.

Aggregate fruit is composed of several ovaries of a single flower. Each unit of these fruits may be a berry, drupe, or nutlet. Examples: strawberry, raspberry, blackberry.

Simple fruit is derived from a single pistil.

A.   Fleshy fruits have a fleshy or leathery pericarp.
    1.   *Berry* has a fleshy pericarp. Examples: grape, tomato, gooseberry, huckleberry.

2. *Pepo* has a hard rind but without internal separations, or septa. Examples: watermelon, cantaloupe, squash, cucumber.
3. *Pome* has a floral cup that forms a thick outer fleshy layer, and a papery inner pericarp (endocarp) forming a multiseeded core. Examples: apple, pear, quince
4. *Drupe* is also called stone fruit, and has a stony endocarp, a thick, leathery, or fleshy mesocarp, and a thin exocarp. The pit is usually one-seeded, but occasionally several one-seeded pits are present. Examples: cherry, coconut, walnut, peach, plum, olive.
5. *Hesperidia* are berrylike fruits with papery internal separations, or septa, and a leathery, separable rind. Examples: orange, lemon, lime, grapefruit.

B. Dry fruit has a thin pericarp that is dry at maturity.
1. *Dehiscent fruit* splits open at maturity and releases mature seed.
   a. *Legume* has a simple (single) pistil that splits open at maturity along two sutures. Examples: bean, pea, soybean, locust.
   b. *Follicle* has a simple (single) pistil that splits open at maturity along one suture. Examples: milkweed, larkspur, spirea.
   c. *Capsule* has a compound pistil that splits open at maturity in one of four ways:
      Loculicidal—splitting open through the midrib of the carpel into the locules. Examples: iris, tulip.
      Circumscissle—splitting open at the middle so that the top comes off like a lid (also called *pyxis*). Examples: plantain, portulaca.
      Septicidal—splitting along the septa. Examples: yucca, azalea.
      Poricidal—splitting open at pores near the top, releasing mature seeds. Example: poppy.
   d. *Silique* and *Silicle* are characteristic of the mustard family, with two valves which at maturity split away from a persistent central partition. A fruit that is several times longer than wide is termed *silique,* while a *silicle* is broad and short.
2. *Indehiscent fruits* do not open at maturity to release the seeds.
   a. *Achene* is a small one-seeded fruit in which the seed is attached to the pericarp at only one point and may be rather loose inside the pericarp. Examples: dandelion, buttercup, sunflower, dock.
   b. *Utricle* is similar to an achene except that it has an inflated papery pericarp. Example: Russian thistle.
   c. *Caryopsis* is similar to an achene except that the entire seed coat is tightly fused with the pericarp. Example: grasses.
   d. *Samara* is similar to an achene except that the pericarp develops a thin, flat, winglike appendage. This is a characteristic of some woody species. Examples: ash, elm, tree of heaven. Double samaras occur in the fruit of maple.
   e. *Nut* is a dry one-seeded fruit from a compound pistil that has a very hard and tough pericarp and that is usually wholly or partially enclosed in an involucre. Examples: acorn, hazel, filbert, chestnut.

f. *Nutlet* is a small, dry fruit composed of one-half a carpel, enclosing a single seed. It is developed by folding and splitting of the carpels into a compound pistil. Examples: members of *Lamiaceae* (mint family) and *Boraginaceae* (forget-me-not family).
g. *Schizocarp* has two fused carpels separating at maturity to form one-seeded mericarps. Example: members of *Apiaceae* (carrot family).

FLORAL TAXONOMY

The arrangement of the floral axis determines the type of *inflorescence* (structure of a flower), and is a stable species characteristic. The main stalk of the inflorescence is the *peduncle*. Lateral stalks supporting the individual flowers are called *pedicels*. Inflorescences may be *determinate* or *indeterminate*. Determinate inflorescences are those in which the axis terminates as a flower. Indeterminate inflorescences terminate in a vegetative bud, which continues to grow and produce flowers throughout the growing season, resulting in flowers of different maturity within the same inflorescences (see Figure 1.11).

Determinate Flowers

Solitary flower—The simplest expression of a determinate inflorescence. Example: corn cockle.

Simple cyme—The simplest branched determinate inflorescence where the lateral flowers develop later than the terminal flower. Example: chickweed.

Compound cyme—A determinate inflorescence where there is secondary branching and each lateral unit becomes a simple cyme. Example: bouncing bet.

Scorpioid cyme—A determinate inflorescence in which the lateral buds on one side are suppressed during growth, resulting in a curved or coiled arrangement. Example: *Heliotropium curassavicum*.

Glomerule—A very compact compound cyme. Example: saxifrage.

Indeterminate Flowers

Raceme—The basic type of inflorescence in which pedicels arise laterally on a long central peduncle. Example: pennycress.

Panicle—An inflorescence in which the lateral branches arising from the peduncle produce flower-bearing branches instead of single flowers. Example: oats.

Spike—An inflorescence in which the flowers arising along the peduncle are essentially sessile, or stalkless, and are attached to the peduncle. Example: wheat.

Catkin—A modified type of spike with a single unisexual flower arising from the peduncle. Example: red alder.

Spadix—A special kind of spike covered by a spathe. Example: skunk cabbage.

Corymb—An inflorescence in which the lower pedicels arising from the peduncle are successively longer than the upper ones, giving a round or flat-topped appearance. Example: bitter cherry.

Umbel—An inflorescence similar to a corymb except that the lateral branches arising

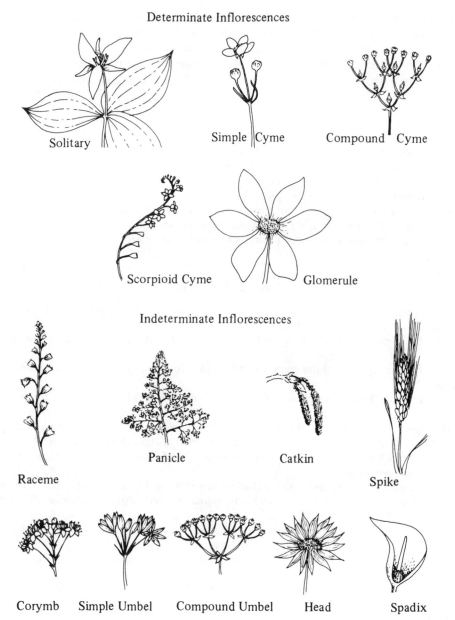

Figure 1.11. *Types of determinate and indeterminate inflorescences. (From Dennis (1967).)*

from the peduncle originate from the same location. Example (simple umbel): pennywart. A *compound umbel* is similar except that each pedicel is branched, bearing multibranched individual flowers. Example: wild carrot.

Head—An inflorescence where the peduncle and the pedicels are tightly clustered, surrounded by a group of flowerlike bracts called *involucre*. Example: sunflower.

Questions

1. What is the difference between floral induction and floral initiation?
2. Do you think that phytochrome and florigen are the same?
3. What is the difference between a complete flower and a perfect flower?
4. Can a dioecious plant have complete flowers?
5. What is the relationship between the peduncle and pedicels?
6. What is the difference between a caryopsis and an achene?
7. What is the difference between a schizocarp and a mericarp?

General References

Boke, N. H. 1947. Development of the adult shoot apex and floral initiation in *Vinca rosea* L. *American Journal of Botany* 34:433–439.

———. 1948. Development of the perianth in *Vinca rosea* L. *American Journal of Botany*. 35:413–423.

———. 1949. Development of the stamens and carpels in *Vinca rosea* L. *American Journal of Botany* 36:535–547.

Bonnett, O. T. 1935. The development of the barley spike. *Journal of Agricultural Research* 51:451–457.

———. 1936. The development of the wheat spike. *Journal of Agricultural Research* 53:445–451.

———. 1937. The development of the oat spike. *Journal of Agricultural Research* 54:927–931.

Dennis, L. J. 1967. *Manual of Introductory Taxonomy and Field Botany*. Corvallis, Ore.: Oregon State University Bookstores.

Garner, W. W. and H. A. Allard, 1920. Effect of the relative length of day and night and other factors of the environment on growth and reproduction in plants. Journal of Agricultural Research 18:553–606.

Lysenko, T. D., 1932: Fundamental results of research on vernalization of agricultural plants, *Bull. Jarovizacci*, No. 4, 1–57. Quoted by Maximow, 1934.

Maheshwari, P. 1950. *An Introduction to the Embryology of the Angiosperms*. New York: McGraw-Hill Book Company.

Maximow, N. A., 1934: The theoretical significance of vernalization, *Imp. Bur. Gen. Aberystwyth* (Wales) *Herbage Pub. Ser. Bull.* 16.

Salisbury, F. B. 1961. Photoperiodism and the flowering process. *Annual Review of Plant Physiology* 12:293–326.

———. 1963. *The Flowering Process*. New York: Pergamon Press.

Searle, N. E. 1965. Physiology of flowering. *Annual Review of Plant Physiology* 16:97–118.

Siegelman, W., and W. L. Butler. 1965. Properties of phytochrome. *Annual Review of Plant Physiology*. 16:383–392.

Stratford, G. A. 1965. Plant hormones II: Florigen and gibberellins. *Essentials of Plant Physiology*. London: Heineman Educational Books, Ltd.

2

Seed Formation and Development

SEED FORMATION

Seed formation begins with the combination of a male and female gamete: a process known as fertilization. Fertilization, or *syngamy*, can occur when both male and female gametophytes are fully mature. This usually occurs in a dual fusion process known as *double fertilization* (Figure 2.1). When the pollen grain lands on the stigma, it germinates by sending out a pollen tube, which grows down the style, through the micropyle and into the embryo sac, with the tube nucleus closely following the tube apex downward. The tube nucleus soon degenerates, but the two pollen sperm cells enter the embryo sac, one fusing with the diploid (2N) polar nucleus to form a triploid (3N) endosperm nucleus and the other fusing with the egg cell to form a diploid (2N) *zygote*, or fertilized egg.

The process of fertilization is very important because it not only results in the formation of a seed but also dictates the level of genetic diversity present in the zygote. Fertilization in angiosperms typically occurs either by self- or cross-fertilization.

Self-Fertilization

Self-fertilization occurs when pollen from the anthers of a flower is transferred to the stigma of the same flower, resulting in fertilization. In most cases, this happens when flowers do not open until the pollination and fertilization of the flower is complete.

Cross-Fertilization

Cross-fertilization occurs when pollen from one flower is transferred to the stigma of another flower to cause fertilization. The flowers can be on the same or different plants. In most agricultural crops, cross-fertilization occurs by two principal methods: wind (anemophily) and insects (entomophily). Unlike self-fertilization, where progeny are genetically similar, cross-fertilization results in progeny that are more dissimilar.

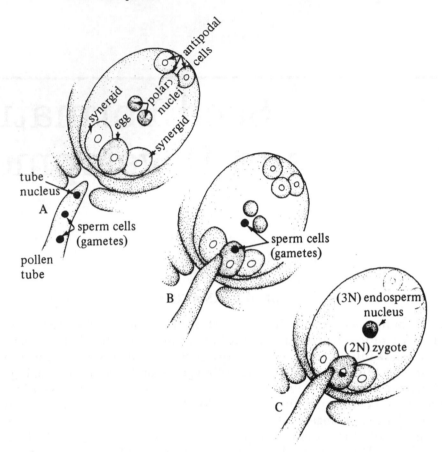

Figure 2.1. *Double fertilization: (A) pollen tube with its two sperm cells and tube nucleus approaching the micropyle, (B) sperm cells approaching egg and polar nuclei, (C) double fertilization has occurred.*

This evolutionary approach produces a population of individuals that are more adaptable to a wide array of environmental conditions.

ASEXUAL REPRODUCTION IN PLANTS

In addition to sexual reproduction, some plants are able to reproduce asexually. There are two types of asexual reproduction in plants: *vegetative propagation* and *apomixis*. Vegetative propagation may be carried out by stolons, rhizomes, tubers, tillers, bulbs, bulbils, or corms. Apomixis is the production of seeds and vegetative propagules by asexual methods. The main features of apomixis are: (1) it substitutes asexual reproduction for sexual reproduction, (2) it occurs in parts of the plant normally concerned with the sexual process (flowering parts), and (3) it occurs without fusion of egg and sperm cells.

There are two types of apomixis: *vegetative proliferation* and *agamospermy*. Vegetative proliferation (also termed *vivipary*) is the conversion of the spikelet, above the glumes, into a leafy shoot. Agamospermy is apomixis through seed production in which substitutions occur for meiosis (reduction division) and fertilization or both. Agamospermy may occur through *adventitious embryony* or *gametophytic apomixis.*

Adventitious embryony is the development of a diploid (2N) embryo from nucellar or integumentary tissue (sporophyte tissue). In gametophytic apomixis, unreduced embryo sacs (gametophytes) are formed without meiosis through the processes of *apospory* or *diplospory.* Reduced embryo sacs may also be developed through *meiosis.*

In apospory, diploid embryo sacs are formed in the nucellus or inner integuments by mitotic divisions. In diplospory, the diploid embryo sac is formed by the megaspore mother cell. Embryos are formed from the diploid egg cell without fertilization through the processes of *parthenogenesis* and *pseudogamy.*

In diploid pseudogamy, endosperm development requires the stimulus of pollination, and the egg cell develops without fertilization; in diploid parthenogenesis, pollination is not required by embryo development. Haploid (1N) embryos are formed from haploid eggs through haploid parthenogenesis.

These relationships are diagrammed in Figures 2.2 and 2.3. Figure 2.2 shows the general relationships between the various types of asexual reproduction and Figure 2.3 contrasts sexual seed production with agamospermy, showing substitutions for meiosis and fertilization.

SEED DEVELOPMENT

In most species, seed formation follows the normal pattern already described. It begins within the minute embryo sac which, with certain exceptions, is about the same in shape, size, and arrangement. In spite of initial similarities, the seed develops according to the genetic specifications for each species, which are coded in the nucleus (chromosomes) of each cell.

The embryo sac may be ellipsoidal, elongated, or variously bent in shape. The longitudinal axis extends from the chalaza to the micropyle and through the antipodal cells, the endosperm nucleus, and the egg apparatus. Morphologically, the micropyle is at the upper end of the embryo sac.

The embryo sac is a biochemical and biophysical system of considerable complexity. As a growing, differentiating structure, it requires a constant nutritive supply, which is provided through the chalaza, establishing a polar gradient from the antipodal to the micropylar end. Nutrition is also obtained from the nucellus and integumentary layers directly through the wall of the embryo sac.

EMBRYOGENY

After sexual fusion, or *syngamy,* a brief period of reorganization occurs, during which the large vacuole adjacent to the zygote gradually disappears, with the zygote cytoplasm becoming more homogeneous and the nucleus larger. The duration of this period varies among species, but it is usually about four to six hours before the zygote begins to divide (see Table 2.1). Lines of polarity in preparation for future division and growth already exist in the embryo sac, having been established in the unfertilized

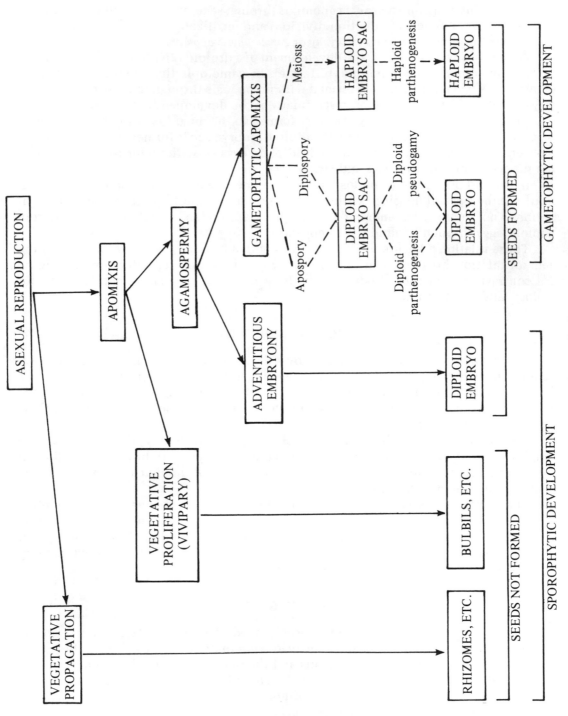

Figure 2.2. Mechanisms of asexual reproduction. (Courtesy of D. F. Grabe.)

Steps in Seed Formation

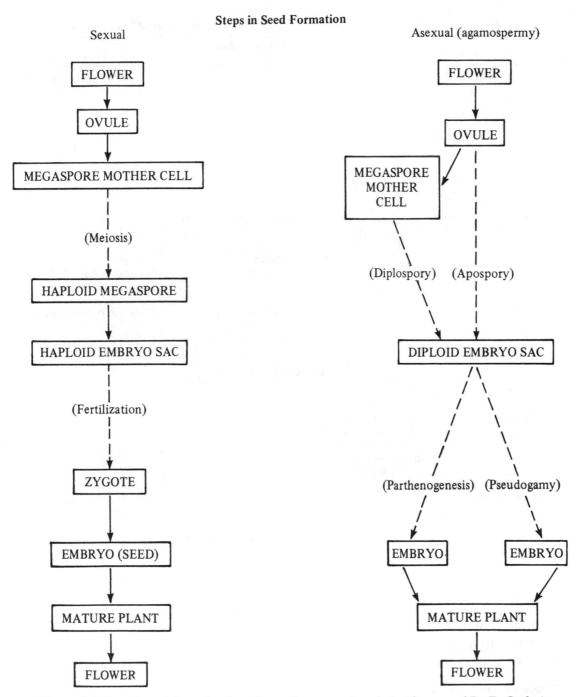

Figure 2.3. *Steps in seed formation through sexual vs. asexual methods. (Courtesy of D. F. Grabe.)*

Table 2.1. Examples of Time Sequence of Embryo Development of Two Species

A. Hordeum distichon palmella

Time elapsed after pollination	Growth of pollen tube	Development of embryo	Development of endosperm
5 min.	Pollen germinated		
10 min.	Male gametes inside pollen tube		
45 min.	Entry of pollen tube into embryo sac	One male gamete in contact with egg	Other male gamete in contact with polar nuclei
5 hr.		Male nucleus forming a sector of the egg nucleus	Male nucleus and polar nuclei in process of fusion
6 hr.		Male sector of zygote nucleus becoming more and more diffuse	First division of primary endosperm nucleus
10 hr.			Second division of primary endosperm nucleus
13 hr.		Prophase of first division of zygote	Four endosperm nuclei
15 hr.		First division of zygote nearing completion	Eight endosperm nuclei

B. Taraxacum kok-saghys

Time elapsed after pollination	Growth of pollen tube	Development of embryo	Development of endosperm
15 min.	Entry of pollen tube into embryo sac		
45 min.	Discharge of pollen tube and approach of male gametes toward egg and secondary nucleus		
1 hr., 15 min.		Syngamy	Triple fusion
3 hr., 50 min.			First division of primary endosperm nucleus
5 hr.		First division of zygote	
6 hr., 15 min.		Two-celled proembryo	Two-nucleate endosperm
8 hr., 15 min.		Four-celled embryo	Four-nucleate endosperm
24 hr., 45 min.		Several-celled proembryo	Multicellular endosperm

Data from Maheshwari (1950).

egg. The still undivided zygote typically elongates along the horizontal axis, and small vacuoles become evenly distributed through the cytoplasm.

Types of Embryo Development

The first few cell divisions from the zygote form the *proembryo*. Plant species may be classified according to the pattern of cell division, which results in different proembryo types. The first division almost always occurs at right angles to the longitudinal axis, resulting in a terminal cell next to the micropyle and a basal cell at the distal end. Depending on the pattern of subsequent divisions, proembryos are classified as *crucifer, asterad, solanad, caryophyllad, chenopodiad,* or *pipered*.

1. The first division of zygote is transverse.
    A. Terminal cell of proembryo divides by a longitudinal wall.
        a. *Crucifer*—basal cell plays only a minor role (or none) in embryo development.
        b. *Asterad*—both the basal and terminal cells contribute to embryo development.
    B. Terminal cell of the proembryo divides by a transverse wall.
        a. *Solanad*—basal cell plays only a minor part (or none) in the development of the embryo.
        b. *Caryophyllad*—basal cell undergoes no further division, and the suspensor, if present, is always derived from the terminal cell.
        c. *Chenopodiad*—basal cell and terminal cell both contribute to embryo development.
2. The first wall of the zygote is longitudinal—or nearly so—*Pipered*.

Although the mature embryos of monocotyledons and dicotyledons appear considerably different, their patterns of embryogeny are similar. The proembryo is divided into the *suspensor* and embryo proper. The suspensor forms into a chain of cells, pushing the embryo proper up into the center of the ovule in contact with the available food supply. The expression *constancy of destination* suggests that each part of the mature embryo inevitably arises from special parts of the proembryo; however, proembryos may vary greatly in size and shape.

Figure 2.4 shows the embryo growth of typical dicotyledonous and monocotyledonous species.

Laws of Embryony

Four laws of embryony (embryogeny) have been established to describe embryo development:

1. *Law of Parsimony:* No more cells are produced than are absolutely necessary.
2. *Law of Origin:* In any species the sequence of cell formation is established in such a way and with such regularity that the origin of any cell can be specified in terms of, or related to, the earlier units of the sequence.
3. *Law of Numbers:* The number of cells produced by different cell generations varies with the species and depends on the rapidity of the segmentation in the cells of the same generation.
4. *Law of Destination:* In the course of normal embryonic development, the cells are formed by divisions in clearly determined directions, and most appear to occupy positions in accordance with the role they must play.

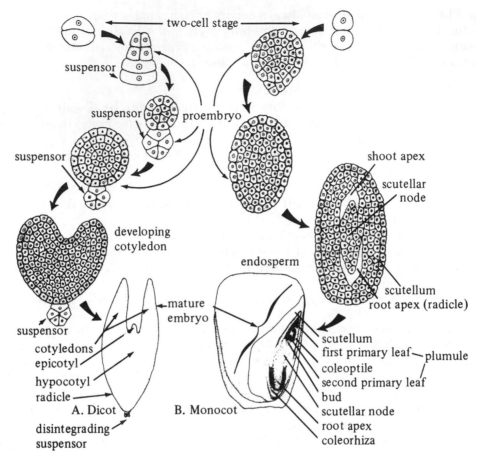

Figure 2.4. *Typical embryo development: (A) dicotyledonous species, (B) monocotyledonous species.*

ENDOSPERM DEVELOPMENT

The endosperm serves as the principal nutritive support for the embryo of many species (especially monocotyledons) during both seed development and germination. In angiosperms, the endosperm normally originates from triple fusion of a sperm cell nucleus from the pollen tube with the diploid polar nucleus (following fusion of the two polar nuclei) of the embryo sac; therefore, its nuclear complement is triploid (3N). In gymnosperms, the endosperm is normally haploid (1N), since it develops from one cell of the female gametophyte.

In seeds of many species, especially dicotyledons, the endosperm develops only a few cells, while in others it may be highly modified and hardly recognizable. In *Orchidaceae,* it is completely suppressed. Triple fusion occurs in *Orchidaceae,* but the products soon degenerate after one or two cell divisions. In most dicotyledonous species, the endosperm is formed but is almost completely consumed during seed development so that the mature seed is composed almost entirely of embryo. Considerable speculation exists about the status of the endosperm. It has been called an anomalous embryo, since the ovum and the two polar nuclei are genetically identical. Regardless of their genetic similarity, the fusion of the two polar nuclei and subsequent fusion with one sperm cell yield the endosperm, while the fusion of the egg with the other sperm cell nucleus yields the zygote.

One of the principal endosperm functions is to provide nutrition for the developing embryo; therefore, its composition is compatible with the embryo's needs. But the endosperm must also draw its nutritive support from the embryo sac and surrounding tissues. The net effect is to surround the embryo with a rich nutritive tissue from which it can draw for development and growth. This creates competition for nutrients, both within and outside the embryo sac.

Types of Endosperm Development

Division of the primary endosperm nucleus yields micropylar and chalazal chambers, one or both of which may contribute to the mature endosperm. When only one develops, the other is crushed and soon degenerates. Endosperm development may be one of three types, depending on the sequence of nuclear division and cell wall formation (see Figure 2.5).

Cellular Endosperm. In this type of endosperm, each nuclear division is accompanied by cell wall formation.

Nuclear Endosperm. This endosperm type is characterized by nuclear divisions unaccompanied by cell wall formation. The nuclei may remain free or may later be separated by cell walls that form in one of three ways: (1) one to three layers of cell wall may form around the periphery, with free nuclei inside, (2) a cell wall may form in the micropylar area, with the rest remaining in a free-cell state, or (3) the entire endosperm may be filled with walled cells. All three endosperm conditions may exist in the same family.

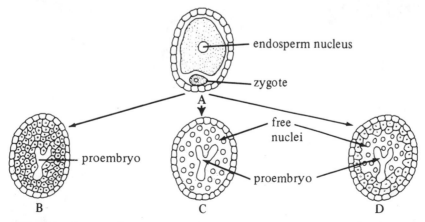

Figure 2.5. *Diagram of types of endosperm development: (A) ovule after fertilization, showing primary endosperm nucleus and zygote, (B) cellular endosperm, (C) nuclear endosperm, (D) helobial endosperm.*

Helobial Endosperm. The helobial endosperm is intermediate between the nuclear and cellular types. Free nuclear divisions occur, but cell wall formation accompanies nuclear division in some parts of the endosperm as well.

Endosperm Haustoria

A remarkable feature of the developing endosperm is its capacity to take nutrients from surrounding tissue for its own development. Nutrient-gathering outgrowths, called *haustoria* (Figure 2.6), may develop at either the micropylar and chalazal ends and reach into the nuclear, integumentary, or even ovary tissue. Haustoria often branch into several prominent lobes, or *diverticulae*. When local food supply is exhausted, the haustoria lobes terminate and become crushed by further endosperm and embryo growth.

The Mature Endosperm

Monocotyledonous endosperms usually reach their maximum morphological development at physiological maturity and remain to comprise a major part of the seed. In dicotyledonous species, the endosperm may not develop or may be used up by the developing embryo and comprise none or only a small part of the mature seed.

Seeds with little or no endosperm are *exalbuminous,* while those with a well-developed endosperm (or perisperm) are known as *albuminous*. Some species have a well-

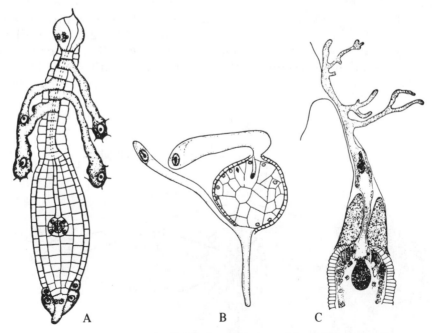

Figure 2.6. *Haustoria arrangements in the developing endosperm of three different species: (A) Centranthera hispida, (B) Nemophila, (C) Impatiens roylei. (From Maheshwari (1950).)*

developed *chalazosperm,* in which both the nucellus and endosperm disappear during development and chalazal tissue proliferates and forms storage tissue.

The outermost layers of the endosperm are known as the aleurone layer. During endosperm development, aleurone cells become thickened and filled with protein granules. These layers function both as storage tissue and for secretion of hydrolytic enzymes, which upon activation during germination help break down storage tissues.

OVERALL SEED DEVELOPMENT

Seed development can be illustrated by the changes that occur in barley, which is typical of most grasses and cereal grains. Endosperm development of barley is of the cellular type, in which the first few divisions of the primary endosperm nucleus give free nuclei. Cell walls form about two days after fertilization, beginning at the periphery of the endosperm which later becomes the aleurone. During early endosperm growth, the proembryo also begins to grow and differentiate; however, its contribution to the overall seed morphology is overshadowed by that of the endosperm.

Cell organelles—plastids, mitochondria, ribosomes, and golgi complexes—become recognizable immediately after initial cell formation, followed by the endoplasmic reticulum. After about three weeks, starch and protein granules completely dominate endosperm composition.

Morphological Development. Morphological development of the seed occurs concurrently with cytological, chemical, and weight changes noted below. Such morphological changes can be illustrated by those occurring in barley. These have been described by Briggs (1978) and are shown in Figures 2.7–2.10, beginning with development of the reproductive meristem and culminating in the development of the mature caryopsis.

Changes in Weight. After sexual fusion, the developing seed begins to increase in weight as a result of nutrient and water intake associated with rapidly accelerating cell division and elongation. Typically in monocots, the developing endosperm accounts for most of the weight increase, with the testa-pericarp weighing somewhat less, and the embryo's weight almost negligible. The developing barley seed undergoes a sharp increase in dry weight until about 35–40 days after fertilization. Immediately after fertilization, most of the dry weight is in the seed coat; however, after about eight days, its weight is surpassed by the endosperm, which later becomes the major seed component.

Chemical Changes. Immediately after fertilization, seed development begins and the seed becomes the primary recipient (sink) of the assimilates from the plant. There are three general stages that can be characterized during seed formation. The first stage is when 80% of the seed growth occurs. It is characterized by numerous cell divisions and elongation and dramatic increases in seed weight as nutrition is supplied through the funiculus by the parent plant. The second stage occurs when the funiculus degenerates and the seed is separated from the parent plant. At this point, the seed possesses its maximum dry weight and seed quality; a stage known as *physiological maturity*. The third stage is when the seed undergoes further desiccation after physiological maturity. This stage is influenced by a variety of weather conditions such as rainfall, high temper-

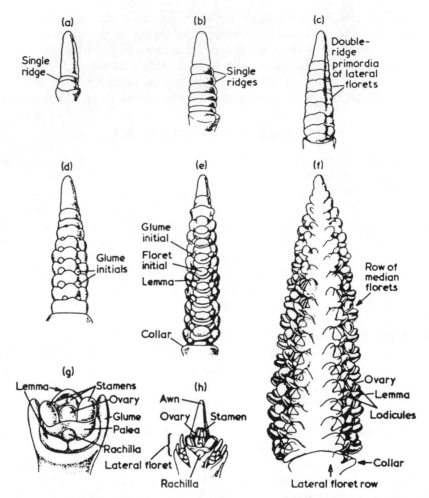

Figure 2.7. *Stages in the differentiation of the ear of a two-rowed barley: (A) single ridges present; these will form leaf primordia. The apex is lengthening; (B) the apex elongated further, but still with single ridges; (C) double ridges forming. Initials of the lateral florets are detectable; (D) the initials of the sterile glumes associated with the median floret are visible; (E) a further stage; the collar is visible; (F) a more advanced ear, viewed from the side; (G) a developing median floret, viewed from the rachis side; (H) a more advanced triad of spikelets from a two-rowed barley." (From Briggs (1978).)*

atures, and exposure to field pathogens that increase and decrease seed moisture content and cause reductions in seed quality. Eventually, seeds reach *harvest maturity,* which is the moisture content (usually 15–20%) at which mechanical harvesting of the seed is possible. Figure 2.11 illustrates these three growth stages.

In monocotyledonous seeds, the major carbohydrate in the endosperm and the entire seed is starch. The carbohydrate content increases rapidly as the endosperm develops, somewhat at the expense of the testa-pericarp tissue, where it decreases slightly. Sucrose and reducing sugar levels, initially high in the young endosperm,

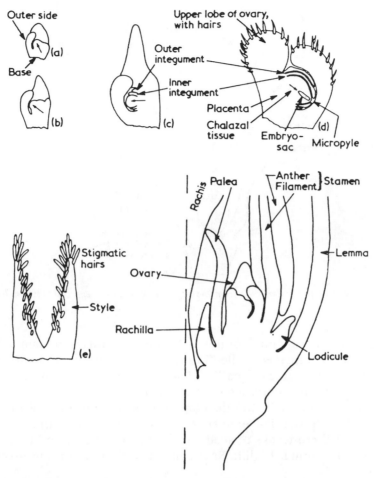

Figure 2.8. *Successive stages in the development of the barley ovary: (A–D) successive stages in median, vertical section. The arrow indicates the approximate position of the embryo-sac mother-cell as the tissue rotates; (E) a view of the ovary apex (at 90° to sections A–D), showing the developing stigmatic hairs on the underdeveloped styles; (F) a longitudinal section of half of the ear, indicating the relative positions of the various parts of the spikelet. (From Briggs (1978).)*

decrease rapidly as the starch content rises. However, both sucrose and reducing sugars increase in the testa-pericarp during early seed development and then decrease rather sharply as the seed matures.

Immediately after fertilization, the endosperm nitrogen of the barley seed is about 50% protein in form. As development proceeds, the protein nitrogen increases rapidly for about 20 days, after which it remains constant. Amide form of nitrogen increases slightly: so its relative proportion in the endosperm remains constant. The testa-pericarp nitrogen content follows a similar trend, although at a slower rate, since the total growth rate of these tissues is slower.

Negligible change in the DNA and RNA of the testa-pericarp occurs during seed

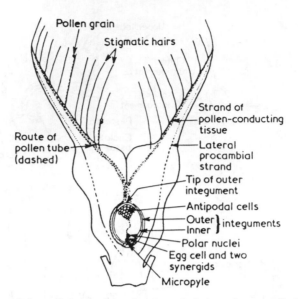

Figure 2.8a. *The route of the pollen tube from the stigmatic hair to the egg cell. (From Briggs,* Barley, *p. 47.)*

development, since they are nucleotides of the nucleus and cytoplasm, and any marked increase in their occurrence is a reflection of cell division. In contrast, DNA and RNA increase rapidly with increased cell division during early embryo and endosperm growth, but level off with increased cell expansion.

Nitrogen is also found in the developing seed in the form of amino acids and protein-bound phosphorus. The endosperm amino acid content increases rapidly during the first two or three weeks of seed development. This period corresponds to the time when the endosperm is high in RNA content, which directs amino acid synthesis.

ENVIRONMENTAL EFFECTS

The environment in which the seed forms affects its development. This is often illustrated by changes in seed size and weight. Components of the environment that influence seed size and weight include soil fertility, moisture, temperature, light, and position on the plant.

Soil Fertility

Generally, plants that have been fertilized with the three major elements (N, P, K) produce larger seeds than those which have not been fertilized. The increase in seed size is due to a greater seed development rate during the seed filling period as a consequence of increased nutrient availability. This is true for soybean (Boswell and Anderson 1976) and tomato (Varis and George 1985) seeds. When examining the influence of individual elements on seed development, nitrogen clearly has the greatest effect. It increases seed size in perennial ryegrass (Ene and Bean 1975), soybean (Ham

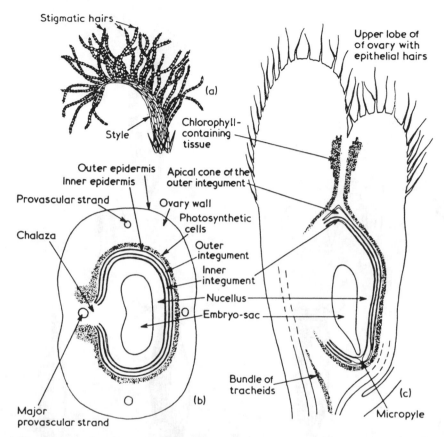

Figure 2.9. *The mature ovary, before anthesis: (A) a style with the apical stigmatic hairs, each composed of four columns of cells around a central lumen, and the simple basal hairs; (B) the ovary in transverse section (C) the ovary in longitudinal section. (From Briggs (1978).)*

et al. 1975), and corn (Eck 1984), although it has also been shown to decrease seed size in wheat (Frederick and Marshall 1985) and tomato (George et al. 1980). These differences in response to nitrogen might be attributed to the time at which the nitrogen was applied. Earlier applications of nitrogen produced greater seed weight in wheat (Langer and Liew 1973) and rice (Humphreys et al. 1987).

Production factors also can influence seed development. Increased competition for limited nutrients by weeds or from crops as a consequence of narrow row spacings and increased numbers of seed per row result in decreased seed size. The intensity of the competition can vary within (Westermann and Crothers 1977) and among species (Elmore and Jackobs 1984).

Moisture

Prolonged droughts and reduced soil moisture content decrease seed size (e.g., soybean and flax), particularly when these effects occur during flowering and seed fill

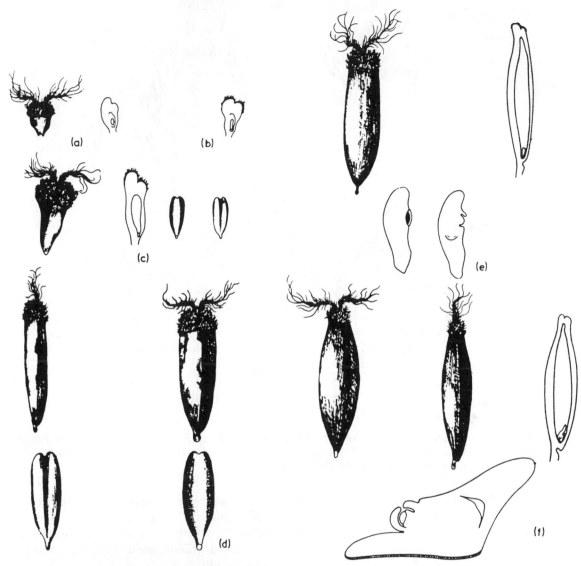

Figure 2.10. *The appearance of the developing grain at various stages after fertilization. The diagrams span the first 10 days' development. Beyond this time, growth in width and depth continues, but first the palea then the lemma adhere to the pericarp (ovary wall). (A) The ovary in surface view and median longitudinal section (L-S) one day after fertilization; (B) median L-S of ovary two days after fertilization; the embryo-sac is lengthening; (C) four days after fertilization; surface view, median L-S. and dorsal and ventral views of the separated embryo-sac. Note the tiny size of the embryo; (D) six days after fertilization—four views; (E) the 8-day-old grain in surface view and median L-S and a surface view and section of the embryo; (F) the 10-day-old grain in plan and side view, and in median L-S and the embryo (L-S, greatly enlarged). Note the decreasing size of the ovary tip. (From Briggs (1978).)*

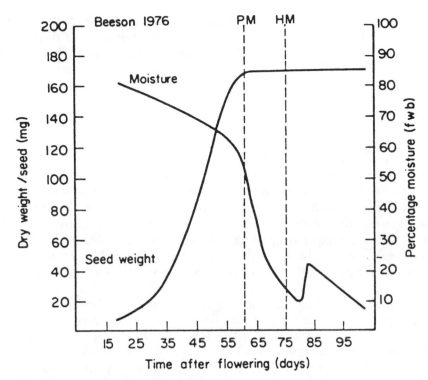

Figure 2.11. *Changes in seed dry weight and moisture content fresh weight basis during seed development and maturation. PM = physiological maturity; HM = harvest maturity. From: Tekrony, D. M., D. B. Egli, and J. Balles. 1980. The effect of the field production environment on soya bean seed quality. In* Seed Production *(ed. P. D. Hebblethwaite), Butterworths, London.*

(Meckel et al. 1984). If drought occurs only before flowering, its primary effect is on a reduction in seed number, while seed size is unchanged. This has been observed in soybean (Wright et al. 1984), corn (Eck 1986), clover (Andrews et al. 1977), and wheat (Sionit et al. 1980). The lack of soil moisture may reduce photosynthesis, which shortens the seed filling period, thereby reducing seed size.

Temperature

High temperatures during seed development produce smaller seeds, while low temperatures favor larger seeds in orchardgrass and perennial ryegrass (Shimzu et al. 1979), bean (Siddique and Goodwin 1980), wheat (Wardlaw et al. 1989), lupin (Downes and Gladstones 1984), flax (Green 1986), and sorghum (Kiniry and Musser 1988). Annual variations in the environment also influence seed size in soybean (Egli et al. 1987), sugar beet (Wood et al. 1980), and birdsfoot trefoil (McGraw et al. 1986).

Light

In general, reduced light to the parent plant results in smaller seeds. This effect has been found in carrot (Gray et al. 1986), pea (Gubbels 1981), soybean (Egli et al.

1987), corn (Kiniry and Richie 1985), and clover (Collins et al. 1978). Partial shading decreases seed weight (Jenner 1979; Peet and Kramer 1980). Short days also reduce seed size in pea (Reid 1979) and perennial ryegrass (Bean 1980). These findings are probably attributed to the lack of light which decreases photosynthesis and results in smaller seeds.

Position on the Plant

The position of the seed in the inflorescence can affect seed development rate. For example, distal seeds in a wheat spike have slower growth rates and shorter seed filling periods than proximal seeds (Simmons and Crookston 1979). Corn seeds at the tip of the ear are smaller than those at the base which has been attributed to inadequate photosynthate supply (Hanft et al. 1986). Sorghum seeds at the base of the panicle are smaller and have a slower growth rate than those elsewhere (Muchow 1990). Physical removal of other reproductive sinks such as flowers or seeds also increases the size of the remaining seeds on/in the inflorescence as in soybean (Egli et al. 1987), wheat (Radley 1978), carrot (Gray et al. 1986), corn (Kiniry et al. 1990), and clover (Rincker et al. 1977).

The inflorescence structure can also affect seed size. In the *Asteraceae,* the composite flowering head produces both ray and disk flowers, with ray flowers producing larger seeds. Interestingly, the environment alters the ratio of ray to disk flowers and thus modifies the seed size of the crop (Venable and Levin 1985; McGinley 1989). In some cases, these two seed types have differing germination requirements (Forsyth and Brown 1982). Smaller seeds are also produced from smaller fruit or fruit that mature later in the growing season. Such trends have been reported in cotton (Leffler et al. 1977), sugar beet (Malik and Shakara 1977), rape (Clarke 1979), carrot (Verkaar and Schenkeveld 1984), and geranium (Roach 1986).

The physiological mechanism(s) governing seed development remain unknown. Some studies have attempted to relate seed development with hormonal levels. Endogenous abscisic acid (ABA) levels, for example, have been correlated with seed weight. Large-seeded soybean genotypes have 50% more ABA than small-seeded genotypes (Schussler et al. 1984). Injecting ABA into wheat seeds increased their absorption of photosynthate (Dewdney and McWha 1979). ABA increases the import of sucrose into barley seeds (Teitz and Dingkuhn 1981), and adding ABA to bean seeds increased their seed size (Clifford et al. 1987). While these studies suggest that ABA may have a direct role in regulating seed development, further work is still required to better understand this complex phenomenon.

Questions

1. In diploid plants is the nucellar tissue of diploid or haploid origin?
2. What is the archesporial cell?
3. Explain why the megagametophyte is sometimes called a macrospore.
4. What is the origin of the perisperm in some seeds? Name some seeds that have a well-developed perisperm.

5. What are some other names for the seed integuments?
6. Name the structure or tissue from which the integuments originate?
7. Can you cite two types of epistase?
8. What is anthesis?
9. Name at least one apomictic species?
10. What is the function of the endosperm in seeds?
11. Why do some seeds have no endosperm tissue at maturity?
12. What is the function of the albumin layer of the endosperm?
13. Why does the DNA content of the embryo increase faster during early seed development, and level off later?
14. Would you expect the DNA content of pericarp and seed coat tissues to increase as rapidly as endosperm and embryo? Why not?
15. How long does the zygote remain at rest before it starts preparing for cell division?

General References

Andrews, P., W. J. Collins, and W. R. Stern. 1977. The effect of withholding water during flowering on seed production in *Trifolium subterraneum* L. *Australian Journal Agriculture Research* 28:301–307.

Bean, E. W. 1980. Factors affecting the quality of herbage seeds. In: *Seed Production,* ed. P. D. Hebblethwaite, pp. 593–604, London: Butterworth.

Bhatnager, S. B., and B. M. Hohri. 1972. Development of angiosperm seeds. In: *Seed Biology,* Vol. I, pp. 77–149. New York: Academic Press.

Boswell, F. C., and D. E. Anderson. 1976. Long-term residual fertility and current N-P-K application effects on soybeans. *Agronomy Journal* 68:315–318.

Bradbury, D., I. M. Cull, and M. M. MacMasters. 1956a. Structure of the mature wheat kernel. I. Gross anatomy and relationship of parts. *Cereal Chemistry* 33(6): 329–342.

———. 1956b. Structure of the mature wheat kernel. II. Microscopic structure of the pericarp, seed coat, and other coverings of the endosperm and germ of hard red winter wheat. *Cereal Chemistry* 33(6): 342–360.

Bradbury, D., M. M. MacMasters, and I. M. Cull. 1956a. Structure of the mature wheat kernel. III. Microscopic structure of the endosperm of hard red winter wheat. *Cereal Chemistry* 33(6): 361–373.

———. 1956b. Structure of the mature wheat kernel. IV. Microscopic structure of the germ of hard red winter wheat. *Cereal Chemistry* 33(6): 373–391.

Briggs, D. E. 1978. *Barley.* London: Chapman and Hall.

Brink, R. A., and D. C. Cooper. 1947. The endosperm in seed development. *Botanical Review* 13:423–541.

Brown, W. V. 1960. The morphology of the grass embryo. *Phytomorphology* 10:215–223.

Buttrose, M. S. 1963. Ultrastructure of the developing aleurone cells of wheat grain. *Australian Journal of Biological Sciences* 16:768–774.

Chute, H. M. 1932. The morphology and anatomy of the achene. *American Journal of Botany* 17:703–723.

Clarke, J. M. 1979. Intra plant variation in number of seeds per pod and seed weight in *Brassica napus* 'Tower.' *Canadian Journal of Plant Science* 59:959–962.

Clifford, P. E., C. E. Offler, and J. W. Patrick. 1987. Injection of growth regulators into seeds growing in situ on plants of *Phaseolus vulgaris* with a double fruit stack system. *Canadian Journal of Botany* 65:612–615.

Collins, W. J., R. C. Rossiter, and A. R. Monreal. 1978. The influence of shading on seed yield and hard-seededness in swards of subterranean clover. *Australian Journal of Agricultural Research* 32:783–792.

Corner, E. J. H. 1951. The leguminous seed. *Phytomorphology* 1:117–150.

Dewdney, S. J., and J. A. McWha. 1979. Abscisic acid and the movement of photosynthetic assimilates towards developing wheat (*Triticum aestivum* L.) grains. *Z. Pflanzenphysiol.* 92:183–186.

Downes, R. W., and J. S. Gladstones. 1984. Physiology of growth and seed production in *Lupinus augustifolius* L. II. Effect of temperature before and after flowering. *Australian Journal of Agricultural Research* 34:501–509.

Eck, H. V. 1984. Irrigated corn yield responses to nitrogen and water. *Agronomy Journal* 76:421–428.

Eck, H. V. 1986. Effects of water deficits on yield, yield components, and water use efficiency of irrigated corn. *Agronomy Journal* 78:1035–1040.

Egli, D. B., R. A. Wiralaga, and E. L. Ramseur. 1987. Variation in seed size in soybean. *Agronomy Journal* 79:697–700.

Elmore, R. W., and J. A. Jackobs. 1984. Yield and yield components of sorghum and soybeans of varying plant heights when intercropped. *Agronomy Journal* 76:561–564.

Ene, B. N., and E. W. Bean. 1975. Variations in seed quality between certified seed lots of perennial rye grass and their relationships to nitrogen supply and moisture status during seed development. *Journal of British Grassland Society* 30:195–199.

Forsyth, C., and N. A. C. Brown. 1982. Germination of the dimorphic fruits of *Bidens pilosa* L. *New Phytology* 90:151–164.

Foulds, F. E. 1929. A study of the comparative morphology of the seeds of *Agropyron*. *Scientific Agriculture* (Ottawa) 10:200–219.

Frederick, J. R., and H. G. Marshall. 1985. Grain yield and yield components of soft red winter wheat as affected by management practices. *Agronomy Journal* 77:495–499.

George, R. A. T., R. J. Stephens, and S. Varis. 1980. The effect of mineral nutrients on the yield and quality of seeds of tomato. In: *Seed Production*, ed. P. D. Hebblethwaite, pp. 561–567, London: Butterworth.

Gray, D., J. R. A. Steckel, and J. A. Ward. 1986. The effect of cultivar and cultural factors on embryo-sac volume and seed weight in carrot (*Daucus carota* L.). *Annals of Botany* 58:737–744.

Green, A. G. 1986. Effect of temperature during seed maturation on the oil composition of low-linolenic genotypes of flax. *Crop Science* 26:961–965.

Gubbels, G. H. 1981. Quality, yield and seed weight of green field peas under conditions of applied shade. *Canada Journal of Plant Science* 61:213–217.

Ham, G. E., I. E. Liener, S. D. Evans, R. D. Frazier, and W. W. Nelson. 1975. Yield and composition of soybean seed as affected by nitrogen and sulfur fertilization. *Agronomy Journal* 67:293–297.

Hanft, J. M., R. J. Jones, and A. B. Stumme. 1986. Dry matter accumulation and carbohydrate concentration patterns of field-grown and in vitro cultured maize kernels from the tip and middle ear positions. *Crop Science* 26:568–572.

Harlan, J. R. 1946. The development of buffalo grass seed. *Journal of the American Society of Agronomy* 38:135–141.

Haupt, A. W. 1934. Ovule and embryo sac of *Plumbago capensis*. *Botanical Gazette* 95:649–659.

Humphreys, E., W. A. Muirhead, F. M. Melhuish, and R. J. G. White. 1987. Effects of time of urea application on combine-sown calrose rice in south-east Australia. 1. Crop response and nitrogen uptake. *Australian Journal of Agricultural Research* 38:101–112.

Jenner, C. F. 1979. Grain-filling in wheat plants shaded for brief periods after anthesis. *Australian Journal of Plant Physiology* 6:629–641.

Johnsen, D. A. 1945. A critical survey of the present status of plant embryology. *Botanical Review* 11:87–107.

Kiniry, J. R., and R. L. Musser. 1988. Response of kernel weight of sorghum to environment early and late in grain filling. *Agronomy Journal* 80:606–610.

Kiniry, J. R., and J. T. Richie. 1985. Shade-sensitive interval of kernel number of maize. *Agronomy Journal* 77:711–715.

Kiniry, J. R., C. A. Wood, D. A. Spanel, and A. J. Bockholt. 1990. Seed weight response to decreased seed number in maize. *Agronomy Journal* 54:98–102.

Langer, R. H. M., and F. K. Y. Liew. 1973. Effects of varying nitrogen supply at different stages of the reproductive phase on spikelet and grain production and on grain nitrogen in wheat. *Australian Journal of Agricultural Research* 24:647–656.

Leffler, H. R., C. D. Elmore, and J. D. Hesketh. 1977. Seasonal and fertility-related changes in cottonseed protein quantity and quality. *Crop Science* 17:953–956.

Leininger, L. N., and A. L. Urie. 1964. Development of safflower seed from flowering to maturity *Crop Science* 4:83–87.

Loewenberg, J. R. 1955. The development of bean seeds (*Phaseolus vulgaris* L.). *Plant Physiology* 30:244–250.

Maheshwari, P. 1950. *An Introduction to the Embryology of the Angiosperms*. New York: McGraw-Hill Book Company.

Malik, K. B., and S. A. Shakara. 1977. Effect of growth regulators on seed development and indeterminate type of growth in sugar beet. *Agriculture Pakistan* 28:65–75.

Martin, J. N. 1937. The strophiole in sweet clover and alfalfa seeds. *Proceedings of the Iowa Academy of Sciences* 44:104.

McGinley, M. A. 1989. Within and among plant variation in seed mass and pappus size in *Tragopogon dubious. Canadian Journal of Botaony* 67:1298–1304.

McGraw, R. L., P. R. Beuselinck, and R. R. Smith. 1986. Effect of latitude on genotype x environment interactions for seed yield in birdsfoot trefoil. *Crop Science* 26:603–605.

Meckel, L., D. B. Egli, R. E. Phillips, D. Radcliffe, and J. E. Leggett. 1984. Effect of moisture stress on seed growth in soybeans. *Agronomy Journal* 76:647–650.

Muchow, R. C. 1990. Effect of high temperature on the rate and duration of grain growth in field-grown *Sorghum bicolor* (L.) Moench. *Australian Journal of Agricultural Research* 41:329–337.

Peet, M. M., and P. J. Kramer. 1980. Effects of decreasing source/sink ratio in soybeans on photosynthesis, photorespiration, transpiration and yield. *Plant Cell and Environment* 3:201–206.

Povilaitis, B., and J. W. Boyles. 1960. Ovule development in diploid red clover. *Canadian Journal of Botany* 38:507–532.

Radley, M. 1978. Factors affecting grain enlargement in wheat. *Journal of Experimental Botany* 29:919–934.

Randolph, L. F., 1936. Developmental morphology of the caryopsis in maize. *Journal of Agricultural Research* 53:881–916.

Rao, V. S. 1959. Nuclear endosperm or noncellular endosperm? *Annals of Botany* (London) 23:364.

Reed, E. L. 1924. Anatomy, embryology, and ecology of *Arachis hypogea. Botanial Gazette* 78:289–310.

Reeves, R. G. 1932. Development of the ovule and embryo sac of alfalfa. *American Journal of Botany* 17:239–246.

———. 1936. Comparative anatomy of the seeds of cotton and other malvaceous plants, I. *Malveae* and *Ureneae. American Journal of Botany* 23:291–296.

Reid, J. B. 1979. Flowering in *Pisum:* Effect of the parental environment. *Annals of Botany* 44:461–467.

Rincker, C. M., J. G. Dean, C. S. Garrison, and R. G. May. 1977. Influence of environment and clipping on the seed-yield potential of three red clover cultivars. *Crop Science* 17:58–60.

Roach, D. A. 1986. Timing of seed production and dispersal in *Geranium carolinianum:* Effects on fitness. *Ecology* 67:572–576.

Schussler, J. R., M. L. Brenner, and W. A. Brun. Abscisic acid and its relationship to seed filling in soybeans. *Plant Physiology* 76:301–306.

Shimzu, N., T. Komatsu, and F. Ikegaya. 1979. Studies on seed development and ripening in temperate grasses. II. Effects of temperature on seed development and ripening and germination behavior in orchard grass (*Dactylis glomerata*) and Italian rye grass (*Lolium multiflorum*). *Bulletin of National Grassland Institute* 15:70–87.

Siddique, M. A., and P. B. Goodwin. 1980. Seed vigor in bean (*Phaseolus vulgaris* L. cv. Apollo) as influenced by temperature and water regime during development and maturation. *Journal of Experimental Botany* 31:313–323.

Simmons, S. R., and R. K. Crookston. 1979. Rate and duration of growth of kernels formed at specific florets in spikelets of spring wheat. *Crop Science* 19:690–693.

Singh, B. 1953. Studies on the structure and development of seeds of *Cucurbitaceae. Phytomorphology* 3:224–239.

Singh, D. 1961. Development of embryos in the *Cucurbitaceae. Journal of the Indian Botanical Society* 40:620–623.

Singh, H., and B. M. Johri. 1972. Development of gymnosperm seeds. In: *Seed Biology,* Vol. I. ed. T. T. Kozlowski, pp. 21–75. New York: Academic Press.

Singh, R. P. 1954. Structure and development of seeds in *Euphorbiaceae: Ricinus communis* L. *Phytomorphology* 4:118–123.

Sionit, N., H. Hellmers, and B. R. Strain. 1980. Growth and yield of wheat under CO_2 enrichment and water stress. *Crop Science* 20:687–690.

Sripleng, A., and F. H. Smith. 1960. Anatomy of the seed of *Convolvulus arvensis. American Journal of Botany* 47:386–392.

Teitz, A., and M. Dingkuhn. 1981. Regulation of assimilate transport in barley by the abscisic acid content of young caryopses. *Z. Pflanzenphysiol.* 104:475–479.

Thompson, R. C. 1933. A morphological study of flower and seed development in cabbage. *Journal of Agricultural Research* 47:215–232.

Varis, S., and R. A. T. George. 1985. The influence of mineral nutrition of fruit yield, seed yield and quality in tomato. *Journal of Horticultural Science* 60:373–376.

Varner, J. E. 1965. Seed development and germination. In: *Plant Biochemistry,* ed. J. Bonner and J. E. Varner. pp. 763–792. New York: Academic Press.

Venable, D. L., and D. A. Levin. 1985. Ecology of achene dimorphism in *Heterotheca latifolia.* I. Achene structure, germination and dispersal. *Journal of Ecology* 73:133–145.

Verkaar, H. J., and A. J. Schenkeveld. 1984. On the ecology of short-lived forbs in chalk grasslands: Semelparity and seed output of some species in relation to various levels of nutrient supply. *New Phytology* 98:673–682.

Wardlaw, I. F., I. A. Dawson, and P. Munibi. 1989. The tolerance of wheat to high temperatures during reproductive growth. II. Grain development. *Australian Journal of Agricultural Research* 40:15–24.

Westermann, D. T., and S. E. Crothers. 1977. Plant population effects on the seed yield components of beans. *Crop Science* 17:493–496.

Winter, D. M. 1960. The development of the seed of *Abutilon theophrasti,* I. Ovule and embryo. *American Journal of Botany* 47:8–14.

Wood, D. W., R. K. Scott, and P. C. Longden. 1980. The effects of mother-plant temperature on seed quality in *Beta vulgaris* L. (sugar beet). *In: Seed Production,* ed. P. D. Hebblethwaite, pp. 257–270, London: Butterworth.

Wright, D. L., F. M. Shokes, and R. K. Sprenkel. 1984. Planting method and plant population influence on soybeans. *Agronomy Journal* 76:921–924.

3

The Chemistry
of Seeds

A knowledge of the chemical composition of seeds is essential for several reasons: (1) seeds are a basic source of food for both man and animals, (2) they are an important source of medicine and drugs, (3) they contain various antimetabolites that adversely affect human and animal nutrition, and (4) they contain reserve food supplies and growth substances that influence seed germination and seedling vigor, seed storage and longevity, as well as industrial and agricultural uses of seeds.

Most of our knowledge of the chemical composition of seeds concerns cultivated species, since these comprise a large share of the food supply and also provide many industrial raw materials. Information about the seeds of wild species is relatively scarce; however, the search for new sources of food and raw materials is gradually yielding more information about seeds of wild plants. Moreover, since seeds are a challenging subject for scientific study, much information about both domestic and wild species is being accumulated simply because of our thirst for knowledge.

Aside from the normal chemical constituents found in all plant tissue, seeds contain extra amounts of chemical substances stored as food reserves to accommodate germination. These reserve foods are stored primarily as carbohydrates, fats (or oils), and proteins. In addition, seeds contain other chemical substances, some of which play minor storage roles, but most of which serve as growth substances and metabolism controls. Compared to other plant parts, the mineral content of most seeds is remarkably low and tends to be centered in the hulls and structural tissue. Seeds with a relatively high mineral composition include bean, cotton, sunflower, soybean, and cereal grains with the hulls intact.

THE INFLUENCE OF GENETIC FACTORS

The chemical composition of seeds is basically determined by genetic factors and varies among different species and seed parts (Tables 3.1 and 3.2). However, it is influenced by environmental and cultural practices.

Table 3.1. Average Chemical Composition of Seeds

Plant	% Protein	% Fat (Lipid)
Acorn (red oak)	3.2	10.7
Barley (Pacific coast states)	8.7	1.9
Bean, Mung	23.6	0.2
Bean, Navy	22.9	1.4
Bean, Pinto	22.5	1.2
Beechnuts	15.0	30.6
Buckwheat	10.3	2.3
Chickpeas	20.3	4.3
Cottonseed kernel (without hull)	38.4	33.3
Flaxseed	24.0	35.9
Kafir grain	11.0	2.9
Mustard, Wild	23.0	38.8
Oats	12.0	4.6
Peas	23.4	1.2
Peanut (without hulls)	30.4	47.7
Rape	20.4	43.6
Rice (rough grain)	7.9	1.8
Rye	12.6	1.7
Soybean	37.9	18.0
Sunflower	16.8	25.9
Vetch	29.6	0.8
Wheat	13.2	1.9

From Morrison (1961).

Table 3.2. Chemical Composition of Different Parts of Corn Seed*

Chemical	Entire Seed	Endosperm	Embryo	Pericarp-Testa
Starch	74.0	87.8	9.0	7.0
Sugars	1.8	.8	10.4	.5
Oil (Lipid)	3.9	.8	31.1	1.2
Protein	8.2	7.2	18.9	3.8
Ash	1.5	.5	11.3	1.0

*Includes only selected chemical components.
From Earle et al. (1956).

Because of genetic factors, the chemical composition of seeds varies widely among species and even among varieties. Through crossing and selection, plant breeders are able to manipulate the chemical composition of many domesticated crops and increase their usefulness as food, fiber, or raw material. Modern varieties of rapeseed, soybean, high-lysine corn, sorghum, and wheat have been bred and developed for higher content of oils, protein, or carbohydrates, and represent a significant improvement over earlier varieties. The effectiveness of breeding and selection techniques in increasing the protein and oil content of crop seeds is shown by work with soybean (Shannon et al. 1972) and corn (Dudley and Lambert 1968).

ENVIRONMENTAL INFLUENCES

Many environmental factors influence the chemical composition of seeds, and because of the interrelationships of these factors, it is sometimes difficult to determine the cause of variability. A two-year study of eight corn hybrids in three Michigan locations revealed a range in protein content from 7.44 to 12.88% within hybrids (Norden et al. 1952). Similar environmental influence on the protein content of chickpeas has been reported in the U.S.S.R. (Ivanov 1933). Wheat protein content varies depending on geographic location of the crop (Baenziger et al. 1985). Environmental conditions from year to year similarly influence the chemistry of peanut (Ketring et al. 1978) and pea (Gubbels 1981) seeds. Among the environmental factors that modify seed chemistry are water availability, temperature, soil fertility, and cultural practices.

Water

Availability of water influences chemical composition of seed. For example, we have long recognized that the protein nitrogen content and quality of seed are lower in years of high rainfall than in drier years and in irrigated land compared to dry land. A study by Greaves and Carter (1923) showed that high rates of irrigation decreased the nitrogen content of wheat, barley, and oat seeds in plants grown in Utah (Table 3.3). This decrease in nitrogen content occurred despite observed increases in phosphorus, potassium, calcium, and magnesium, which are not readily soluble in water. Similar studies for both wheat and sorghum have shown that the nitrogen content of the mature seed decreases linearly with the amount of water supplied during seed development (Stone and Tucker 1968; Stone et al. 1964; Mathers et al. 1960). These studies serve as clear examples of the ways in which high-moisture environments, whether created by rainfall or irrigation, can influence the mineral composition of seeds. Yet, we know relatively little about why these changes occur. It is not clear, for example, whether the primary effect of excess moisture is on mineral absorption by the roots, or on the rate of grain fill with carbohydrates and the concomitant dilution of the basic cell constituents. Further studies are needed to answer these questions. In contrast, plants exposed to low soil moisture or drought conditions have an increase in seed protein content. This has been found in wheat (Karathanasis et al. 1980), perennial ryegrass

Table 3.3. Effect of Irrigation on the
Mineral Content of Wheat, Barley, and Oat Seed

Element	Increase (+) or decrease (−) over controls, %		
	Wheat	Barley	Oats
Nitrogen	− 21	−19	−40
Phosphorus	+ 55	+30	+35
Potassium	+ 35	+14	+31
Calcium	+155	+41	+22
Magnesium	+ 32	+ 9	+65

From Greaves and Carter (1923).

(Ene and Bean 1975), corn (Francois et al. 1986), soybean (Pikaard and Cherry 1984), and bean (Robinson 1983) seeds.

Temperature

Although few studies have been conducted to demonstrate the influence of temperature on seed structure and composition, those studies that have been done show an association. Howell and Carter (1958) found that the oil content of soybean seeds depended on the temperature during pod development (see Figure 3.1). Seeds that matured at 21°C contained 19.5% oil, whereas those that matured at 30°C contained 22.3% oil. Soybeans also develop a higher oil content when they are planted early in the season and (therefore) mature under a warmer temperature than when they are planted later and mature during cooler weather. Canvin (1965) examined the influence of temperature on fatty acid content of developing rapeseeds and showed that as the temperature increased, oleic acid levels increased and erucic acid levels decreased. This study has special significance since erucic acid has a bitter taste and is an undesirable component of rapeseeds that are to be used as feed.

In sunflowers, low temperature during seed development favors the production of the preferred linoleic acid, and high temperatures enhance the quantity of oleic acid in the oil (Harris et al. 1980). As a result, sunflower seeds that are grown for high oil

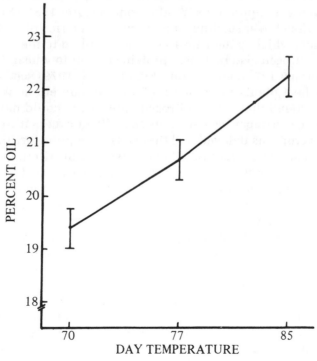

Figure 3.1. *Oil content of soybean seed produced with day temperatures as shown during periods of pod filling; night temperature was 65°F in all cases. (From Howell Carter (1958).)*

content are planted late so that flowering occurs later in the season (Owen 1983; Jones 1984). Unger (1986) showed that the best-quality sunflower oil was produced during the cooler conditions of late summer. Similar responses have been reported for flaxseed (Green 1986). Beyond oil content, temperature of seed development also influences protein and carbohydrate quality. For example, high temperatures increase seed protein content as in wheat (Campbell et al. 1981), and cool temperatures lower protein content in soybean (Radford et al. 1977) seeds. Other studies have shown that high night temperatures accelerate kernel development and maturation in rice seeds, producing "chalky" kernels. At low night temperatures, the seeds develop the desired "milky white" kernels.

Soil Fertility

Perhaps the easiest environmental parameter to control in an evaluation of factors that affect seed chemical composition is the mineral nutrition that the mother plant receives. In most instances, seeds that are mineral deficient will perform poorly compared to normal seeds unless they are planted in a soil that is nutritionally adequate and provides the missing essential element or elements. Many studies have been conducted to evaluate the influence of nitrogen, phosphorus, and potassium on seed quality. Some of these are cited below.

Corn plants grown under either high nitrogen fertilization or low plant populations produce seed with a higher protein content than those produced under low nitrogen availability or high plant populations (Wolfson and Shearer 1981). Similar results with rice indicate that when lower seeding rates are used, each rice plant has greater access to available nitrogen, which is then absorbed and supplied to the seeds (DeDatta et al. 1972). Added soil nitrogen also increases protein content in wheat (Glenn et al. 1985), rice (Allen and Terman 1978), and cotton (Elmore et al. 1979) seeds. The application of foliar urea to wheat plants increased seed protein content as well (Altman et al. 1983). Scott (1969) found that excess nitrogen application could have an indirect and detrimental influence on sugar beet seed quality. When excess nitrogen was applied, the ripening of the crop was delayed and the seeds from plots fertilized with nitrogen were less mature than seeds from plots that received no nitrogen. In this instance, the reduction in germination was a nitrogen-induced maturity effect.

Other studies indicate the importance of phosphorus for production of high-quality seeds. For example, Harrington (1960) demonstrated that phosphorus nutrition of parent plants failed to influence seed performance of the progeny, Austin (1966) showed that freshly harvested seed from phosphorus-deficient watercress plants had lower (and slower) germination than seeds from nondeficient plants. Other studies have shown that added phosphorus to the parent plant increases seed phosphorus content in pea (Peck et al. 1980), soybean (Cassman et al. 1981), and wheat (Porter and Paulsen 1983). Low-phosphorus seed produces smaller plants than nondeficient seed.

Few studies have been conducted on the influence of potassium on seed quality and composition. Harrington (1960) reported that potassium-deficient plants of *Capsicum frutescens* gave a higher proportion of abnormal seeds with dark-colored embryos and seed coats. Both normal and abnormal seeds from such plants had a lower percentage of germination than control seeds, and their viability declined more rapidly in storage.

Other essential elements added to the parent plant also will be found in increased concentrations in the seed. This is true for calcium in peanut (Coffelt and Hallock 1986); manganese, zinc, and boron in soybean (Parker et al. 1981; Raboy and Dickinson 1984; Touchton and Boswell 1975); copper in wheat (Loneragan et al. 1980); and cadmium and selenium in lettuce and wheat (Cary 1981).

Cultural Practices

Still other environmental factors associated with plant morphology and production practices are known to modify seed chemical content. For example, chemical composition of the seed can be influenced by its position on the plant. The relative proportions of seven fatty acids have been shown to vary dependent on the position of the seed in the pod of rape (Diepenbrock and Geisler 1979). First and second seed in a wheat spikelet have higher nitrogen concentration than third and fourth seeds (Simmons and Moss 1978). The time during the season when the seeds develop also influences their chemistry such as the oil content in rape seed (Auld et al. 1984). Soybean seeds which mature later in the growing season have higher protein levels (Gbikpi and Crookston 1981). Competition within the field also modifies seed chemistry content since there is more competition among plants for limited nutrients. As competition increases, protein content declines while oil content increases in sunflower (Robinson et al. 1980; Majid and Schneiter 1987), rice (Nandisha and Mahadevappa 1984), and wheat and barley (Read and Warder 1982). These results suggest that dense plantings for oil crops may be an advantage as long as no reduction in yield is obtained. Interestingly, there appears to be an inverse relationship in oil-containing seeds between oil and protein content: As the protein content increases, oil content declines. This is true in soybean (Poole et al. 1983); sunflower (Mathers and Stewart 1982), and in meadowfoam (Crane et al. 1981).

CARBOHYDRATE STORAGE IN SEEDS

Carbohydrates are the major storage substance in seeds of most cultivated plants. Cereals and grasses are especially rich in carbohydrates and low in fats and proteins. Peas and beans are moderately high in carbohydrates, with lesser amounts of proteins and a low fat content. Seeds from many trees, such as chestnut, buckeye, and some oak species have a relatively high carbohydrate content and are quite low in fats and proteins.

Starch and hemicellulose are the major forms of carbohydrate storage in seeds. Other carbohydrates that occur in nonstorage forms are pectins and mucilages.

Starch

Seeds are composed largely of metabolically inactive food reserves that are stored until needed during germination. Starch is stored in two related forms, amylose and amylopectin, which are two polymers of D-glucose, one linear and the other branched (Figure 3.2). Starch is the principal, most widespread storage carbohydrate of seeds.

Amylose is composed of 200 to 1000 glucose units linked by $\alpha1,4$ glucosidic bonds, with a molecular weight ranging from 10,000 to 100,000. The molecule has a helical structure with six glucose rings to one revolution. Amylose stains blue when exposed to iodine and is 100% digestible by β-amylase.

Amylopectin, in contrast, is a much larger molecule consisting of 20–25 glucose units per branch, linked by both $\alpha1,4$ and $\alpha1,6$ glucosidic bonds with a molecular weight ranging from 50,000 to 1,000,000. Amylopectin is only about 50% digestible by the enzyme β-amylase and stains purplish-red when exposed to iodine.

Both amylose and amylopectin are hydrolyzed by the enzymes α-and β-amylase during normal metabolism and germination. Alpha-amylase possesses a molecular weight of 60,000 and requires the cation Ca^{++} for activation and stabilization against proteolytic destruction and thermal denaturation of the enzyme. This enzyme attacks both amylose and amylopectin randomly, cleaving both the $\alpha1,4$ and $\alpha1,6$ linkages. Although the action pattern varies, there is a general amylolysis (amylase catalyzed breakdown) of the starch molecule to large amounts of dextrin molecules to maltose and finally glocuse. The enzyme is formed *de novo* in the aleurone of barley and the

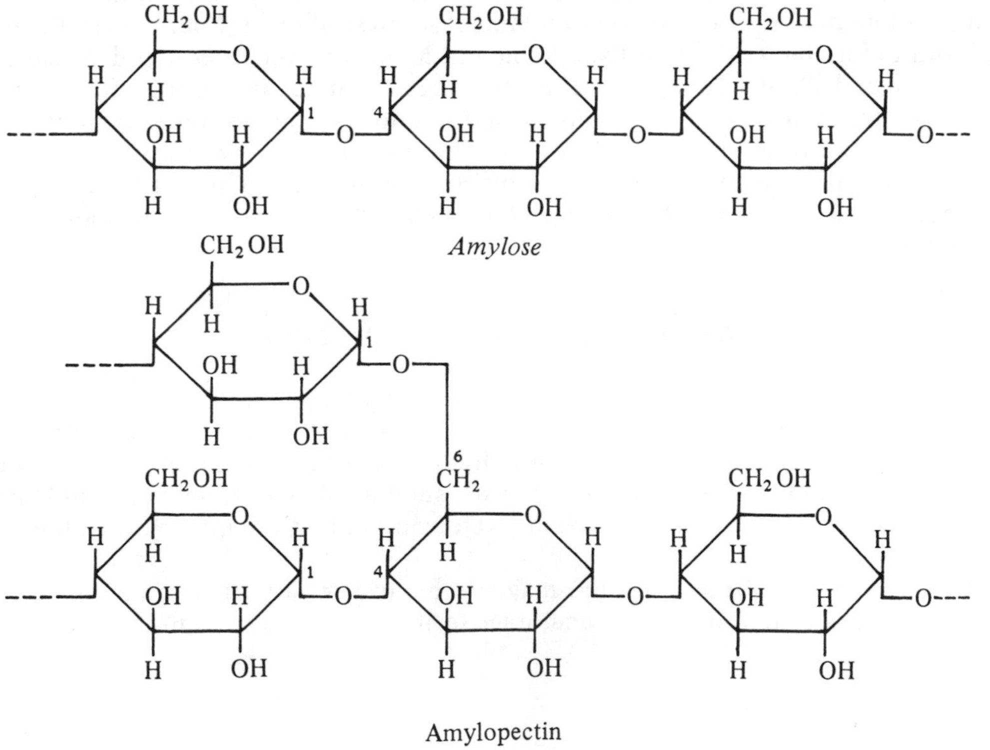

Figure 3.2. *Chemical structure of amylose and amylopectin showing linear ($\alpha1,4$ and 1,6 links) configuration.*

scutellum of corn. In corn, α-amylase accounts for 90% of the amylolytic activity and β-amylase for the remaining 10%.

Beta-amylase hydrolyzes both amylose and amylopectin from the nonreducing end by breaking the α1,4 linkage and forming maltose. In the case of amylose, almost complete hydrolysis occurs. However, since β-amylase is unable to bypass the branched chain point or α1,6 linkage of amylopectin, this molecule is only partially hydrolyzed, yielding the residual β-*limit dextrin*. This means that only the outer portions of the amylopectin molecule are attacked by β-amylase. This enzyme is already formed and is present *in situ* in seeds. In wheat kernels, β-amylase is present in an active latent form, apparently chemically bound to the wheat glutenin by disulfide linkages. During germination, the secretion of a substance capable of releasing the glutenin-bound enzyme may occur, accounting for the observed β-amylase activity.

In seeds, most starch is laid down in discrete subcellular bodies called *starch grains* that range from 2 to 100μ in size within the endosperm. Many starch grains appear to form around a central point, the hilum, around which shells of the polysaccharide are deposited. These shells probably reflect a diurnal periodicity in the synthesis and deposition of starch, since they are absent from starch grains of seeds developing in continuous light under experimental conditions. The shape of the starch grain appears to depend on the amylose content—the less angular, rounded grains having relatively high amylose levels.

Most starch grains are composed of about 50–75% amylopectin and 20–25% amylose. However, some seeds such as rice may be high in amylose (up to 37%) and are classified as starchy. The difference in appearance between amylopectin and amylose starch forms is illustrated in waxy corn (almost 100% amylopectin) versus normal corn (50–75% amylopectin) (Kerr 1950).

Hemicellulose

Other than starch, the major form of carbohydrate storage in seeds is hemicellulose. Hemicellulose is a widely used but poorly defined class of polysaccharides usually found in the cell walls of plants, although in certain seeds they are found as reserve food materials. This definition includes xylans, mannans, and galactans, plus a number of similar but less common polysaccharides. They are usually found in the thickened tertiary layers of cell walls of the endosperm or cotyledons instead of in the interior region of the endosperm. They are composed principally of mannins with small amounts of sugar (glucose, galactose, arabinose) as side chains on the main linear polymers of mannose residues. Hemicelluloses are particularly characteristic of seeds of many of the palm species, such as date palm and South American and Polynesian ivory nut palms. Seeds of the date palm have a small cylindrical embryo embedded in a sizable horny endosperm of hemicellulose. Hemicellulose has also been reported in the endosperm or cotyledon of several other species (Mitchell 1930).

Other Seed Carbohydrates

Mucilages. Apart from starch and hemicellulose, the amount of carbohydrates found in seeds is similar to that found in other parts of the plant. An exception might be the mucilages, which infrequently may be found in rather large amounts. A well-known example is the seed of buckhorn plantain, which is covered with a thick coating of mucilage that becomes sticky when wet and tends to cling to material that it touches. Obviously, this is a seed dispersal mechanism. However, seed conditioners also use the sticky coat characteristic when cleaning buckhorn seed from certain crop seed (see Chapter 10). Mucilages from buckhorn plaintain seed may be separated commercially for industrial use.

Mucilages are complex carbohydrates consisting principally of polyuronides and galacturonides that chemically resemble the pectic compounds and hemicelluloses. Physically, they are similar to the gums found in the bark and stems of many plants.

Pectic Compounds. These compounds occur in seeds and in other plant parts mainly as components of the cell wall and the middle lamella. The three principal pectic compounds are pectic acid, pectin, and protopectin. Pectic acid is a long, straight-chain substance composed of about 100 galacturonic acid molecules. Pectin differs from pectic acid by esterification of many of the carboxyl groups and has a greater number of galacturonic residues per chain. Pectins form viscous colloidal sols in water that set into firm gels under the proper conditions; they are used as setting agents in jams and jellies. Protopectins differ from pectins in their larger molecular chain. They occur in the primary cell wall and the middle lamella where they bind the cell walls together. When protopectins are converted into pectins, they are instrumental in the softening of ripening fruits.

LIPID STORAGE IN SEEDS

The value of seed-borne lipids (see Table 3.4) for food and industrial uses has contributed greatly to our knowledge of the composition of oily seeds. Seed oils have

Table 3.4. Percentage of Fats and Oils in the Dry Matter of Different Plant Species

Species	Percentage fat or oil	Species	Percentage fat or oil
Coconut	65	Spurge	35–45
Brazil nut	70	Rapeseed	33–43
Castor bean	60	Sesame seed	50–55
Sunflower seed	45–50	Colza seed	43–53
Flaxseed	30–35	Madia sativa seed	32
Cottonseed	15–20	Kafir seed	2.5
Peanut	40–50	Feterita seed	2.4
Hemp	30–35	Milo seed	2.3
Walnut	50–65	Corn seed	2.1
Cacao bean	40–50	Wheat seed	1.8
Poppy seed	40–50	Field pea	1.5
Pumpkin seed	41	Bean	2.8
Almond	40–50	Rye seed	1.9
Soybean	15–20	Rice seed	2.5
Cantaloupe seed	30	Buckwheat seed	1.1

From Miller (1931).

$$
\begin{array}{lllll}
R_1\,COOH & CH_2\,OH & CH_2\,OCOR_1 & \\
R_2\,COOH \quad + & HOCH & \longrightarrow \quad R_2\,COOCH & + & H_2\,O \\
R_3\,COOH & CH_2\,OH & CH_2\,OCOR_3 & \\
\\
\textit{3 fatty acids} & \textit{glycerol} & \textit{lipid molecule} & & \textit{water} \\
& & \textit{(triglyceride)} &
\end{array}
$$

Figure 3.3. *How three fatty acids combine with glycerol to form a lipid.*

tremendous versatility for industrial uses. In contrast to animal fats, their highly unsaturated chemical nature has caused increased interest in them for health purposes (e.g., canola).

Except for certain fruits, the occurrence of high lipid concentrations differentiates seeds from other plant organs. High lipid content is usually associated with decreased protein content (for example, in soybeans, rapeseed, cotton). However, in some species, such as certain oaks, it is associated with high carbohydrate levels.

Lipids are plant or animal substances that are insoluble in water, but soluble in ether, chloroform, benzene, or other organic solvents. They are esters of either fatty acids or glycerol or their various hydrolytic products. They are known as glycerides, or, more specifically, *triglycerides,* because each glycerol molecule is combined with three fatty acid molecules (Figure 3.3). Aside from those described by Bloor (1928), the characteristics of lipids are: (1) total fatty acid complement, (2) their glyceride structure, and (3) certain other substances that may be associated with the glycerides as impurities or as part of the lipid molecules.

The term *lipid* is often used interchangeably with *fat* and *oil,* although it is actually the generic term for both. Oils are distinguished from fats only in that they remain in liquid form at ordinary room temperatures, whereas fats remain in solid form. Oils are often designated as fatty oils to differentiate them from essential oils, which are chemically unrelated.

Fatty Acids

The fatty acids are so named because they are a constituent of natural fats, and in the free state they resemble fats in physical properties. Free fatty acids are seldom found in plant parts other than in germinating or deteriorating seeds as a result of fat hydrolysis. Fatty acids are saturated or unsaturated, depending on the type of carbon linkage in the molecule. The unsaturated fatty acids contain one or more double-bond links, which means the compound can take up hydrogen atoms to form a saturated compound. The unsaturated fatty acids are most common in seeds with oleic acids (one double bond) and linoleic acids (two double bonds), accounting for 60% of the weight of all the lipids present in most oilseed crops (Table 3.5). Saturated fatty acids are also present and contain an even number (n) of carbon atoms (n usually being between 4 and 24). Palmitic acid (n = 14) is the most common saturated fatty acid in oilseeds.

Table 3.5. Fatty Acid Content (%) of Seeds and Seed Products of Several Plant Species

Fat or oil	Lauric	Myristic	Palmitic	Stearic	Oleic	Linoleic	Linolenic
Coconut	45	20	5	3	6		
Palm kernel	55	12	6	4	10		
Olive	—	—	14.6	—	75.4	10	
Peanut	—	—	8.5	6	51.6	26	
Cottonseed	—	—	23.4	—	31.6	45	
Rapeseed/canola	—	—	2	4	60	17	10
Maize	—	—	6.0	2	44.0	48	
Linseed	—	3	—	—	—	77	17.0
Soybean	—	—	11.0	2	20.0	64	3.0
Cantaloupe seed oil	—	0.3	10.2	4.5	27.2	56.6	
Sunflower oil	—	—	3.5	2.9	33.4	57.5	
Hubbard squash seed oil	—	—	13.0	6.0	37.0	44.0	

Modified from Miller (1931).

Glycerol and Other Alcohols

Glycerol and other alcohols are combined with fatty acids to form different kinds of lipids. Of these, trihydroxy alcohol and glycerol or glycerine are most often involved and form esters (glycerides) with many different fatty acids.

Classification of Lipids

Lipids may be classified as (a) simple, (b) compound, or (c) derived. The simple lipids include esters of fatty acids and glycerol or various other alcohols. Among these simple lipids are fats and fatty oils. Compound lipids are esters of the fatty acids containing additional chemical groups. Phospholipids are compound lipids in which one of the three fatty acid units is replaced with phosphoric acid combined with choline. Derived lipids are derived from simple and compound lipids by hydrolysis and are soluble in fat solvents. They include various fatty acids and large molecular alcohols that are soluble in the fat solvents and may combine with the fatty acids. Among these are cholesterol.

The great majority of seed lipids are simple lipids, which include fats, fatty oils, and waxes. Waxes are simple lipids that are fatty acid esters of some alcohol other than glycerol. The term is used only in a chemical sense, since they occur both as liquids and as solids. Waxes are less soluble in ordinary fat solvents than fats. They can be saponified (see below), but only with difficulty. They are usually found as coverings for the leaves, fruits, and seeds of many plants.

Hydrolysis of Lipids

When oilseeds with a high lipid content germinate, there is a rapid disappearance of lipid with a concomitant rise in sucrose content. Concurrently, lipase activity rises sharply and participates in the stepwise hydrolysis of triglycerides to diglycerides, monoglycerides, and finally free glycerol and free fatty acids. Fatty acids are further oxidized by both α- and β-oxidation during the course of germination. These processes are further discussed in Chapter 4. Most lipids are stored in oil storage bodies called *spherosomes* that range in size from 0.2 to 6.0μ in diameter. Many of the enzymes essential for fatty acid biosynthesis or for triglyceride hydrolysis are present in the spherosomes.

PROTEIN STORAGE IN SEEDS

Proteins are nitrogen-containing molecules of huge size and exceedingly complex structure, the greater part of which yield amino acids upon hydrolysis of the peptide bond (Figure 3.4). Proteins are so important to both plant and animal life that all physiological reactions of living cells revolve around their physical and chemical properties and those of related compounds. Aside from water, they are the principal component of all protoplasm of both plant and animal cells.

Proteins comprise a valuable food storage component in seeds of most plant species. Soybeans are one of the few species known in which proteins comprise more of the reserve food supply than do fats or carbohydrates. Most of the plants known to have high-protein seed are legumes having nitrogen-fixing capacity. However, many high-protein species are nonlegumes. Seed storage proteins are less complex than protoplasmic proteins and less likely to be tied up with lipids and other prosthetic groups, although they are structurally similar.

The great majority of seed proteins are metabolically inactive and serve merely as food reserves for use by the growing embryo during germination. Metabolically active proteins constitute a small amount of the total, but they are extremely important to the developing and germinating seed. As enzymes, they catalyze all metabolic processes in digestion, translocation, and utilization of stored food reserves; no growth can occur without them. The *nucleoproteins* are another extremely important form of active protein and are molecules of enormous size (the molecular weight may number in the millions). They are formed by protein-nucleic acid linkage and are a huge conglomerate molecule composed of a protein, a pentose sugar, a cyclic nitrogen compound (purine or pyrimidine), and phosphoric acid. If the pentose sugar is deoxyribose, the resulting nucleoprotein is called *deoxyribonucleic acid* (DNA); if the pentose sugar is D-ribose, the nucleoprotein is called *ribonucleic acid* (RNA). Both these forms have been emphasized for their function in protein synthesis and their crucial role in the structure and function of chromosomes, genes, and life itself.

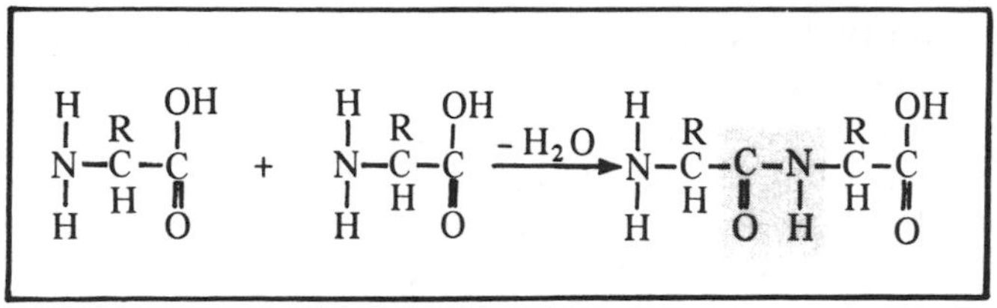

Figure 3.4. *Formation of a peptide bond. Two amino acids combine with loss of water, resulting in a dipeptide. The peptide link in the dipeptide is indicated within the tinted area at right.*

Proteins are stored in the seed in units known as protein bodies that are 1–20μ in diameter bounded by a lipoprotein unit membrane. These are somewhat like starch grains in size and shape and are usually a mixture of different proteins. They may be visualized as the kind of units that are deposited in the albuminous aleurone layers of cereal seeds and that, during seed germination, play an important role both as food reserves and as hydrolytic enzymes to aid in starch breakdown.

An insight into seed proteins has been provided by the work of Osborne (1924) with wheat seeds. He purified four proteins from wheat and classified them on the basis of their solubility, two of which were metabolically active (*globulin* and *albumin*), and two that were nonactive (*glutelin* and *prolamine*). Osborne considered these to be the only proteins occurring in seeds. Crocker and Barton (1957) described them as follows:

Seed albumins are soluble in water at neutral or slightly acid pH and are coagulated by heat. Examples are leucosins of cereal grains, legumelin of various pulse seeds, and ricin of rice. These are mainly enzymes.

Seed globulins are soluble in saline solutions, but not in water and are generally more difficult to coagulate with heat than are animal globulins. Their solubility is modified by combined acids and concentration of saline solutions. Globulins are found predominately in dicot seeds such as legumes. Examples are legumin, vignin, glycinin, vacilin, and arachin.

Seed glutelins are soluble in aqueous or saline solutions or ethyl alcohol, but can be extracted with strong acid or alkaline solutions. Glutelins are found in most cereal seeds. Examples are glutenin of wheat and oryzenin of rice.

Seed prolamines are soluble in 70–90% ethyl alcohol, but not in water; however, the salts with acids and bases are soluble in water. They are found only in cereal seeds: gliadin of wheat and rye and zein in maize are examples. Upon hydrolysis they yield proline, glutamic acid, and ammonia.

In general, glutelins and prolamines form the major components of seed protein that are metabolically inactive and associated with structural architecture. Seed proteins generally have a high nitrogen and proline content and are low in lysine, tryptophan, and methionine.

OTHER CHEMICAL COMPOUNDS FOUND IN SEEDS

Tannins

Although one usually thinks of tannins as occurring in other plant parts, especially the bark, they also occur in seeds, particularly in the seed coat structures. They are found in the testa of cocoa and bean seed (Bonner and Varner 1965).

Tannins have been used since the dawn of civilization for removing hair from animal skins during the tanning process. They are naturally occurring compounds of high molecular weight (500–3000), containing enough phenolic, hydroxyl, or other suitable groups to enable them to form effective cross-links between proteins and other macromolecules. This characteristic gives them a unique ability to tie up proteins and inhibit their enzymatic activity, and is the basis for their use in the tanning process.

Alkaloids

The death of Socrates gives us a clue about the nature of alkaloids: Socrates died from drinking a cup of hemlock containing the alkaloid poison coniine. Better-known alkaloids that occur in plants or their seeds are morphine from poppy fruits, strychnine from seeds of *Strychnos nux vomica,* atropine from the deadly nightshade, and colchicine from meadow saffron. Other extremely common alkaloids are caffeine from coffee and tea, nicotine from tobacco leaves, and theobromine from cacao.

Alkaloids are complex cyclic compounds containing nitrogen. Most are white solids; however, nicotine is a liquid at room temperatures.

Glucosides

While most glucosides are found in the vegetative organs of plants, some do appear in seeds. Examples of glucosides in seeds and vegetative plant parts are salicin in bark and leaves of willows; amygadlin in seeds of almonds, peaches, and plums; sinigrin in black mustard; aesculin in horse chestnut seed; and quercitron in the bark of oak (Bonner and Varner 1965).

Glucosides are formed by a reaction between a sugar (usually glucose) and one or more nonsugar compounds. In their pure state, they are mostly crystalline, colorless, bitter, and soluble in either water or alcohol. Some of the glucosides, such as saponin (which comes from tung seed), are highly poisonous to both man and animals.

Phytin

Phytin, the insoluble mixed potassium, magnesium, and calcium salt of myoinositol hexaphosphoric acid, is the major form of phosphorus storage in seeds. In cereal grains, phytin is generally associated with the protein bodies in the aleurone layer and is more or less absent from the protein bodies of the cotyledons. During the germination of seeds, phosphatases that hydrolyze phytin increase severalfold. The phytase activity is highest in the scutellum and aleurone layers. Since a large proportion of the seed's phosphate, magnesium, and potassium are present in phytin, much of the seed's subsequent metabolism is dependent on the hydrolysis of phytin and the concomitant release of magnesium and potassium ions. In lettuce seeds, 50% of the total phosphorus is tied up in phytin.

Hormones

The term *hormone* is used to designate certain organic compounds that, when present in minute quantities, exert important regulatory effects on the metabolism of either plants or animals. They are of enormous importance in both plants and animals. A well-known animal hormone is adrenalin, secreted by the adrenal glands, which exerts a marked influence on both the heart and vascular system. Many important plant hormones also occur in seeds. They are variously designated as phytohormones, growth hormones, growth substances, and growth regulators.

Gibberellins. The presence of gibberellins in higher plants was demonstrated by Radley (1956), who was able to induce tall growth in dwarf peas by giving them an extract from normal tall plants. It is now agreed that gibberellins are a normal constituent of green plants, as well as seeds. They appear to have a major physiological role in both seed development and seed germination. They also exert an influence on floral induction and initiation.

The best-known gibberellin is gibberellic acid (GA_3), although at least 50 others have been identified. Gibberellins are produced commercially by fermentation of fungal cultures (*Gibberella* spp.).

Cytokinins. Another group of compounds that occurs in seeds and exerts hormonal influence is the cytokinins (kinins). They were first discovered in coconut milk (Van Overbeek et al. 1941). Fifteen years later, kinetin was purified and the chemical structure identified. The first naturally occurring cytokinin to be extracted from seeds was zeatin.

Cytokinins appear to be necessary for cell growth and differentiation; perhaps this is the basis for their influence in promoting seed germination. They have also been shown to inhibit senescence in leaves (Richmond and Lang 1957) and to regulate the flow of chemicals through the plant system (Mothes et al. 1959).

Inhibitors. There is growing conviction among plant physiologists that dormancy—whether in seeds, buds, tubers, or other plant parts—is regulated by a balance, or interaction, between endogenous inhibitors and growth promoters such as gibberellins and auxins. Dormin (Cornforth et al. 1965; Thomas et al. 1965) and abscisin II (Ohkuma 1965) were preliminary names given to abscisic acid (ABA), which is reportedly instrumental in inhibiting the formation of the leaf abscission layer and (particularly) the onset of winter dormancy in deciduous plants. Another inhibitor that influences seed dormancy is coumarin. Ethylene gas can both inhibit and promote seed germination, and is considered a hormone (Crocker et al. 1935), as is the growth regulator, maleic hydrazide (Meyer et al. 1960). Chapter 6 contains a comprehensive discussion of germination inhibitors.

Vitamins

From the standpoint of human physiology, vitamins qualify as growth regulators and cannot be sharply distinguished from hormones, since they are necessary for the diet but are required in small, often minute, quantities. Although the role of vitamins in animal life and their almost universal occurrence in green plants have long been recognized, their role in plant growth is not so well known. Chemically, vitamins represent a very heterogeneous group.

In contrast to animals, who are dependent on green plants for their vitamin requirements, green plants are self-sufficient for their vitamin needs (although certain parts of a plant may depend on other parts for their vitamin supply). All known vitamins or their immediate precursors are synthesized in higher plants. Whether all are necessary in a plant's metabolism is not certain. However, specific roles for certain vitamins have been determined. Thiamine appears to be necessary for the developing embryo and endosperm of seeds in some species. It is also needed for normal root development. The basis for both needs seems to be the role of thiamine in continued cell division, which is rapidly occurring in these parts. In the case of both roots and the developing seed, thiamine is produced in the vegetative plant parts or by the cotyledon and translocated into the areas where needed. Biotin and ascorbic acid are apparently involved in respiration processes in seeds. The role of biotin is not known, but ascorbic acid appears to be responsible for regulating the oxidation-reduction potential during germination.

Questions

1. What types of chemical compounds are found in seeds?
2. List the factors that influence the chemical composition of seeds.
3. Do you think seeds are more important as food or as raw materials for industrial products?
4. Make a list of seeds that are useful in the production of (a) drugs and (b) industrial products.
5. How does the nutritional value of seeds differ from the food value of other sources of protein, carbohydrates, and oils?
6. Why is the presence of phytates in seeds important to the role of seeds as human and animal food?
7. Describe three environmental factors that can contribute to alterations in seed chemical composition.
8. Detail the differences between amylose and amylopectin and emphasize how the enzymes, α- and β-amylase contribute to their hydrolysis.
9. What are the two most common unsaturated fatty acids found in most oil seeds?
10. Name the four principal classifications for seed proteins and list the differences between each group.

General References

Allen, S. E., and G. L. Terman. 1978. Yield and protein content of rice as affected by rate, source, method, and time of applied N. *Agronomy Journal* 70:238–242.

Altman, D. W., W. L. McCuistion, and W. E. Kronstad. 1983. Grain protein percentage, kernel hardness, and grain yield of winter wheat with foliar applied urea. *Agronomy Journal* 75:87–91.

Auld, D. L., B. L. Bettis, and M. J. Dial. 1984. Planting date and cultivar effect on winter rape production. *Agronomy Journal* 76:197–200.

Austin, R. B. 1966. The growth of watercress *Rorippa nasturtium-aquaticum* L. (Hayek) from seed as affected by the phosphorus nutrition of the mother plant. *Plant and Soil* 24:113–120.

Baenziger, P. S., R. L. Clements, M. S. McIntosh, W. T. Yamazaki, T. M. Sterling, D. J. Sammons, and J. W. Johnson. 1985. Effect of cultivar, environment, and their interaction and stability analyses on milling and baking qualities of soft red winter wheat. Crop Science 25:5–8.

Bloor, W. R. 1928. Biochemistry of fats. *Chemical Review* 2:243–300.

Bonner, J., and J. E. Varner, eds. 1965. *Plant Biochemistry*. New York: Academic Press.

Campbell, C. A., H. R. Davidson, and G. E. Winkleman. 1981. Effect of nitrogen, temperature, growth stage and duration of moisture stress on yield components and protein content of Manitou spring wheat. *Canadian Journal of Plant Science* 61:549–563.

Canvin, D. T. 1965. The effect of temperature on the oil content and fatty acid composition of the oils from several oil seed crops. *Canadian Journal of Botany* 43:63–69.

Cary, E. E. 1981. Effect of selenium and cadmium additions to soil on their concentration in lettuce and wheat. *Agronomy Journal* 73:703–706.

Cassman, K. G., A. S. Whitney, and R. L. Fox. 1981. Phosphorus requirements of soybean and cowpea as affected by mode of N fertilization. *Agronomy Journal* 73:17–22.

Coffelt, T. A., and D. L. Hallock. 1986. Soil fertility responses of Virginia-type peanut cultivars. *Agronomy Journal* 78:131–137.

Cornforth, J. W., B. V. Milborrow, G. Ryback, and P. F. Wareing. 1965. Identity of sycamore "dormin" with abscisin II. *Nature* 205:1269–1270.

Crane, J. M., W. Calhoun, and T. A. Ayres. 1981. Seed and oil characteristics of fertilized meadowfoam. *Agronomy Journal* 73:255–256.

Crocker, W., and L. V. Barton. 1957. *Physiology of Seeds*. Waltham, Mass.: Chronica Botanica Company.

Crocker, W., A. E. Hitchcock, and P. W. Zimmerman. 1935. Similarities in the effects of ethylene and the plant auxins. *Contributions from Boyce Thompson Institute* 7:231–248.

DeDatta, S. K., W. N. Obcemea, and R. K. Jana. 1972. Protein content of rice grain as affected by nitrogen fertilizer and some triazines and substituted ureas. *Agronomy Journal* 64:785–788.

Diepenbrock, W., and G. Geisler. 1979. Compositional changes in developing pods and seeds of oilseed rape (*Brassica napus* L.) as affected by pod position on the plant. *Canadian Journal of Plant Sciences* 59:819–830.

Dudley, J. W., and R. J. Lambert. 1968. Genetic variability after 65 generations of selection in Illinois high oil, low oil, high protein, and low protein strains of *Zea mays* L. *Crop Science* 9:179–181.

Earle, F. R., J. J. Curtice, and J. E. Hubbard. 1956. Composition of the component parts of the corn kernel. *Cereal Chemistry* 23:507.

Elmore, C. D., W. I. Spurgeon, and W. O. Thom. 1979. Nitrogen fertilization increases N and alters amino acid concentration of cottonseed. *Agronomy Journal* 71:713–716.

Ene, B. N., and E. W. Bean. 1975. Variations in seed quality between certified seed lots of perennial rye grass and their relationships to nitrogen supply and moisture status during seed development. *Journal of British Grassland Society* 30:195–199.

Francois, L. E., E. V. Maas, T. J. Donovan, and V. L. Youngs. 1986. Effect of salinity on grain yield and quality, vegetative growth, and germination of semi-dwarf and durum wheat. *Agronomy Journal* 78:1053–1058.

Gbikpi, P. J., and R. K. Crookston. 1981. Effect of flowering date on accumulation of dry matter and protein in soybean seeds. *Crop Science* 21:652–655.

Glenn, D. M., A. Carey, F. E. Bolton, and M. Vavra. 1985. Effect of N fertilizer on protein content of grain, straw and chaff tissues in soft white winter wheat. *Agronomy Journal* 77:229–232.

Greaves, J. E., and E. G. Carter, 1923. The influence of irrigation water on the composition of grains and the relationship to nutrition. *Journal of Biological Chemistry* 58:531–541.

Green, A. G. 1986. Effect of temperature during seed maturation on the oil composition of low-linolenic genotypes of flax. *Crop Science* 26:961–965.

Gubbels, G. H. 1981. Quality, yield and seed weight of green field peas under conditions of applied shade. *Canadian Journal of Plant Science* 61:213–217.

Harrington, J. F. 1960. Germination of seeds from carrot, lettuce, and pepper plants grown under severe nutrient deficiencies. *Hilgardia* 20:219–55.

Harris, H. C., J. R. McWilliam, and V. J. Bofinger. 1980. Prediction of oil quality of sunflower from temperature probabilities in eastern Australia. *Australian Journal of Agricultural Research* 31:477–488.

Howell, R. W., and J. L. Carter. 1958. Physiological factors affecting composition of soybeans. II. Responses of oil and other constituents of soybeans under controlled conditions. *Agronomy Journal* 50:664–667.

Ivanov, N. N. 1933. Cause of the chemical variation of chickpea seeds. *Bulletin of Applied Botanical Genetics, Plant Breeding Series* 3(1):3–11. (*Chemical Abstracts* 27:5370, 1933).

Jones, O. R. 1984. Yield, water use efficiency, and oil concentration and quality of dryland sunflower grown in the southern high plains. *Agronomy Journal* 76:229–235.

Karathanasis, A. D., V. A. Johnson, G. A. Peterson, D. H. Sander, and R. A. Olsen. 1980. Relation of soil properties and other environmental factors to grain yield and quality of winter wheat grown at international sites. *Agronomy Journal* 72:329–336.

Kerr, R. W., ed. 1950. *Chemistry and Industry of Starch.* 2d ed. New York: Academic Press.

Ketring, D. L., C. E. Simpson, and O. D. Smith. 1978. Physiology of oil seeds. VII. Growing season and location effects on seedling vigor and ethylene production by seeds of three peanut cultivars. *Crop Science* 18:409–413.

Loneragan, J. F., K. Snowball, and A. D. Robson. 1980. Factors influencing variability in manganese content of seeds, with emphasis on barley (*Hordeum vulgare*) and white lupins (*Lupinus albus*). *Australian Journal of Agricultural Research* 41:29–37.

Majid, H. R., and A. A. Schneiter. 1987. Yield and quality of semidwarf and standard-height sunflower hybrids grown at five plant populations. *Agronomy Journal* 79:681–684.

Mathers, A. C., and B. A. Stewart. 1982. Sunflower nutrient uptake, growth, and yield as affected by nitrogen or manure, and plant population. *Agronomy Journal* 74:911–915.

Mathers, A. C., F. G. Viets, Jr., M. E. Jensen, and W. H. Sletten. 1960. Relationship of nitrogen and grain sorghum yield under three moisture regimes. *Agronomy Journal* 52:443–446.

Meyer, B. S., D. B. Anderson, and R. H. Bohning. 1960. *Introduction to Plant Physiology.* New York: D. Van Nostrand Company.

Miller, E. C. 1931. *Plant Physiology.* New York: McGraw-Hill.

Mitchell, E. M. 1930. A microchemical study of hemicelluloses of endosperms and cotyledons. *American Journal of Botany* 17:117–138.

Morrison, F. B. 1961. *Feeds and Feeding.* Ithaca, N.Y.: Morrison Publishing Company.

Mothes, K., L. E. Engelbrecht, and O. Kulajewa. 1959. Uber die Wirkung des Kinetins auf Stickstoffverteilung und Eiweiss-synthese in isolierten Blattem. *Flora: Oder Allgemeine Botanische Zeitung* 147:445–464.

Nandisha, B. S., and M. Mahadevappa. 1984. Influence of mother-plant nutrition and spacing on planting value of rice seeds (*Oryza sativa* L.). *Seed Research* 12:25–32.

Norden, A. J., E. C. Rossman, and E. J. Benne. 1952. Some factors that affect protein content of corn. *Michigan Agricultural Experiment Station Bulletin* 34:210–225.

Ohkuma, K. 1965. Synthesis of some analogs of abscisin II. *Agricultural and Biological Chemistry* 29:962–964.

Osborne, T. B. 1924. *Monographs on Biochemistry: The Vegetable Proteins.* London: Longmans, Green, and Company.

Owen, D. F. 1983. Differential response of sunflower hybrids to planting date. *Agronomy Journal* 75:259–262.

Parker, M. B., F. C. Boswell, K. Ohki, L. M. Shumand, and D. O. Wilson. 1981. Manganese effects on yield and nutrient concentration in leaves and seed of soybean cultivars. *Agronomy Journal* 73:643–646.

Peck, N. H., D. L. Grunes, R. M. Welch, and G. E. MacDonald. 1980. Nutritional quality of vegetable crops as affected by phosphorus and zinc fertilizers. *Agronomy Journal* 72:528–534.

Pikaard, C. S., and J. H. Cherry. 1984. Maintenance of normal or supranormal protein accumulation in developing ovules of *Glycine max* L. Merr. during PEG-induced water stress. *Plant Physiology* 75:176–180.

Poole, W. D., G. W. Randall, and G. E. Ham. 1983. Foliar fertilization of soybeans. I. Effect of fertilizer sources, rates and frequency of application. *Agronomy Journal* 75:195–200.

Porter, M. A., and G. M. Paulsen. 1983. Grain protein response to phosphorus nutrition of wheat. *Agronomy Journal* 75:303–305.

Raboy, V., and D. B. Dickinson. 1984. Effect of phosphorus and zinc nutrition on soybean seed phytic acid and zinc. *Plant Physiology* 75:1094–1098.

Radford, R. L., C. Chavengsaksongkram, and T. Hymowitz. 1977. Utilization of nitrogen to sulphur ratio for evaluating sulphur-coating amino acid concentration in seed of *Glycine max* and *G. soya. Crop Science* 17:273–277.

Radley, M. 1956. Occurrence of substances similar to gibberellic acid in higher plants. *Nature* 178:1070–1071.

Read, D. W. L., and F. G. Warder. 1982. Wheat and barley responses to rates of seeding and fertilizer in southwestern Saskatchewan. *Agronomy Journal* 74:33–36.

Richmond, A. E., and A. Lang. 1957. Effect of kinetin on protein content and survival of detached *Xanthium* leaves. *Science* 125:650–651.

Robinson, R. G. 1983. Yield and composition of field bean and adzuki bean in response to irrigation, compost, and nitrogen. *Agronomy Journal* 75:31–35.

Robinson, R. G., J. H. Ford, W. E. Lueschen, D. L. Rabas, L. J. Smith, D. D. Warnes, and J. V. Wiersma. 1980. Response of sunflower to plant population. *Agronomy Journal* 72:869–871.

Scott, R. K. 1969. The effect of sowing and harvesting dates, plant population and fertilizers on seed yield and quality of direct-drilled sugar beet seed crops. *Journal of Agricultural Science* 70:373–385.

Shannon, J. G., J. R. Wilcox, and A. H. Probst. 1972. Estimated gains from selection for protein and yield in the F4 generation of six soybean populations. *Crop Science* 12:824–826.

Simmons, S. R., and D. N. Moss. 1978. Nitrogen and dry matter accumulation by kernels formed at specific florets in spikelets of spring wheat. *Crop Science* 18:139–143.

Stone, J. F., and B. B. Tucker. 1968. Nitrogen content of grain as influenced by water supply to the plant. *Agronomy Journal* 61:76–78.

Stone, J. F., R. H. Griffin, II, and B. J. Ott. 1964. Irrigation studies of grain sorghum in the Oklahoma Panhandle, 1958 to 1962. *Oklahoma Agricultural Experiment Station Bulletin* B-619.

Thomas, T. H., P. F. Wareing, and P. M. Robinson. 1965. Action of the sycamore "dormin" as a gibberellin antagonist. *Nature* 205:1270–1272.

Touchton, J. T., and F. C. Boswell. 1975. Effects of boron application on soybean yield, chemical composition and related characteristics. *Agronomy Journal* 67:417–420.

Unger, P. W. 1986. Growth and development of irrigated sunflower in the Texas high plains. *Agronomy Journal* 78:508–515.

Van Overbeek, J., M. E. Conklin, and A. F. Blakeslee. 1941. Factors in coconut milk essential for growth and development of very young *Datura* embryos. *Science* 94:350–351.

Wolfson, J. L., and G. Shearer. 1981. Amino acid composition of grain protein of maize grown with and without pesticides and standard commercial fertilizers. *Agronomy Journal* 73:611–613.

4

Seed
Germination

In the germination process, the seed's role is that of a reproductive unit; it is the thread of life that assures the survival of all plant species. Furthermore, because of its role in stand establishment, seed germination remains a key to modern agriculture. Thus, especially in a world acutely aware of the delicate balance between food production and world population, a fundamental understanding of germination is essential for maximum crop production.

DEFINITION OF GERMINATION

Various definitions of seed germination have been proposed, and it is important to understand their distinctions. To the seed physiologist, germination is defined as the emergence of the radicle through the seed coat. To the seed analyst, germination is "the emergence and development from the seed embryo of those essential structures which, for the kind of seed in question, are indicative of the ability to produce a normal plant under favorable conditions" (AOSA 1991). Others consider germination to be resumption of active growth by the embryo resulting in the rupture of the seed coat and the emergence of a young plant. This definition presumes that the seed has been in a state of quiescence (see Chapter 6), or rest, after its formation and development. During this period of rest, the seed is in a relatively inactive state and has a low metabolic rate. It can remain in this state until environmental conditions trigger the resumption of growth. Regardless of which definition is preferred, it should be emphasized that one cannot actually see the process of germination unfold. Therefore, all definitions include some measure of seedling development, even though this occurs subsequent to the germination event.

Some seeds are capable of germination only a few days after fertilization and long before their normal harvesting time; others are dormant and require an extended rest period or additional development before germination can occur. Depending on the species, this period may last for only a few days or for as long as several years. Regardless of the length of time between maturity and resumption of growth, seed germination may be characterized by several general processes that are discussed in this chapter.

MORPHOLOGY OF GERMINATION

Based on the fate of the cotyledons or storage organs, two kinds of seed germination occur, and neither appears to be related to seed structure. These two types are illustrated by the germination of bean and pea seeds. Although these seeds are similar in structure, their germination patterns are quite different.

Epigeal Germination

Epigeal germination (Figure 4.1) is characteristic of bean and pine seeds and is considered evolutionarily more primitive than hypogeal germination (described below). During germination, the cotyledons are raised *above the ground* where they continue to provide nutritive support to the growing points. During root establishment, the hypocotyl begins to elongate in an arch that breaks through the soil, pulling the cotyledon and enclosed *plumule* through the ground and projecting them into the air. Afterward, the cotyledons open, plumule growth continues, and the cotyledons wither and fall to the ground.

Hypogeal Germination

Hypogeal germination (Figure 4.1) is characteristic of pea seeds, all grasses such as corn, and many other species. During germination, the cotyledons or comparable storage organs remain *beneath the soil* while the plumule pushes upward and emerges above the ground. In hypogeal germination, the epicotyl is the rapidly elongating structure. Regardless of their aboveground or belowground locations, the cotyledons or comparable storage organs continue to provide nutritive support to the growing points throughout germination.

The *coleoptile,* a temporary sheath enclosing the plumule, is associated with hypogeal germination of many species (e.g., grasses). It provides protection and rigidity to the emerging plumule as it pushes through the soil and is exposed to light. Then it stops growing and disintegrates as the plumule breaks through and continues to grow.

REQUIREMENTS FOR GERMINATION

Seed Maturity

Seeds of most species are capable of germinating long before physiological maturity (Holmes 1953; Harrington 1959; Bowers 1958; Giri 1967; Williams, 1972; Hill and Watkin 1975, Pegler 1976). For example, smooth bromegrass seeds are able to germinate only a few days after fertilization (Grabe 1956). In other cases, maximum seed germination can only be obtained if the seed is allowed to dry down slowly as it matures. This is true, for example, in soybeans where the ability to synthesize germination requiring enzymes such as malate synthase and isocitrate lyase occurred during the slow matura-

A. Epigeal Germination of Bean Seed

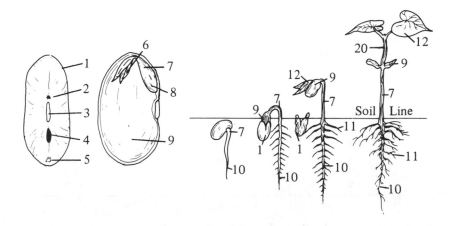

B. Hypogeal Germination of Corn Seed

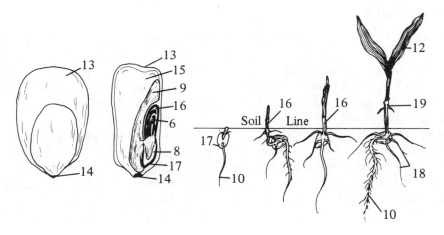

1. Seed coat	8. Radicle	15. Endosperm
2. Micropyle	9. Cotyledon (scutellum)	16. Coleoptile
3. Hilum	10. Primary root	17. Coleorhiza
4. Raphe	11. Secondary root	18. Seminal root
5. Chalaza	12. Leaf	19. Adventitious roots
6. Plumule	13. Pericarp	20. Epicotyl
7. Hypocotyl	14. Point of caryopsis attachment	

Figure 4.1. *Two patterns of seed germination and germination structures present in a dicot (bean) and a monocot (corn).*

Table 4.1. Germination of Perennial Sow Thistle and Canada
Thistle Seeds Harvested at Different Stages of Maturity
(From Kinch and Termunde (1957))

Days After Flower Opening	Percent of Germination Perennial Sow Thistle	Canada Thistle
2	0	0
3	0	0
4	4	0
5	—	0
6	34	19
7	66	37
8	70	76
9	83	88
10	—	90
11	—	80

tion process (Adams et al. 1983). Table 4.1 shows the relationship between seed maturity and germination capability of sow thistle and Canada thistle.

Environmental Factors

Water. Water is a basic requirement for germination. It is essential for enzyme activation, breakdown, translocation, and use of reserve storage material. In their resting state, seeds are characteristically low in moisture and relatively inactive metabolically. That is, they are in a state of quiescence. Thus, quiescent seeds are able to maintain a minimum level of metabolic activity that assures their long-term survival in the soil and during storage.

Moisture availability is described in various ways. *Field capacity* moisture is about optimum for germination in soil; however, germination varies among species (see Table 4.2) and may occur at soil moistures near the *permanent wilting point*. In other situations, soils are either very high in salt concentrations or are located in semiarid regions making water less available for seed germination. Even under such conditions, however, a number of crops such as basin wild rye and tall wheat grass among others are still

Table 4.2. The Germination Response of Seeds of Different Species Subjected to 14 Days of Moisture Stress (From Eslick and Vogel (1959))

Species Exhibiting Increased Germination	Species Not Affected	Species Exhibiting Decreased Germination
western wheat grass	crested wheat grass	big bluegrass
standard crested wheat grass	tall wheat grass	orchard grass
Ruby Valley milk vetch	Russian wild rye	mountain bromegrass
broadleaf bird's-foot trefoil	timothy	switchgrass
	tall oat grass	Kentucky bluegrass
	red fescue	alsike clover
	smooth bromegrass	white sweet clover
	ladino clover	sickle milk vetch
	red clover	

able to germinate when many other species cannot (Young and Evans 1981; Ries and Hoffman 1983; Roundy 1985; Dudeck and Peacock 1985).

In other situations, the initial stages of germination may even proceed with moisture available from a high-humidity environment, although such conditions are not adequate for complete germination. This is often demonstrated by a phenomenon known as *precocious germination* or sprouting when seeds actually germinate in the head or pod following rains or high-humidity conditions (Figure 4.2). The physiological factors that regulate precocious seed germination are being intensively investigated because this phenomenon causes annual soft white winter wheat crop yield losses that result in millions of dollars of lost crop revenue in the Pacific Northwest, as well as the soft white wheat areas of Michigan, New York, and Ontario, Canada.

High moisture levels may inhibit germination. For example, when moisture content

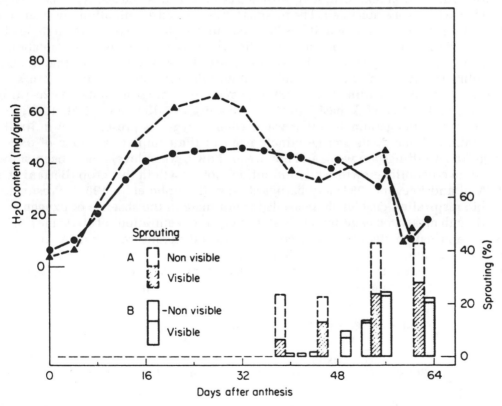

Figure 4.2. *Degree of sprouting in two lines of wheat (A and B), grown in the field, in 1977. Curves show changes in water content during the development and maturation of the grain. Vertical columns show the percentage of grains sprouting on the ear. Sprouting is visible when the coleoptile emerges through the pericarp, and is invisible when it does not emerge. Water content: A-line . . .▲ . . .; B-line • (From Mitchell et al. (1980).)*

was increased from 20% to 40%, dwarf bean germination was reported to decrease significantly (Ensor 1967). Excess moisture or flooding is also known to retard the germination of sugar beet seed (Snyder 1975), corn (Fausey and McDonald 1985), and several other species (Heydecker et al. 1969).

Air (Oxygen and Carbon Dioxide). Air is composed of about 20% oxygen, 0.03% carbon dioxide, and about 80% nitrogen gas. If one provides different proportions of these three gases under experimental conditions, it soon becomes clear that oxygen is required for germination of most species. Carbon dioxide (CO_2) concentrations higher than 0.03% retard germination, while nitrogen gas has no influence (Table 4.3).

Respiration increases sharply during seed germination. Since respiration is essentially an oxidative process, an adequate supply of oxygen must be available. If the oxygen concentration is reduced substantantially below that of air, germination of most seeds is retarded. For example, Van Toai et al. (1988) demonstrated that corn seed germination declined with decreasing oxygen concentrations below that of ambient air. Al-Ani et al. (1985) indicated that maximum seed germination for most crops including wheat, sorghum, corn, soybean, and sunflower occurred at oxygen concentrations close to those of ambient air. The reasons for decreased germination under anaerobic conditions remain unknown. It has been postulated that anoxic conditions lead to the production of increased ethanol in cells which is toxic to normal metabolism (Thomson and Greenway 1991). Other studies which have examined the ethanol toxicity hypothesis have been unable to substantiate this physiological process in seeds (Van Toai et al. 1985; Martin et al. 1991). However, there are some notable exceptions to this postulate. Rice (Takahashi 1985), barnyard grass (Kennedy et al. 1987), and other aquatic plants germinate under water where oxygen is present only in limited concentrations. Rice seeds can germinate even in the complete absence of oxygen although the seedlings are weak and abnormal. Low oxygen levels have been shown to stimulate coleoptile growth while inhibiting root growth in this crop (Bertani et al. 1981; Alpi and Beevers 1983) and barnyard grass (Rumpho et al. 1984). Presumably anaerobic respiration enables these seeds to germinate in the absence of oxygen.

Although most species germinate best in oxygen concentrations of air, some species actually germinate better at gaseous concentrations different from that in air. Cattail (*Typha latifolia*) and Bermuda grass (*Cynodon dactylon*) germination is aided by oxygen concentrations below that of air (Morinaga 1926). Since cattail is an aquatic plant, its response might be expected; however, the reaction of Bermuda grass is somewhat

Table 4.3. Effects of CO_2/O_2 Ratios on the Germination of Oat Seeds (From Forward (1958))

Gas Mixture		
Percent CO_2	Percent O_2	Percent Germination
0.0	20.9	100
16.9	17.4	93
30.0	14.7	50
35.0	13.6	31
36.8	13.2	10
38.7	12.8	1

surprising. Lettuce and onion seeds have been reported to require less oxygen for germination than most agronomic crop seeds (Siegel and Rosen 1962). On the other hand, seeds of carrot, curly dock, sunflower, cocklebur, and various cereals germinate better under oxygen concentrations above that of air (Albaum et al. 1942; Barton 1941; Morinaga 1926). The influence of carbon dioxide on seed germination is usually opposite that of oxygen. Most seeds fail to germinate if the CO_2 partial pressure is increased over the 0.03% of air; however, a decrease usually does not hinder germination. Several seeds reportedly have a minimum CO_2 requirement, notably seeds of lettuce and timothy (Mayer and Poljakoff-Mayber 1989; Thornton 1936). It should also be emphasized that these relationships can be attributed to seed dormancy. Some studies have shown that oxygen is able to decrease endogenous levels of inhibitors (Simpson 1990). When this occurs, germination responses which are not directly attributable to increased levels of respiration will be enhanced.

Temperature. Seed germination is a complex process involving many individual reactions and phases, each of which is affected by temperature. The effect on germination can be expressed in terms of cardinal temperatures; that is, *minimum, optimum,* and *maximum* temperatures at which germination will occur. The minimum temperature is sometimes difficult to define since germination may actually be proceeding but at such a slow rate that determination of germination is often made before actual germination is completed. The optimum temperature may be defined as the temperature giving the greatest percentage of germination within the shortest time. The maximum temperature is governed by the temperature at which denaturation of proteins essential for germination occurs. Not only does germination have cardinal temperatures, but each stage has its own cardinal temperature; therefore, the temperature response may change throughout the germination period because of the complexity of the germination process.

The response to temperature depends on a number of factors, including the species, variety, growing region, quality of the seed, and duration of time from harvest. As a general rule, temperate-region seeds require lower temperatures than do tropical-region seeds, and wild species have lower temperature requirements than do domesticated plants. High-quality seeds are able to germinate under wider temperature ranges than low-quality seeds.

The optimum temperature for most seeds is between 15 and 30°C. The maximum temperature for most species is between 30 and 40°C. Some species will germinate at temperatures approaching the freezing point. Certain flower, alpine, and rock garden species are notable for their low-temperature germination. Seeds of Russian pigweed are reportedly able to germinate in frozen soil or even on ice (Aamodt 1935). Figure 4.3 shows two groups of seeds—one with low optimum germination temperatures and the other with higher optimum temperature requirements.

Germination at Alternating Temperatures. Seeds of many species require daily fluctuating temperatures for optimum germination. Such diurnal periodicity is common and seems to be more prevalent in species that have not had a long history of domestication. For example, seeds of many tree and native grass species germinate best under alternating temperature conditions (Figure 4.4). The need for fluctuating temperatures during germination seems to be associated with dormancy, but alternating temperatures may accelerate germination of nondormant seeds as well. There is an adaptive advan-

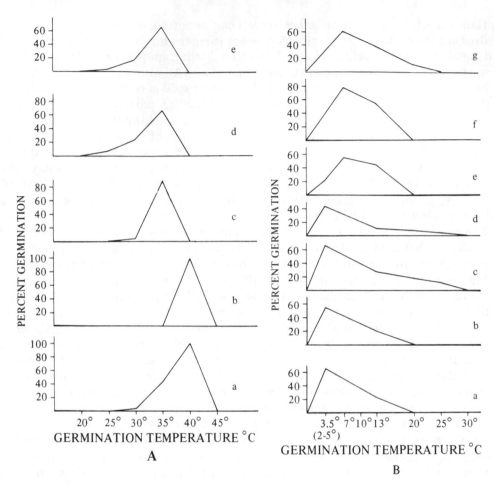

Figure 4.3. *Maximum germination of some arable weed seeds at constant temperatures: (A) species in which the non-after-ripened seeds have maximum germination at high temperatures; (a) Chenopodium rubrum, (b) Chenopodium filicifolium, (c) Datura stramonium, (d) Polygonum persicaria, (e) Gnaphalium uliginosum; (B) species in which the non-after-ripened seeds have maximum germination at low temperatures; (a) Juncus bufonius, (b) Veronica hederifolia, (c) Polygonum convolvulus, (d) Campanula rapunculoides, (e) Delphinium consolida, (f) Fumaria officinalis, (g) Arenaria serpyllifolia. (From Vegis (1963).)*

tage to the requirement for alternating temperatures. For example, vegetation insulates the soil surface against large diurnal temperature fluctuations and, in open areas, the soil serves as an insulator with deeper portions of the profile maintaining more constant temperatures. Thus, the necessity for alternating temperatures ensures that most seeds germinate on or near the soil surface in the absence of surrounding vegetation. Where alternating temperatures are needed, the range between high and low seems to be more important than the actual temperatures (Murdoch et al. 1989). McDonald et al. (1994a) reported that this range was 10°C for germination of most cool-season grass seeds.

The reason for the effects of alternating temperatures on germinating seeds is not

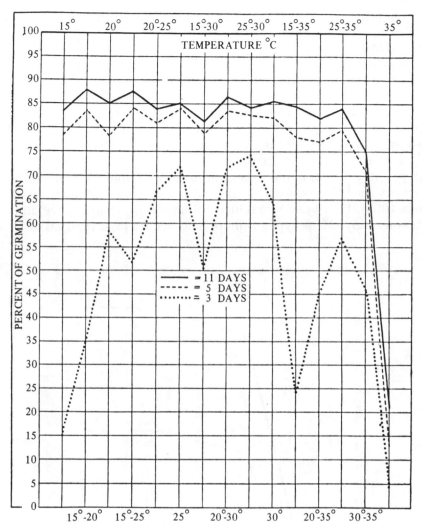

Figure 4.4. *Germination of Kentucky bluegrass under alternating versus constant temperature conditions. (From Harrington (1923).)*

known. The most commonly cited effect has been their differential influence on the sequential steps of germination. Evidence indicates that alternating temperatures cause a change in the macromolecular structure of components in the seed which, in the original form, prevents germination (Cohen 1958). It has also been suggested that alternating temperatures may create a balance of the intermediate products of respiration at the high-temperature cycle, which, though unfavorable for germination at that temperature, may promote germination at a lower one (Toole et al. 1958). Perhaps the most likely explanation is that alternating temperatures create a shift in the inhibitor-promoter balance where the inhibitor is decreased during the low-temperature cycle and the promoter increased during the high-temperature phase, which ultimately leads to germination.

   Stratification or Prechilling. The practice of preconditioning imbibed seeds in cool, moist conditions to promote germination is well known (Figure 4.5). This process is called *stratification,* a term used by the nursery industry in which propagating stocks have been stored between layers of moist sand and sawdust prior to planting in the field the next spring. Today, the term is used to denote any combination of moisture and low- (or sometimes high-) temperature preconditioning; this is a common practice in seed testing laboratories and is called *prechilling.* Stratification is discussed more extensively under dormancy in Chapter 6.

   Chilling Injury. Seeds of lima bean and cotton are subject to chilling injury during imbibition if dry seeds are exposed to temperatures of 5–15°C (Pollock and Toole 1966; Christiansen 1968). Injury can be avoided if imbibition occurs above 20°C, followed by lower temperatures. This type of injury also occurs in several other species. The

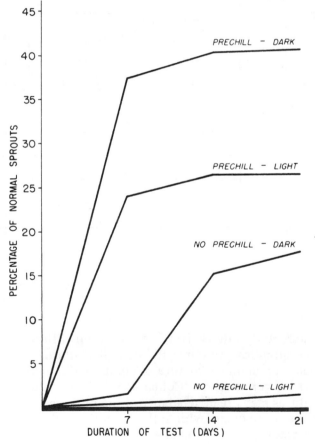

Figure 4.5. *Average rate of germination of 41 green needlegrass accessions under four conditions. (From Niffenegger and Schneiter (1963).)*

basis of low-temperature imbibition injury is not known, but the first measurable effect is that the seeds leach out more organic nutrients than uninjured seeds.

It has generally been assumed that low temperatures created a stress condition in cell walls that caused increased cell leakage during imbibitional chilling. However, Cohn et al. (1979) showed that imbibitional chilling injury was a consequence of stelar lesions in radicles of corn. Willing and Leopold (1983) proposed that low-temperature injury during imbibition interfered with membrane expansion, possibly by lowering elasticity and hindering incorporation of lipid material into the expanding cell membrane. Other studies have shown that imbibitional chilling injury is not a consequence of reduced energy metabolism in corn (Cohn and Obendorf 1976, 1978) and soybean (Ashworth and Obendorf 1980). It has been shown, however, that seeds experiencing imbibitional chilling injury leak more than those that do not (Leopold and Musgrave 1979) and that there is a correlation between the degree of chilling injury and soybean maturity class. Early-maturing cultivars are more susceptible to imbibitional chilling damage than later-maturing cultivars (Bramlage et al. 1979). The practical implications of overcoming such injury, either by presoaking or by other means, could be important for earlier planting at lower temperatures.

Light. While moisture, oxygen, and favorable temperature are essential for germination of all seeds, certain species also require light. The influence of light on germination of seeds has long been recognized. The light response of seeds of several hundred species has been studied to determine those whose germination was promoted by light, by darkness, or were indifferent to light. Almost half of the species investigated responded to light.

The mechanism of light control in seed germination is similar to that controlling floral induction, stem elongation, pigment formation in certain fruits and leaves, radicle development of certain seedlings, and unfolding of the epicotyl of bean seedlings. Both light intensity (lux or candlepower) and light quality (color or wavelength) influence germination. The nature of light quality can be illustrated by visible light, which is composed of different wavelengths of different colors (Figure 4.6). By the use of a prism, the visible light that appears colorless to the naked eye can be broken down into its several colors. A rainbow is such a phenomenon, with the water in the clouds

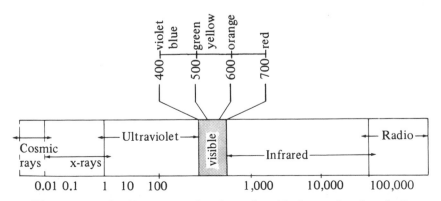

Figure 4.6. *The spectrum of radiant energy plotted on a logarithmic wavelength scale (in nanometers).*

acting as a prism and exposing the component parts of the spectrum as separate bands. Wavelengths shorter than 400 nm (in the ultraviolet region) and those longer than 700 nm (in the infrared or far-red region) cannot be seen by the naked eye.

Light Intensity. The influence of light intensity on different species varies greatly. Germination of some seeds that require light has been reported to be promoted by moonlight representing 100 lux at the most, whereas lettuce seeds need much higher light intensities. Light intensities of 1080–2160 lux (100–200 foot-candles) from indirect light in the average seed testing laboratory are probably adequate for germination of most species. Most germinators used for light-requiring seeds are equipped with supplemental light providing several thousand lux. In comparison, a bright, sunny day provides up to 108,000 lux (10,000 foot-candles) and a cloudy overcast day about 16,200 lux (1500 foot-candles). Figure 4.7 shows the effect of light intensity on seed of New Zealand browntop.

Light Quality. Figure 4.8 shows the influence of light quality on seed germination. The greatest promotion of germination occurs in the red area (660–700 nm) with a peak at 660 nm, followed by an inhibition zone in the far-red area above 700 nm. Wavelengths below 290 nm inhibit germination, with a second inhibition zone in the blue region (440 nm), whereas between 290 and 400 nm a clear-cut effect is lacking.

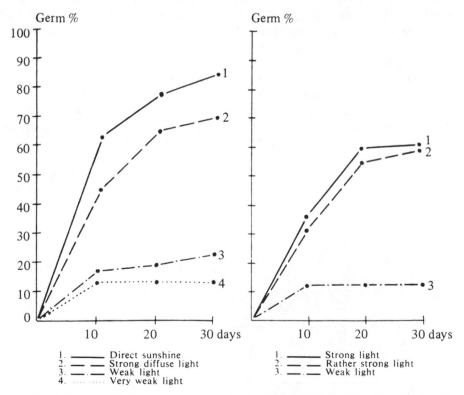

Figure 4.7. *The influence of light intensity on germination of New Zealand browntop seeds. (From Gadd (1955).)*

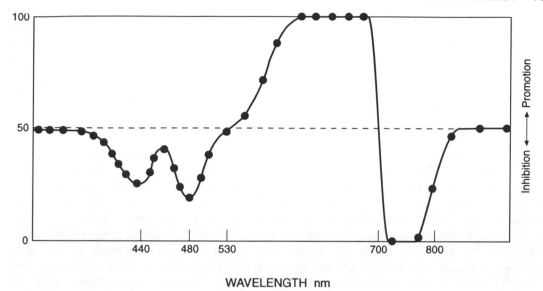

Figure 4.8. *The action of radiation of specific wavelengths (nm) in relation to the germination of light-sensitive lettuce seed. The percentages of germination in the spectrum (immediately following an exposure to red light sufficient to effect 50% germination) are indicated as ordinates, the wavelengths of the spectral light being indicated as abscissae (From Flint and McAlister (1937).)*

In 1952, a research group at the U.S. Department of Agriculture in Beltsville, MD, first reported the photoreversibility of lettuce seed germination. By alternately exposing imbibed seeds to red and far-red light, seed germination could be alternately promoted and inhibited—the ultimate effect which was independent of temperature depended on the last wavelength given. Table 4.4 illustrates that reversibility. Since 1952, this phenomenon has been discovered in many other species such as tobacco, birch, peppergrass, pine, elm, and shepherd's purse.

Table 4.4. Reversal of Light Effect by Repeated Alternations of Red (R) and Infrared (I) Irradiations on Grand Rapids Lettuce Seed. (From Toole et al. (1953))

Irradiation	Seeds germinating when exposures were made at	
	27° C Percent	7° C Percent
None (dark controls)	14	11
R	70	72
R + I	6	13
R + I + R	74	74
R + I + R + I	6	8
R + I + R + I + R	76	75
R + I + R + I + R + I	7	11
R + I + R + I + R + I + R	81	77
R + I + R + I + R + I + R + I	7	12

Photoreversible germination in seeds appears to be due to a single coupled photoreversible chemical reaction. The chemical reaction is controlled by the wavelength of light absorbed in plant cells. The photoreversible substance was first isolated from corn seedlings grown in the dark (Butler et al. 1964). It is a protein pigment known as *phytochrome*. In its most concentrated form, it appears blue, is faded by red light, and reintensified by far-red light. The far-red absorbing form (induced by exposure to red light) is believed to be the biologically active form that functions by inducing the synthesis of enzymes essential for germination.

Day Length. Seeds of several species exhibit a photoperiodically controlled germination response. The mechanism seems to be controlled by the activation of phytochrome similar to that associated with floral induction. As was the case for floral induction, there are long-day species (e.g., *Begonia evansiana* (Nagao et al. 1959) and *Betula* (Black and Wareing 1955) and short-day species (e.g., *Amaranthus retroflexus* (Kadman-Zahavi 1960) and *Atriplex dimorphostegia* (Koller 1971). Other categories are listed (Koller 1972) according to their response to photoperiod, such as seeds inhibited by prolonged irradiation (e.g., *Raphanus* (Isikawa 1962), which germinate in the total absence of light but are inhibited by continuous light. In some cases, there may be an interaction of a high-energy light reaction with phytochrome, which determines the photoperiodic effect (Ellis et al. 1986). It is further pointed out that the photoperiodic response, especially that of long-day species, is temperature-dependent (Koller 1972). For example, long-day species are especially sensitive to day length and are almost absolute in their long-day requirement. The need for long days decreases progressively as temperatures increase and, at high temperatures, germination can proceed in the absence of light.

Factors that Influence Light Sensitivity. Light sensitivity during seed germination depends on the species and variety, as well as on environmental factors before and during germination.

Age of Seed. Light influence is strongest immediately after harvest and diminishes with age of the seed and eventually disappears (Toole et al. 1957; Fujii and Yokohama 1965; Hagon 1976). This may be one reason why differing accounts of light requirements for seeds from the same species exist in the literature.

Period of Imbibition. Light sensitivity of Grand Rapids lettuce seeds decreased during the imbibition period to 10 hr, remained constant for another 10 hr, and again increased sharply (Borthwick et al. 1954). A similar response has been observed for peppergrass and tobacco seeds (Table 4.5). Tobacco seeds that ordinarily require high light intensities are able to germinate under lower intensities after four days of imbibition (Kincaid 1935). The sensitivity of loblolly pine (*Pinus taeda*) seeds to light also increases during storage (Toole et al. 1958b). Other studies have shown that light promoted the germination of *Avena fatua* and Grand Rapids lettuce seeds when seed moisture content was high but was inhibitory if seed moisture content was low (Hsiao and Simpson 1971). Holm and Miller (1972) demonstrated that restriction of water with mannitol from the seed resulted in a reduction of germination which could be overcome by a red light treatment. One general conclusion is that less light is needed to stimulate seed germination with increasing time of imbibition.

Table 4.5. Germination of Seeds of Tobacco (*Nicotiana*) and Peppergrass (*Lepidium*) Seeds Exposed to Red Light for Various Periods After 4 and 23 Hours Imbibition in Dark (From Toole et al. (1953))

Period of exposure to red light (minutes)	Germination percentage[1] after indicated hours of imbibition			
	4-hour imbibition		23-hour imbibition	
	Nicotiana	Lepidium	Nicotiana	Lepidium
0	4	0	6	11
¼	6	14	22	62
1	6	46	37	65
4	14	61	58	64
16	29	66	88	76
64	69	64	95	76

[1]*Nicotiana* germinated at 25°C. and *Lepidium* at 20° to 30° C.

Imbibition Temperature. *Lepidium* seeds, which had imbibed water at 20°C, germinated only 31% in response to light; however, when the seeds imbibed at 35°C, germination increased to 98% (Toole et al. 1955b). In addition to the fact that increased temperature during imbibition hastens metabolic events associated with germination, it also enhances water uptake, resulting in more rapid hydration and increased sensitivity to light.

Stratification. Seeds of loblolly pine (*P. taeda*) and Eastern white pine (*P. strobus*) became more responsive to light following stratification, and the increased sensitivity was roughly proportional to the duration of stratification (Toole et al. 1962).

Germination Temperature. Many data exist showing the interaction of light and temperature on germination; however, it is not possible to reduce these to a unified picture. It is clear that the response to each can sometimes be increased, decreased, or changed qualitatively by the other, while in other cases it cannot. Germination of peppergrass (*Lepidium*) (Toole et al. 1955a) and Virginia pine (*P. virginiana*) (Toole et al. 1961) seeds is increased when red irradiation is preceded by a low temperature and followed by a high one. Low-temperature treatment could substitute for the red light requirement for germination of dock (*Rumex*) seeds; and weak red light, while insufficient to promote germination alone, always promoted it in combination with a high-temperature treatment (Isikawa and Fujii 1961).

In some instances, the dependence of seed germination on light is reduced when seeds experience alternating temperatures. For example, *Agrostis tenuis* seeds have their light requirement either completely or partially removed when exposed to alternating temperatures (Toole and Koch 1977). Light can influence germination of downy brome the least at optimum temperatures and most at less suitable germination temperatures (Thill et al. 1984). In most cases, total germination is enhanced when seeds are exposed to both alternating temperatures and light (McDonald et al. 1994a).

The temperature treatment that increases the sensitivity to red light reduces the sensitivity to far-red light (Toole et al. 1956). Virginia pine (*P. virginiana*) seeds germinated well when exposed to light at 25°C but germinated weakly at either 20° or 30°C (Toole et al. 1961). Peppergrass (*Lepidium*) and tobacco (*Nicotiana*) seeds that failed to germinate at 20°C became responsive to light when temperatures were alternated between 15° and 25°C (Toole et al. 1957). These few examples illustrate the complex nature of the light-temperature interaction on germination. However, the exact mechanism of the light and temperature interaction remains unknown.

Phytochrome. The action spectrum of light-sensitive lettuce seed was discovered in the mid-1930s (Flint 1934; Flint and McAlister 1935, 1937) when it was shown that the greatest promotive response from a given irradiance occurs in the red region (600–700 nm) and the far-red (720–760 nm) portion of the spectrum causes the greatest inhibition (Figure 4.8).

The light and dark response of phytochrome is illustrated by the scheme in Figure 4.9. Red light (660 nm) exposure converts phytochrome to the physiologically active, far-red absorbing form, and germination proceeds. Exposure to far-red light (730 nm) reconverts phytochrome to the physiologically inactive red-absorbing form, and germination is blocked. Only a brief flash of either red or far-red light is necessary to convert phytochrome to the opposite form. In seeds of the princess tree (*Paulownia tomentosa*), a prolonged dark period also permits conversion of P_{FR} to P_R (Toole et al. 1958a). This conversion can also be promoted by prolonged or repeated brief exposures to high temperatures or far-red light.

We now know that all of the P_R is never completely converted to P_{FR} by red irradiation because of the overlap of the action spectra of P_{FR} and P_R (Butler et al. 1964). In far-red irradiation, phytochrome is driven strongly toward P_{FR} with about 2% remaining in the active P_R form (Figure 4.10).

Light-requiring seeds need a certain number of P_{FR} molecules to germinate and this may vary among species. In some cases, the P_{FR} requirement for germination may be so low that the seeds germinate irrespective of the light treatment. Such seeds are referred to as light-insensitive.

Despite more than 40 years of intense research into phytochrome action, we still do not know precisely how it functions. Recent advances in molecular and genetic techniques, however, are beginning to increase our understanding of phytochrome action. It is now known that phytochrome is synthesized in the seed as P_R, which is the biologically inactive form and most seeds therefore require a light stimulus to convert the P_R to P_{FR}, the biologically active form that leads to germination. The location of phytochrome in seeds has also been a matter of debate. It is generally assumed that it is associated with cellular membranes. But is there a specific part of the seed that contains the phytochrome? It has been reported that light sensitivity is confined to

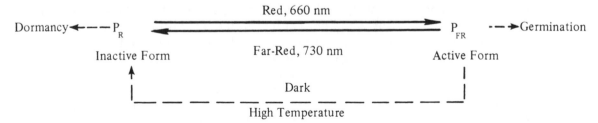

Figure 4.9. *Photoreversible phytochrome response.*

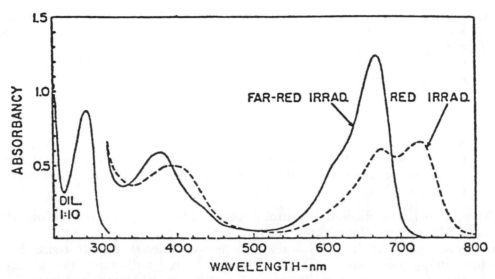

Figure 4.10. *Absorption spectra of phytochrome; measured spectra. The solution was diluted tenfold for measurements below 300 nm. (From Butler et al. (1964).)*

only one portion of the seed in certain species. For example, in *Phacelia* seeds light sensitivity is localized at the chalazal and micropylar ends. In lettuce seeds, only the micropylar end is light sensitive. These, of course, are the thinnest portion of the seed coats and may simply allow more light to penetrate to the embryo. Evidence exists (Bewley and Black 1978) that the phytochrome pigment is actually located in the embryonic axis. In addition, there now appear to be at least five differing isoforms of phytochrome based on molecular studies of phytochrome mutants (Viestra 1993). What role these have, if any, in mediating seed germination must still be determined.

While the physiological role of P_{FR} is still unclear, the following four mechanisms have been proposed:

1. P_{FR} influences gibberellin synthesis. The observation that gibberellins can substitute for light in breaking the dormancy of many light-sensitive seeds (Chen 1970) has led to the suggestion that P_{FR} increases gibberellin synthesis. It has been shown that inhibitors of GA biosynthesis delay the light-induced germination of a number of species, suggesting that gibberellins play some part in the germination of light-sensitive lettuce seeds (Jones and Stoddart 1977). An alternative hypothesis is that P_{FR} stimulates the release of endogenous gibberellins from a bound form.
2. P_{FR} selectively activates specific genes. Following activation of P_{FR}, a number of hydrolytic enzymes are known to increase (Chen and Varner 1973). It has been suggested that P_{FR} selectively exposes portions of the genome, which promotes the rapid synthesis of enzymes essential for germination. For example, cloning of phytochrome genes from moss plants has shown these to have kinase enzymatic activity (Algarra et al. 1993). The same may be true for germinating seeds.

3. P_{FR} alters membrane permeability (Bewley and Black 1978). Phytochrome is localized in cell membranes. It has been suggested, therefore, that it may have a specific influence on cell membrane permeability. If, for example, gibberellins bound to sugars are hydrolyzed and released during imbibition, they may not exhibit a physiological function until they are isolated from the sequestered form. This release may occur following a photoinductive reaction, causing P_{FR} to alter membrane permeability, resulting in the release and transfer of gibberellins to their site of action (such as the aleurone layer).
4. Kinase activity has been reported to increase (Algarra et al. 1993), which may cause a change in the balance between the pentose phosphate shunt and the glycolytic pathway, thus altering the capacity to germinate (Taylorson and Hendricks 1976).

The phytochrome-mediated control of germination has immense ecological significance. This system allows buried seeds to remain dormant (although fully imbibed) until they are exposed to light. It has also been demonstrated that light passing through a leafed canopy possesses a high percentage of far-red radiation. This retards the germination of seeds under heavy tree canopies where the light intensity is not adequate for seedling establishment. Thus, phytochrome has evolved as a survival mechanism that promotes seed germination only under conditions in which seedling establishment is most likely to succeed.

PATTERN OF SEED GERMINATION

Most seeds undergo a specific sequence of events during germination. The major events are imbibition, enzyme activation, initiation of embryo growth, rupture of the seed coat, and emergence of the seedling.

Imbibition

The early stages of imbibition or water uptake into a dry seed are a crucial period for seedling germination. Seeds are sensitive to rapid imbibition, chilling, and anoxia; common events that occur with the increasing emphasis on early planting and conservation tillage. Imbibition is an essential process initiating seed germination. It is the first key event that moves the seed from a dry, quiescent, dormant organism to the resumption of embryo growth. Consequently, an orderly transition of increased hydration, enzyme activation, storage product breakdown, and resumption of embryo growth must occur. Imbibition is not merely an uncontrolled physical event: it is now recognized that chemical conformation events, seed coat effects, and seed quality factors govern the directed flow of water into the seed (Leopold and Vertucci 1989). Thus, any consideration of seed germination physiology and its resultant impact on stand establishment should initially focus on water uptake. The extent to which water imbibition occurs is dependent on three factors: (1) composition of the seed, (2) seed coat permeability, and (3) water availability.

Composition of the Seed. Seeds typically possess extremely low water potentials attributed to their osmotic and matric characteristics. These potentials may be as low as −400 MPa in the dry seed of rape, wheat, and corn (Shakeywich 1973). The low water potentials are a consequence of the relationship of water with components of the seed. Water in a soybean seed, for example, basically exists in three forms (Figure 4.11) dependent on its hydrational status (Leopold and Vertucci 1986). Below 8% moisture (region 1), water is chemisorbed to macromolecules through ionic bonding, has limited mobility, and acts as a ligand rather than a solvent. Between 9% and 24% moisture (region 2), water is weakly bound to macromolecules and begins to have solvent properties, and diffusion gradually becomes evident as water takes on the properties of a bulk solution. Above 24% moisture (region 3), water is bound with negligible energy, and its properties are similar to bulk water. The macromolecular surface of the seed is considered fully wetted at 35% moisture (Leopold and Vertucci 1986).

The Q_{10} value of imbibition is 1.5–1.8, indicating that imbibition is a physical process not dependent on metabolic energy and related to the properties of the colloids present in seed tissues. This is supported by the observation that imbibition occurs equally in both dead and live seeds (Figure 4.12). The principal component of seeds that is responsible for the imbibition of water is protein. Proteins are zwitterions that exhibit both negative and positive charges that attract the highly charged polar water molecules. The difference in imbibition of protein-containing seeds compared with

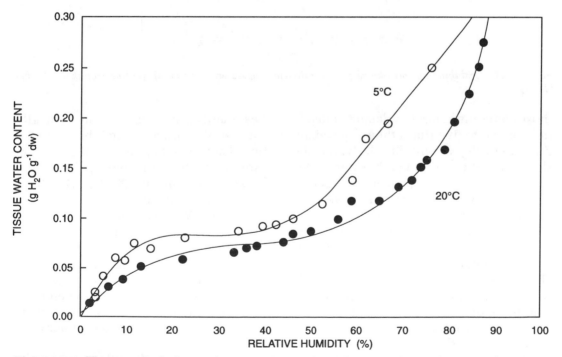

Figure 4.11. *Water sorption isotherms of ground soybean pellets at 5° and 20°C, showing the reverse sigmoidal curve typically found for seeds. Relative humidity was controlled by saturated salt solutions. (From Leopold and Vertucci (1986).)*

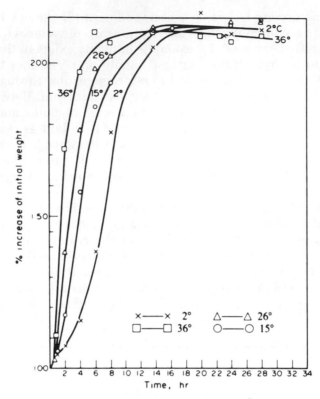

Figure 4.12. *Imbibition of heat-killed peas at different temperatures. (From Mayer and Poljakoff-Mayber (1989).)*

those containing starch is demonstrated by soybeans and corn. Soybean seeds typically imbibe two to five times their dry weight in water, while corn seeds imbibe 1.5 to 2.0 times their dry weight. Similarly, the observation that the embryo of cereal seeds can absorb about twice as much water as the endosperm can be explained by the greater proportion of protein present in the embryonic tissues (Chung and Pfost 1967). Other studies on five legume species have shown that the embryonic axis characteristically had higher water-binding properties than the cotyledons (Vertucci and Leopold 1987). McDonald et al. (1988b) showed that the soybean embryonic axis hydrated more rapidly and completely than any other seed part due to its higher protein composition.

Other chemical constituents of seeds also contribute to imbibition. The mucilages of various seeds increase imbibition as do the cellulose and pectins located in the cell walls. In contrast, starch molecules have little impact on imbibition—even when large quantities of starch are present, as in seeds of many grasses. Starch, because of its uncharged structure, only attracts water at very acid pH or after high-temperature treatment—conditions that do not occur in nature.

Seed Coat Permeability. Entry of water into seeds is greatly influenced by the nature of the seed coat (or pericarp). Water permeability is usually greatest at the micropylar area where the seed coat is ordinarily quite thin. The hilum of many seeds also permits easy water entry. The same appears to be true in many grass seeds which possess a pericarp that completely surrounds the seed except at the pedicel (black lawith the formation of the black layer. This open, porous structure results in a more rapid hydration of the embryo that progressively moves from the radicale to the coleoptile end (McDonald et al. 1994b). A slower, more progressive wetting front simultaneously moves through the seed coat and into the endosperm (Figure 4.13).

Seeds of certain species have special tissues around these natural openings that prevent water entry and contribute to hard seed coat (impermeable to water) dormancy (see Chapter 6). This hard-seededness has been attributed to small elongated pores and a high density of waxy material embedded in the testa epidermis (Calero et al. 1981; Tully et al. 1981; Yaklich et al. 1986; Mugnisjah et al. 1987). In other instances, hard-seededness has been attributed to the presence of lipids, tannins, and pectic substances in the seed coat (Denny 1917). The incidence of hard-seededness is both genetically and environmentally controlled and is greatest when seed maturation occurs under high temperature, high humidity (Dassou and Kueneman 1984; Potts et al. 1978), and dry (Hill et al 1986) conditions. McDonald et al. (1988a) showed that the soybean seed coat delayed water uptake during the first eight hours of soaking. It also directed water movement both tangentially and radially to the embryonic axis. The radial movement of water was attributed to the presence of a radicle pocket that possessed a high incidence of hourglass cells (Figure 4.14). These cells may increase the water storage capacity around the radicle tip, ensuring a ready source of water for turgor pressure essential for germination.

The seed coat acts as a semipermeable membrane, permitting the entry of water and certain solutes while restricting others. For example, the leakage of the inositol pinitol from imbibing soybean seeds is substantially greater than for sugars such as sucrose/raffinose, stachyose, fructose, and glucose (Nordin 1984). Other studies have shown soybean seed leachate to contain high concentrations of K^+ and protein (Seneratna and McKersie 1983) and enzymes such as malate dehydrogenase (Duke et al. 1983). This leakage is more pronounced if the seeds are deteriorated (Duke et al. 1986; Seneratna and McKersie 1983; Schoettle and Leopold 1984). This differential permeability may be the result of the ionization of the acidic and basic groups of the membrane lipids (Weatherby 1943). Such membranes repel ions of similar charge while attracting those of opposite charge. Thus, un-ionized molecules of liquids or bases do not permeate as readily as ionized molecules. The cuticle of wheat seeds has been observed to carry an electric charge that influences the permeability to various solutes (Brown 1932).

Water Availability. The environmental forces that determine the rate of water imbibition by seeds are complex. The ability to imbibe water is dependent on cell water potential and is a result of three forces:

1. Cell wall matric forces. Cell walls and intracellular inclusions such as mitochondria, ribosomes, and spherosomes are characterized by the presence of membranes. These membranes possess charges that attract water molecules and contribute to the total cell water potential.

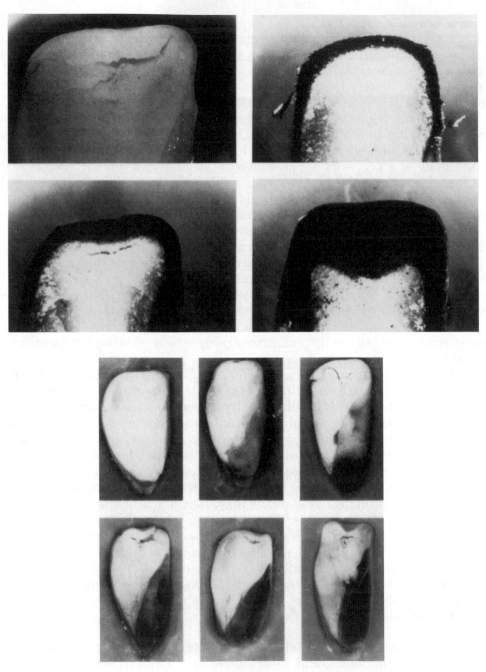

Figure 4.13. *Movement of water through the seed coat and embryo of a corn seed. The top four photographs illustrate the movement of iodine into the corn seed endosperm after 0, 6, 24, and 48 hours. The bottom six photographs illustrate staining of the corn seed embryo with nitroblue tetrazolium chloride after 0, 3, 6, 15, 24, and 48 hours of soaking in water (From McDonald et al. (1994).)*

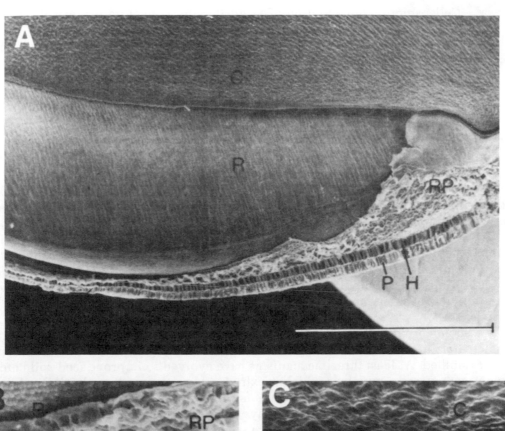

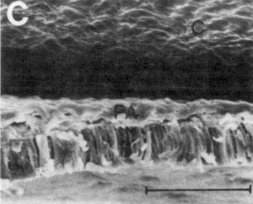

Figure 4.14. *Scanning electron micrograph of* **(A)** *Cotyledon (C), radicle (R), radicle pocket (RP), hourglass cells (H), and palisade cells (P);* **(B)** *Enlarged view of radicle (R), radicle pocket (RP), hourglass cells (H), and palisade cells (P);* **(C)** *The seed coat opposite the hilum (note the absence of hourglass cells showing the palisade cells (P), parenchyma (PA), and cotyledon (C). Bars represent 1.0 mm. (From McDonald et al. (1988b).)*

2. Cell osmotic concentration. The greater the concentration of soluble compounds, the greater the attraction for water.
3. Cell turgor pressure. As water enters a cell, it exerts a swelling force on the cell wall called *turgor pressure*. Unlike the cell wall matric forces and osmotic concentration that attract water molecules into a cell, turgor pressure, which is a result of the restraining force of the cell wall, tends to retard water absorption.

The soils in which seeds are planted also exhibit their own water potentials. The physical properties of soils determine the retention and conductivity of water. For example, it is well known that soils heavy in clays are able to absorb water more vigorously and retain it longer than those possessing high quantities of sand. In effect, the seed water potential must compete with the soil water potential for imbibition to occur. Initially, the difference between seed and soil water potential is quite large. However, as imbibition continues, this difference is reduced in the immediate vicinity of the seed. If it were not for the conductive ability of soils, imbibition would be quickly halted. However, most soils exhibit a high degree of hydraulic conductivity that replenishes the available water surrounding the seed as it continues the process of imbibition. This is important since seeds are sessile and a continuous flow of water is essential for maximum imbibition. Seeds rarely attract water from beyond 10 mm in most soils.

Associated with seed and soil water potential is the degree of seed-soil contact. The greater the intimate contact of the seed with the soil, the greater the amount of water imbibed. At least three mechanisms have evolved to improve seed-soil contact. Some seeds possess mucilage which is extruded from the epidermal cells of the seed as it imbibes water. The mucilage serves to increase the contact of the seed with the soil by increasing the number of pathways through which water may be absorbed by the seed. Another mechanism to enhance seed-soil contact is to increase the amount of seed contact with a specific volume of soil. This can be accomplished by altering seed coat configuration. Seeds possessing textured seed coats are more likely to have a greater seed-soil contact than smooth-coated seeds, and thus they will imbibe water more rapidly. A final factor is seed size. Small seeds possess a greater surface area to volume ratio than large seeds. This greater surface area permits them to have access to a greater amount of water than large seeds, which means that they will hydrate more rapidly.

Enzyme Activation

Dry seeds are characterized by a remarkably low rate of metabolism that is undoubtedly attributable to their low moisture content (which may be as low as 5 to 10% in unimbibed seeds). As soon as the seed becomes imbibed, however, marked changes in its metabolism occur. A triphasic pattern of water uptake has been demonstrated during the germination of most seeds (Figure 4.15). Enzyme activation begins during Phases I and II of imbibition. During Phase II, the seed undergoes many processes essential for germination. Increased respiration and leakage of nutrients from the im-

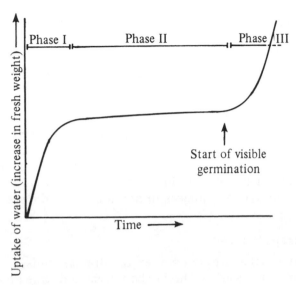

Figure 4.15. *Triphasic pattern of water uptake by germinating seeds. (From Bewley and Black (1978).)*

bibed seed leads to loss of dry weight. Finally, in Phase III, root elongation is observed. The root becomes functional during this phase and is responsible for the increased water uptake noted in Phase III.

The process of enzyme activation during Phase II of water imbibition serves to break down stored tissue, aid in the transfer of nutrients from storage areas in the cotyledons or endosperm to the growing points, and trigger chemical reactions that use breakdown products for the synthesis of new materials.

In monocots, gibberellins are released from the scutellum and move through the endosperm of the aleurone layer. There they trigger the synthesis of hydrolytic enzymes including α-amylase, ribonuclease, endo-β-glucanase, and phosphatase. The release of these hydrolytic enzymes results in a degradation of the endosperm and endosperm cell walls. This process begins near the scutellum, with subsequent hydrolysis occurring to the sides and upwards through the endosperm. The mechanism for this degradation is not fully understood but scientists have offered two hypotheses. One suggestion is that since enzymes are released from the scutellum, they arrive at the aleurone layer closest to the scutellum before they reach those aleurone cells farther removed. This enables the aleurone cells closest to the scutellum to initiate hydrolytic enzyme synthesis first. The other postulate is that a symmetrical release of enzymes from the aleurone occurs. However, the scutellum, which is also rich in other enzymes, simply provides an additional source of hydrolytic enzymes to digest the endosperm tissue. Regardless of the mechanism, the germinating seed now possesses many products that are a consequence of the activity of hydrolytic enzymes released from the aleurone and scutellum.

Trigger Chemical Reactions Used in the Synthesis of New Materials. It has already been demonstrated that the early events that occur during the enzyme activation phase include the synthesis of storage product enzymes such as α-amylase, ribonuclease, and phosphatase. These events are mediated by the hormone gibberellic acid. However, during this lag phase many other hydrolytic enzymes are formed. This synthesis is preceded by an increase in the endoplasmic reticulum, ribosomes, and ribosomal RNAs—essential components of the enzyme-synthesizing machinery. Such enzymes as ATPase, phytase, proteases, lipase, and peroxidase all increase during enzyme activation. In many cases, these enzymes further break down storage compounds or take the hydrolyzed products and resynthesize them into molecules essential for new growth. In other instances, the enzymes are indirectly involved in the synthesis of new materials by ensuring that the energy molecules essential for these reactions are present in the cytoplasm in adequate amounts. Clearly, the enzyme activation phase (which appears to be a metabolically inactive period based on water uptake) is one of the most essential and dynamic phases in preparation of the seed for embryonic axis elongation.

Breakdown of Storage Tissues

Generally, enzymes that break down carbohydrates, lipids, proteins, and phosphorous-containing compounds are the first to be activated during Phase II of water uptake by seeds. The controlling mechanism for directing this storage tissue degradation has not yet been precisely elucidated.

Since the embryonic axis requires energy for growth, storage compounds must be hydrolyzed to soluble forms, translocated from the endosperm to the embryo, and transformed to energy molecules that can be immediately utilized by the embryonic axis. The endosperm initially becomes rich in soluble products such as glucose and maltose. These are then absorbed by the scutellum. In the scutellum, glucose and maltose are transformed by a series of *in situ* enzyme reactions to form sucrose. (Sucrose itself is not hydrolyzed within the scutellum because the essential enzymes are not present.) The sucrose molecule is then transported to the adjacent embryonic axis as the principal energy molecule for growth. This is illustrated in Figure 4.16.

In dicots, the hormonal regulation of storage product degradation is not as clear as in monocots. This may be due to the absence of an aleuronelike tissue that synthesizes hydrolytic enzymes. Additionally, the role of hormones in dicot seed germination has been debated. In some instances, gibberellins are known to trigger hydrolytic enzyme synthesis, but the degree of activation is never as great as that noted in cereals. Some investigators believe that dicot seed germination is mediated by the growing embryonic axis. As the axis continues to grow, it incorporates breakdown products into the synthesis of new compounds. This reduces the concentration of compounds in the cotyledons, which in turn stimulates the hydrolysis of other storage reserves for use by the embryonic axis. Should this stimulation prove to be too great, and hydrolyzed storage products begin to accumulate, a feedback mechanism may be operative that retards further storage reserve hydrolysis. This is illustrated in Figure 4.17.

The mobilization and transfer of nutrients in most dicot seeds is through the conductive tissue of the cotyledons to the growing embryonic axis. In bean seeds, for example, the hydrolysis of protein bodies occurs at the center of the cotyledons and then moves toward the outside of the cotyledons. Like grass seeds, storage compounds must initially be hydrolyzed to a soluble form before they can be translocated.

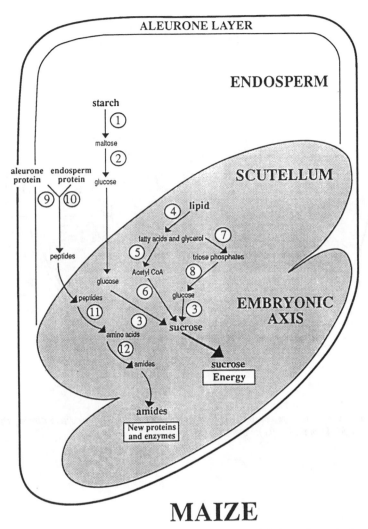

Figure 4.16. *Breakdown of storage compounds and mobilization of their products during the germination of corn seeds. (From McDonald (1994).)*

Carbohydrate Metabolism. Amylopectin and amylose are hydrolyzed by α- and β-amylase enzymes. These enzymes split either starch structure yielding the disaccharide maltose, which is then split into two monosaccharide glucose units. Some glucose units are converted into the highly mobile disaccharide sucrose for translocation to other sites, after which it is reconverted to glucose or used directly in synthesis.

Glucose may be further broken down by respiration. The first step is known as *glycolysis,* which yields two pyruvic acid molecules. These, then, are completely broken down into CO_2 and water by a series of reactions known as the *tricarboxylic acid* (Krebs) *cycle*. The reactions of glycolysis occur in the cytoplasm, while those of the Krebs cycle occur in the mitochondria. Both processes yield energy as ATP (Figure 4.18).

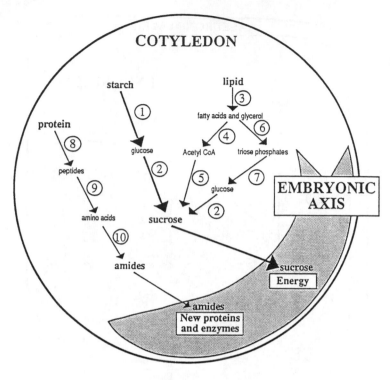

Figure 4.17. *Breakdown of storage compounds and mobilization of their products during the germination of soybean seeds. (From McDonald (1994).)*

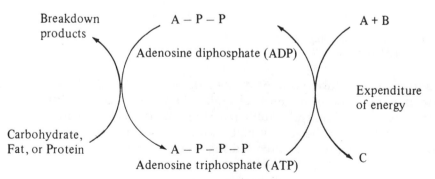

Figure 4.18. *In the breakdown of storage substances by the cells, energy is stored in the form of special "energy-rich" phosphate bonds in the molecule of adenosine triphosphate (ATP). The ATP is then available for energy to transport or to drive reactions as A + B → C during which ATP loses some energy and phosphate and is changed to ADP.*

As early as 1890, Haberlandt, in his study of rye germination, suggested that events occurring in the growing points of the embryo initiate starch hydrolysis by the amylases. The responsible substance, however, was not identified. It is now believed that gibberellins are produced in the growing root-shoot axis and scutellum and migrate to the aleurone layers of the endosperm, where they stimulate synthesis of amylase and other hydrolytic enzymes (Jones and Armstrong 1971; Paleg 1960; Jacobsen and Varner 1967; Briggs 1963; Radley 1967). These enzymes hydrolyze the starch and other storage products of the endosperm into solutes that are transported through the scutellum to the growing points, where they nourish the growing seedling. Limited work with noncereal species indicates that similar systems occur in other kinds of seeds.

Lipid Metabolism. For oil-bearing seeds, the first step in utilization of reserve storage materials also involves a hydrolytic reaction using the enzyme lipase to cleave the ester bonds and yield free fatty acids and glycerol. The free fatty acids are further degraded by one of two processes—either α-or β-oxidation (Figure 4.19).

α-**oxidation.** This form of free-fatty acid breakdown plays a minor role in germinating seeds, but has been observed in peanuts and sunflower. It involves the successive loss of one carbon atom and CO_2 by the aid of fatty-acid peroxidase and aldehyde dehydrogenase enzymes.

β-**oxidation.** The major means of fatty acid breakdown during germination is by β-oxidation, with the aid of β-oxidase, yielding acetyl coenzyme A and energy in the form of ATP. Acetyl coenzyme A may then enter the Krebs cycle for complete oxidation to CO_2, H_2O, and energy (ATP), or it may enter the *glyoxylate cycle* for conversion by a complicated series of reactions to yield sucrose. Sucrose is then available for translocation to growing sites and for use in biosynthesis.

Protein Metabolism. Relatively little is known about the exact nature of reserve protein breakdown during seed germination. However, proteinases, the proteolytic enzymes—endopeptidases, carboxypeptidases, aminopeptidases (Bond and Bowles 1983)—are involved in cleaving the peptide bonds of the protein and releasing the amino acids. Protineases have been observed in many seeds and increase rapidly during germination (Ryan 1973). Proteolytic enzymes differ in their specificity in attacking certain peptide linkages. The specificity seems to be determined by adjacent end groups on the protein molecule, the type of amino acid side chains, and the relationship between the size of the molecule and the number of free end groups. Hydrolysis of protein bodies evidently occurs differently in cereals and legumes. In soybeans, protein bodies are hydrolyzed by internal digestion (Wilson 1987) while in corn the hydrolysis begins at the surface of the protein bodies (Torrent et al. 1989).

As proteins are broken down during seed germination, there is a concomitant rise in amino acids and amides in the cotyledon, followed by protein synthesis in the growing parts of the embryo. Initial evidence of such changes has been observed in germinating lettuce seed as a decrease in the ratio of protein to soluble nitrogen (Klein 1955).

After free amino acids are released from their protein complexes, they may be further broken down by any of three processes: (1) deamination to give ammonia and a carbon skeleton that subsequently enters various metabolic processes, (2) transamination enzymes to yield ketoacids, which enter the Krebs cycle for further breakdown to

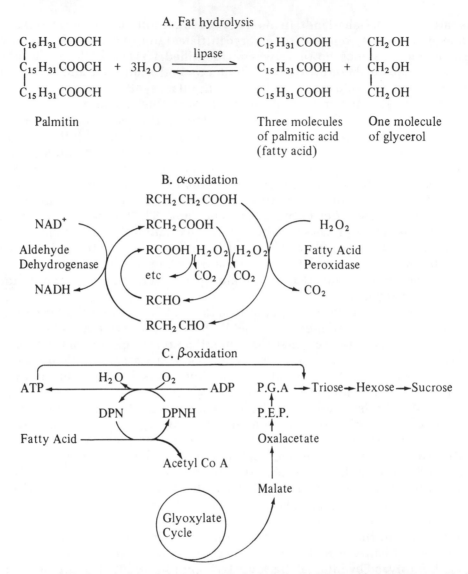

A. Fat hydrolysis

$$\begin{matrix} C_{16}H_{31}COOCH \\ | \\ C_{15}H_{31}COOCH \\ | \\ C_{15}H_{31}COOCH \end{matrix} + 3H_2O \underset{\text{lipase}}{\rightleftharpoons} \begin{matrix} C_{15}H_{31}COOH \\ C_{15}H_{31}COOH \\ C_{15}H_{31}COOH \end{matrix} + \begin{matrix} CH_2OH \\ | \\ CH_2OH \\ | \\ CH_2OH \end{matrix}$$

Palmitin   Three molecules   One molecule
of palmitic acid   of glycerol
(fatty acid)

B. α-oxidation

C. β-oxidation

Figure 4.19. *Pathways of fat degradation: (A) fat hydrolysis, (B) α-oxidation, (C) β-oxidation.*

CO_2, H_2O, and energy (ATP), or (3) direct utilization for synthesis of new proteins in other parts of the germinating seed. Regardless of the pathway followed, the breakdown products are eventually available for use by the developing seedling.

New proteins are synthesized on the surface of ribosomes in the cell cytoplasm. Here messages are received from the DNA of the nucleus by means of messenger RNA. These messages determine the kind and sequence of amino acids to be combined into new proteins.

Phosphorus-Containing Compounds. About 80% of the phosphorus in seeds is stored as calcium, magnesium, or manganese salts of inositol hexaphosphate, or phytin. The other 20% is in organic compounds such as nucleotides, nucleic acids, phospholipids, phosphorylated sugars, phosphoproteins, and a trace of inorganic phosphate. During seed germination, phytin is broken down, releasing inorganic phosphorus for synthesis of other phosphorus-containing compounds. Its breakdown is catalyzed by phytase, a phosphatase enzyme.

Nucleotides, such as ADP and ATP, are complex phosphorus-sugar compounds that have the ability to store and release chemical energy in their phosphate bonds. During germination, these are converted back and forth as energy is released and used again. Each time that energy is released, inorganic phosphorus is also released.

Although the phospholipids are not a major food reserve in the seed, they are present, and are broken down during germination. Their breakdown scheme is similar to that for other lipids, and the catalyzing enzyme is a phospholipase.

Initiation of Embryo Growth

Studies have been conducted on the developmental changes that occur in seeds as they initiate embryo growth. Because monocot and dicot seeds are different in their morphological structure, it is not surprising that the alterations these seeds exhibit are unique. Monocot seeds generally display a germination pattern similar to that exhibited by corn. During the first 120 hours of germination, there is a marked decrease in the dry weight of the endosperm with a concomitant increase in the dry weight of the embryonic axis (Figure 4.20). These changes are, in part, a reflection of decreases in total nitrogen and insoluble protein that occur in the endosperm, and the subsequent translocation of these compounds to the emerging axis. Similar changes would be anticipated following endosperm starch hydrolyzation to maltose and then to glucose, which is then enzymatically altered to sucrose and translocated to the axis. Figure 4.12 illustrates many of the metabolic changes that occur during the early stages of germination.

Similar transitions are also apparent in dicot seeds. In *Vigna sesquipedalis* (cowpea), the major storage tissues, the cotyledons undergo a decrease in dry weight as the hypocotyl and subsequently the epicotyl, show increases (Figure 4.21). Like corn, soluble carbohydrates, soluble nitrogen, and nucleic acid phosphorous levels decrease in the cotyledons and are found in the emerging embryonic organs of the hypocotyl, roots, epicotyl, and plumule. These events show that the storage tissues function primarily as reservoirs from which the emerging axis can draw nutrients for rapid germination and emergence.

Protrusion of the Radicle

The actual protrusion of the radicle, which signals that the germination process is complete, can be accomplished through either cell elongation or cell division. In general, cell elongation precedes cell division. This is true in lettuce, corn, barley,

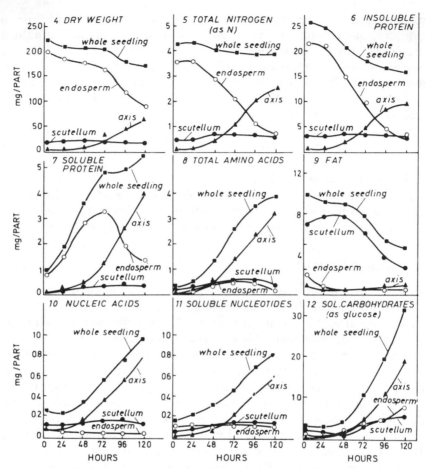

Figure 4.20. *Changes in content of various components in different parts of Zea mays during germination and growth. (From Ingle et al. (1964).)*

beans, and peas (Berlyn 1972; Brown 1932; Foard and Haber 1966; Haber and Luippold 1960). In pine seeds, both cell division and elongation occur simultaneously (Berlyn 1972), while in cherry seeds, cell division precedes cell elongation (Pollock and Olney 1959). Thus, the protrusion of the radicle through the seed coat is initiated by cell elongation, followed by cell division in most seeds. The cell elongation process reportedly occurs in two stages in seeds of peas, barley, and vetch (Rogan and Simon 1975). In the first stage, a slow elongation of the radicle occurs without any increase in dry weight and only a small increase in fresh weight. This stage may represent active cell wall preparation for the synthesis of new wall materials during later elongation. The second phase is a rapid elongation of the radicle with marked increases in both fresh and dry weight accompanied by rapid mobilization of nutrients into the radicle. These events lead to protrusion of the radicle and the seed's change from an autotrophic to a heterotrophic organism.

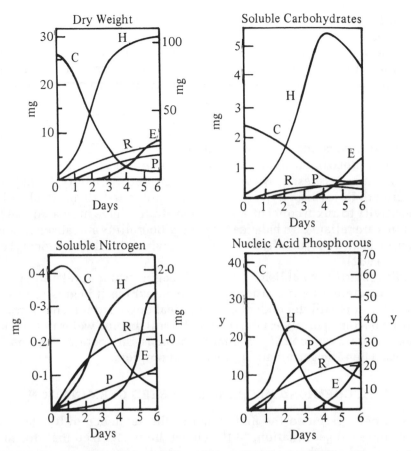

Figure 4.21. *Changes in content of dry weight, soluble carbohydrates, soluble nitrogen, and nucleic acid phosphorus of Vigna sesquipedalis during germination. (From Oota et al. (1953).)*

Seedling Establishment

Seedling roots, of course, grow down and shoots grow up and these gravitropic responses are mediated by auxin (indole-3-acetic acid). Auxin is synthesized in the corn root apical meristem (Feldman 1981), moves forward through the stele to the root cap, and is preferentially redistributed toward the lower side of the cap. At that point, the redistributed auxin moves through the root cortical cells to the zone of elongation. There, the auxin on the lower side of the root inhibits cell growth and causes a downward curvature (Evans et al. 1986). Thus, while the meristem is the site of auxin synthesis, the root cap functions in redistributing auxin asymmetrically so that roots grow downward (Young et al. 1990). With respect to corn shoot growth, corn seedlings must rely upon the elongation of the mesocotyl and coleoptile before the leaves can unfold above the soil surface. While both tissues elongate, it is the mesocotyl that is primarily responsible for the greatest elongation. Auxin is believed to be synthesized in the coleoptile tip and moves basipetally to the mesocotyl where it has its greatest effect (Bandurski et al. 1984; Parker and Briggs 1990).

In legume seeds, rapid growth of the seedling hypocotyl occurs immediately following radicle protrusion. Cavalieri and Boyer (1982) demonstrated that water potentials decreased from the root to the hypocotyl crook and radially from the stele to the cortex. They also showed that water potential in the hypocotyl elongating zone was not uniform and was most negative immediately below the hypocotyl crook. The control of gravitropic responses for elongating seedling hypocotyls is not yet as clear as it appears to be in corn. Traditional concepts suggest that there is a redistribution of auxin in which lower tissues of a horizontal hypocotyl elongate more in response to a higher auxin gradient than upper tissues. However, Rorabaugh and Salisbury (1989) found that differential growth in soybean hypocotyls was due not to redistribution but differences in tissue sensitivity to auxin concentration. Other studies have indicated that hypocotyl elongation rates are reliant on a balance between gibberellins and abscisic acid concentrations (Bensen et al. 1990) or ethylene evolution, which inhibits hypocotyl elongation (Seyedin et al. 1982).

The seedling starts to establish itself when it begins water uptake and photosynthesis. Initially, it undergoes a transition stage during which it begins to produce some of its own food, but is still dependent on food breakdown from reserve storage tissue. As the seedling becomes firmly established in the soil, begins water uptake, and manufactures most of its own food, it gradually becomes independent of the exhausted storage tissues. Then the germination process is complete.

A BIOCHEMICAL MECHANISM OF SEED GERMINATION

In 1968, Amen presented the first model of the sequence of biochemical events that occur during seed germination. At the outset, he recognized that this model must be a generalized scheme because every plant species is unique; some require light while others need cold temperatures to induce germination. His attempt to compile the literature of that time into a workable model was noteworthy and provided the impetus and stimulus for research workers to test the model for its validity. Surprisingly, little new or additional information has been gained to test Amen's hypothetical scheme. His model, however, highlights many of the events that must occur during germination and is, therefore, important to consider in a discussion of seed germination.

According to Amen, germination is governed by a balance between inhibitors and promotors. When inhibitors are present in physiologically greater concentrations than promotors, dormancy ensues. A trigger agent such as light or temperature is necessary to either inactivate or decrease the level of inhibitors in seeds. When this occurs, a germination stimulant such as gibberellic acid can exert its promotive influence and the process of germination is initiated. Figure 4.22 shows that gibberellins increase the de novo synthesis of proteolytic enzymes, α-amylase and ribonuclease, while activating the release of β-amylase. The amylases hydrolyze starches, providing the essential sugars for germination. Ribonucleases are essential for nucleic acid hydrolysis, which subsequently can be used to recode new RNA species for later germination processes. The proteolytic enzymes serve in conjunction with cellulases to degrade

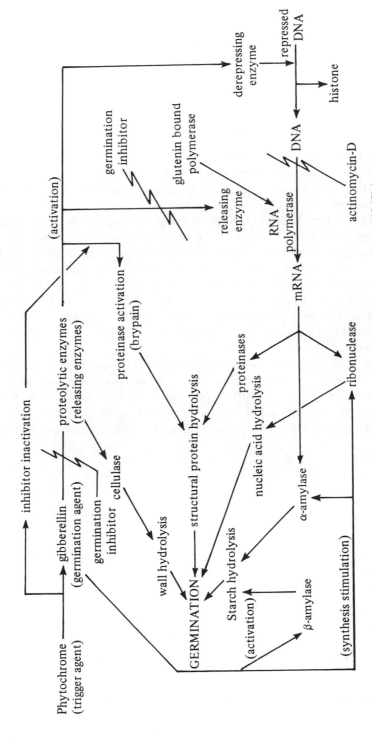

Figure 4.22. *Possible biochemical pathways in the initiation of seed germination. (From Amen (1968).)*

cell walls, which is an essential first step in the loosening of the seed coat prior to radicle protrusion. Other studies have indicated that gibberellins may derepress portions of the DNA molecule, allowing transcription of mRNA which then can be translated to form many of the de novo enzymes responsible for germination. Many of these events have been shown in germinating seeds, and this model serves to integrate these independent processes into a unified concept of germination.

OVERALL CHANGES DURING GERMINATION

It is useful to consider the overall changes that occur during seed germination. Because of the great number of species available and their diversity, innumerable examples could be cited. However, certain patterns of development occur that are common to all seeds.

Dry Weight. During the first few days, the germinating seedling undergoes a net loss in dry weight (Figure 4.23) due to the high respiration rate and some exudation and leakage through the seed coat. As the epicotyl and radicle begin to grow, they gain weight rapidly, at the expense of the cotyledon or endosperm, which undergoes a rapid weight loss as the food reserves are broken down and transported away. In epigeal species, the developing hypocotyl uses most of the initial breakdown products for rapid synthesis and growth. After the third to fifth days, its growth diminishes, and the remaining breakdown products are utilized for synthesis in the epicotyl and developing root system.

The storage carbohydrates, fats, and proteins decrease rapidly in the cotyledon and endosperm during germination. Their degradation products are translocated to the growing points, where they accumulate prior to utilization in further synthesis.

The increase in nucleic acids, especially DNA, closely parallels their increase during cell division. Each cell division results in a doubling of the nuclear material.

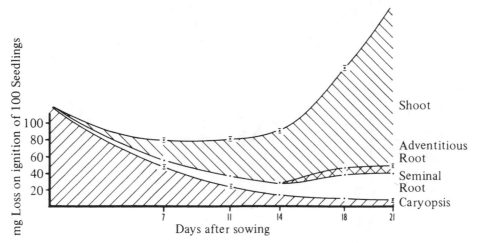

Figure 4.23. *Changes in weight distribution in perennial ryegrass seed through 21 days after planting. (From Anslo (1962).)*

Early hypocotyl growth is largely from cell division; therefore, DNA increases rapidly. After a few days, further growth is primarily from elongation of previously formed cells; therefore, DNA formation decreases. Growth in the plumule and radicle results equally from cell division and cell elongation; consequently, DNA increases steadily in these organs as germination proceeds.

RNA is located in both the nucleus and the cytoplasm; so its occurrence is less closely associated with nuclear division. Therefore, the RNA changes reflect the overall effect of both cell division and cell elongation.

CHEMICAL PROMOTION OF SEED GERMINATION

Gibberellins

Since about 1955, gibberellins have been known to promote seed germination of a great variety of species. Like thiourea, the gibberellins are not used extensively in routine germination tests, but may be useful in certain situations. A number of gibberellins promote germination, but the form most often used is gibberellic acid (GA_3).

Like thiourea, gibberellins can substitute for light and temperature in promoting germination. They can also promote germination of seeds not having these requirements. Formerly, the action of gibberellin and red light on imbibed seeds was assumed to be identical. However, scientists now believe that their mode of action is only partially similar (Ikuma and Thimann 1960; Fujii et al. 1960). Gibberellins are believed to be important in controlling the germination of seeds in nature (Phiney and West 1960; Koller et al. 1962). There is evidence that natural gibberellinlike substances appear during successive stages of after-ripening and germination. Gibberellinlike substances have also been isolated from seeds of beans, lettuce, and many other species. It is now well established that gibberellins play an important role in the regulation of seed germination.

Cytokinins

Another group of endogenous hormones that promote seed germination in some species are the cytokinins. Kinetin (6-furfurylaminopurine) is the best known of these. While cytokinins can break primary dormancy of some seeds, they appear to be more effective in overcoming secondary dormancy (Tilsner and Upadhyaya 1985). The mechanism of cytokinin regulation of seed germination is not known, but three possibilities have been suggested:

Transcriptional Mediation. The binding of cytokinins to ribosomal preparations has been demonstrated in wheat seed embryos (Fox and Erion 1975). This finding may suggest that cytokinins can regulate specific gene expression.

Translational Mediation. Cytokinins have been found in association with wheat embryo tRNA and always with tRNA species which recognize uridine as their initial codon letter (Burrows et al. 1970). It has been suggested that the ribosome assumes a specific configuration in response to the uridine codon which controls which tRNA species are permitted access to the codon, thus mediating the proteins to be synthesized

(Burrows 1975). In this way, cytokinins can regulate some of the initial processes of germination.

Membrane Permeability Mediation. Thomas (1977) showed that cytokinins influence many phytochrome-controlled processes. Since phytochrome is located in cell membranes and can alter membrane permeability through its reorientation, cytokinins may conceivably mediate seed germination through their effect on membrane permeability. Such a system would permit the release of gibberellins from the scutellum to the aleurone during the first stages of germination.

The exact role of cytokinins in promoting seed germination has yet to be resolved. However, cytokinins certainly can break dormancy in seeds with a chilling requirement such as sugar maple, *Leucadendron daphnoides,* and *Protea compacta,* or with a light requirement such as *Rumex obtusifolius,* lettuce, and celery.

Ethylene

Ethylene (C_2H_4) is known to stimulate seed germination of many species (in addition to its influence on fruit ripening, bud dormancy, leaf abscission, and other growth processes). Apparently, it is involved in the regulation of seed dormancy, although its effect is not limited to dormant seed. It has been shown to enhance the germination rate of aged as well as immature seed. Ethylene is thought to be involved in regulating auxin levels in dormant seed, and is known to be released during seed germination of several species. It can also act synergistically with gibberellin and red light in the germination of lettuce seeds. There has also been a suggestion that ethylene and nitrate interact in stimulating seed germination (Saini and Spencer 1986; Egley 1984). Ethylene is now used in seed testing to break the dormancy of peanut seeds.

Hydrogen Peroxide

The stimulating effect of hydrogen peroxide (H_2O_2) on seed germination and subsequent seedling vigor has been observed in a number of species, including many conifers, legumes, tomatoes, and barley. This chemical acts as a respiration stimulant that accelerates the breakdown of reserve food substances, thus providing a more rapid supply of energy and materials for synthesis in the growing points. Hydrogen peroxide also has disinfectant properties and can be used to disinfect seed to prevent mold growth on the seed and germination media. A laboratory test has been developed using the hydrogen peroxide treatment as a rapid viability check of conifer seed.

Auxins

As noted in Chapter 3, auxins and other growth regulators are a universal component of plants and a common constituent of seeds. In view of their presence in seeds and their role as plant growth regulators, it is not surprising that they can influence germination as well. The best-known auxin, indoleacetic acid (IAA), has been shown to increase lettuce seed germination and to be temperature dependent. A definite relationship between IAA concentration and its action has not yet been established, although high concentrations inhibit germination, while low concentrations are generally promotive or ineffective. There is also evidence that IAA interacts with light in influencing seed germination. Most studies have shown that auxins have little promotive effect on seed germination.

Potassium Nitrate

Potassium nitrate (KNO_3) is the most widely used chemical for promoting seed germination. Solutions of 0.1 to 0.2% KNO_3 are common in routine germination testing and are recommended by the Association of Official Seed Analysts and the International Seed Testing Association for germination tests of many species. Figure 4.24 shows the effect of KNO_3 on New Zealand browntop seed germination.

Most seeds that are sensitive to KNO_3 are also sensitive to light. At one time it was assumed that KNO_3 substituted for light, but now it is believed that the light sensitivity is only increased. On the other hand, KNO_3 can completely counteract light inhibition of rice grass (*Oryzopsis*) seed germination.

KNO_3 may interact with temperature in influencing seed germination in some

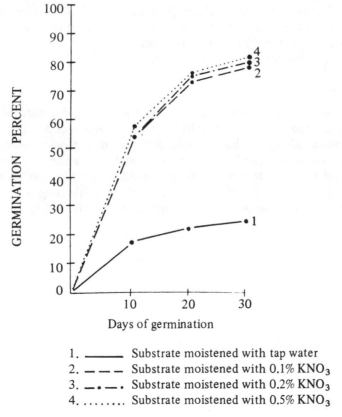

1. ——— Substrate moistened with tap water
2. — — — Substrate moistened with 0.1% KNO_3
3. —.—. Substrate moistened with 0.2% KNO_3
4. .... Substrate moistened with 0.5% KNO_3

Figure 4.24. *The effects of different potassium nitrate concentrations on germination of New Zealand browntop seed. (From Gadd (1955).)*

species. This has been demonstrated in a comprehensive study of several native range grasses (Toole 1938). It is also known to act synergistically, or cooperatively, with gibberellic acid (Hashimoto 1958) and kinetin (Ogawara and Ono 1961) in inducing germination of tobacco seed.

However, KNO_3 may be detrimental to the germination of some seeds. Reported inhibition of lettuce seed germination by KNO_3 is somewhat surprising in view of the light and low temperature requirements of this species (AOSA 1991). Other studies have shown that the stimulatory effects of KNO_3 affect the respiratory system directly (Adkins et al. 1984) and that these are more pronounced in light than in darkness (Hilton 1984). Still others have proposed that nitrate acts to stimulate oxygen uptake (Hilton and Thomas 1986) or serves as a cofactor of phytochrome (Hilhorst 1990). These studies demonstrate that the precise role of KNO_3 in stimulating seed germination remains to be determined.

Thiourea

Although not used in routine germination tests like KNO_3, thiourea does promote germination of many seeds. Unlike KNO_3, thiourea is able to substitute for the light and temperature requirements of germination, perhaps replacing the light and temperature requirements for the physiological processes occurring naturally during after-ripening. Thiourea has been reported to replace the growth promotor that develops naturally during stratification (Villiers and Wareing 1960).

Other Chemicals

Other chemicals may also promote seed germination under natural conditions. For example, the host plants of two parasitic species, witchweed (*Striga*) and mistletoe (*Orobanche*), secrete a substance that stimulates the germination of nearby seeds of these parasitic species. Scopoletin, a phenolic compound, better known as a growth inhibitor, has been observed to promote germination of aged white mustard (*Sinapsis alba*) seed (Libbert and Lubke 1957). The phenomenon of allelopathy suggests that many plants produce chemicals that either delay or prevent seed germination (Karssen and Hilhorst 1992).

OTHER FACTORS THAT AFFECT GERMINATION

Osmotic Pressure

High osmotic pressures of the germination solution make imbibition more difficult and retard germination (Rodger et al. 1957). The ability of seeds to germinate under high osmotic pressure differs with variety as well as with species, but all are affected. Some halophyte species, however, actually have seeds that germinate better in salinities ranging from 0.2 to 1.7 mol l^{-1} (Ungar 1987). Mannitol solutions at varying concentrations have long been used to obtain different osmotic pressures for studying germination responses (Dotzenko and Dean 1959). More recently, other research studies have employed polyethylene glycol (PEG) to subject seeds to varying osmotic pressures. These osmoticums have been used to minimize the effects of salt toxicity (Bliss et al. 1984).

Hydrogen-Ion Concentration (pH)

Germination can proceed over a wide range of hydrogen-ion concentrations. The germination of almost all species occurs readily between pH values of 4.0 and 7.6 (Justice and Reece 1954).

Presoaking

Presoaking seeds in water has been suggested as a means to speed up germination, and it has accelerated germination of several grass species (Chippendale 1934; Johnston 1964). Another approach has been to expose seeds to an osmoticum so that their moisture content is increased but remains below the moisture level needed to initiate germination—a process called *osmoconditioning*. In both cases, the seeds are usually redried prior to germination. The basis for the acceleration is uncertain; however, it is likely that hydrolytic processes begin during presoaking, and the resulting simple sugars that are released can be utilized for synthesis immediately upon germination. It is also possible that membrane repair may occur by enzymes activated during the hydration process.

Another well-known benefit from presoaking at moderate temperatures (20°C) is the protection it provides from chilling injury during the subsequent germination at lower temperatures (5° to 15°C) (Pollock and Toole 1966). In such cases, the seed is not redried but is held moist until germination.

Conversely, soaking seeds may be detrimental to germination capacity and should not be used unless definitely needed. Kidd and West (1918) reported reduction of germination and seedling growth as a result of soaking. Prolonged soaking has been found to cause injury to seeds of many species (Tilford et al. 1924). Presoaking injury to soybeans has been attributed to low oxygen concentration inside the seed (Resuhr 1941), although more often to leaching or to exosmosis of enzymes and nutrients from the seeds. The latter view is supported by leachate conductivity experiments.

Effect of Low Temperature

Frost and cold nights may cause considerable seed injury prior to harvest. After extensive study of low-temperature injury to corn seed, Rossman (1949a) concluded that the extent of injury depended on: (1) temperature level, (2) duration of exposure, (3) moisture content of the seed, (4) physiological maturity of the seed, (5) husk protection, and (6) variety. Of these, seed moisture was most crucial.

If conditions are otherwise favorable, seeds can withstand extremely low temperatures. Air-dried (10–12% moisture) seeds of wheat, barley, and several other species have even withstood temperatures of liquid nitrogen (−192°C) for 24 to 110 hours (Brown and Escombe 1898; Adams 1905) and temperatures of liquid hydrogen (−252°C) for 6 hours (Thiselton-Dyer 1899) without injury. Corn seed at 20% moisture survived temperatures of −7°C, but at 25% moisture considerable injury occurred (Rossman 1949a).

Rossman (1949b) reported several additional responses to freezing in corn seed. These include: (1) viability of soaked seed was decreased more than freshly harvested seed with comparable moisture content, (2) seedlings were killed outright, (3) repeated freezing and thawing cycles were less injurious than continuous freezing when the total times were equal, (4) the first freeze-thaw was more injurious than subsequent ones, (5) rate of thawing had no effect on the viability of frozen seeds, and (6) slow drying of frozen seeds was less injurious than fast drying, presumably because it gives the protoplasm time to recover (Rossman 1949b).

As seed matures in the field, it becomes increasingly hardy. The cold resistance of corn seed with 15 to 20% moisture was adequate to avoid injury from ordinary autumn freezing (Kiesselbach and Ratcliff 1920). When maturing oat seeds were collected from fields before and after frosts, seeds in the milk to soft dough stage were easily injured, but resistance increased rapidly thereafter (Fryer 1919).

Mature soybean seeds are surprisingly resistant to freezing, and can be stored at 18% moisture up to sixteen months at $-10°C$ without significant injury (Robbins and Porter 1946). Soybean seeds with 30 to 32% moisture have been stored at $-29°C$ for extended periods of time without injury, whereas sorghum seeds at 16 to 19% moisture did not survive under the same conditions.

Judd et al. (1982) reported a curvilinear relationship between soybean seed moisture and the temperature at which freezing occurred. Immature seeds (in green pods) at 65% moisture content were not injured by freezing temperatures of $-2°C$; seeds at physiological maturity (in yellow pods) at 55% moisture showed significant reductions in germination following an eight-hour exposure at $-7°C$; while germination of seed in brown pods at 35% moisture was reduced.

The mechanism of low-temperature injury is not completely understood. It is generally attributed to formation of ice crystals in the inter- and intracellular components, which then expand and rupture the physical and functional integrity of the cell membranes. Extreme cell desiccation and precipitation of the protoplasm (when water is withdrawn by intercellular ice formation) may also contribute to freezing injury (Luyet and Gehenio 1938; Stanwood 1986).

Radiation

Exposure to gamma radiation usually retards seed germination. Its effect differs in varieties as well as species of seeds, and is more pronounced at higher temperatures and higher seed moisture contents. Germination damage has been observed in bluegrass, peanuts, and onions at 10 Krad radiation; wheat and sorghum at 20 to 40 Krad; corn at 40 to 80 Krad; and crimson clover and radish at 80 Krad (Justice and Kulik 1970). Germination losses up to 38.1% were reported. Gamma radiation up to 80 Krad caused retardation of shoot and root growth up to one-half to two-thirds that of the untreated seed (Justice 1967).

The studies cited above reflect the effect of gamma radiation above 10 Krad radiation. Sahid and Soemartono (1974) reported results of gamma radiation studies on rice seeds showing that the speed of germination was actually increased by exposures to radiation below 10 Krad. Like other investigators, however, they found that stronger exposures depressed germination.

Mechanical Damage

Mechanical damage during harvesting, conditioning, and handling is a major problem in the seed industry, especially for fragile, large-seeded species. Injury symptoms may be of several different forms: (1) gross damage to the seed coat that is easily visible, such as "splits" and seed coat cracks; (2) internal damage, visible only after seed germination; (3) microscopic breaks, especially of the seed coat, that reduce performance and increase susceptibility to microorganisms; and (4) cryptic (hidden, or internal) injury, which is probably physiological in nature and reduces the germination vigor, lengthens the time to maturity, and reduces yield (McDonald 1985).

Sources of Mechanical Injury. Injury may occur from any kind of physical abuse. While threshing abuse probably is the most serious, seed conditioning and handling also contribute to seed injury. The effect of all individual impacts the seed receives appears to be cumulative in total seed damage; thus, it is essential always to minimize physical impact to assure the best germination.

Influence of Moisture Content. Susceptibility to mechanical injury increases as the moisture content of the seed decreases; however, safe moisture contents vary among species. Large-seeded legumes are particularly sensitive, and excessive injury begins to occur at moisture contents below 15%. Germination of navy beans was raised from 36 to 76% by increasing the moisture before threshing from 11 to 16% (Dexter 1966). Small-seeded grasses and legumes can safely be threshed at moisture levels as low as 8 to 10%. The cereals are more susceptible, and injury occurs at moisture levels below 11 to 12%.

Influence of Genotype. The genetic control of susceptibility to mechanical injury in some crop species has been clearly established (Atkin 1958; Davis 1964). Twelve percent moisture is the reported critical moisture for distinguishing between tolerant and susceptible navy bean lines (Dorrel and Adams 1969).

Questions

1. What similarity do you see between seed germination and seed deterioration?
2. What type of germination is characteristic of the greater number of species—hypogeal or epigeal?
3. What environmental factors are required for seed germination? Which do you consider to be of the greatest importance?
4. How can rice seed germinate in the complete absence of air?
5. How would you define the optimum temperature for seed germination of a given species?
6. What is the origin and meaning of the term *stratification?*
7. Do you think that the average light-supplied seed germinator can deliver optimum light intensity for lettuce seed germination?
8. What factors influence the sensitivity of light-requiring seeds to light intensity? How is the light action on seed germination similar to that on floral induction?
9. What groups of chemicals can promote seed germination? Which are used to promote germination in routine seed testing?
10. Why are soybeans more resistant to low-temperature injury than most seeds?
11. What is imbibition pressure?

12. How does most water enter the seed during imbibition?
13. What is the importance of mitochondria during seed germination?
14. What is the importance of nucleotides (ADP and ATP) during germination?
15. What are the breakdown pathways for starch, proteins, and lipids during germination?
16. What is the relationship between gibberellic acid and amylase during barley seed germination?
17. Cite three potential mechanisms to explain the mode of action of phytochrome in promoting seed germination.
18. List the factors that influence the availability of water necessary for seed germination.
19. Cite at least three processes that occur during the outwardly tranquil enzyme activation stage of seed germination.
20. Contrast and provide examples of trigger and germination agents.

General References

Aamodt, O. S. 1935. Germination of Russian pigweed seeds in ice and on frozen soil. *Scientific Agriculture* 15:507–508.

Adams, C. A., M. C. Fjerstad, and R. W. Rinne. 1983. Characteristics of soybean seed maturation: Necessity for slow dehydration. *Crop Science* 23:265–268.

Adams, J. 1905. The effect of very low temperatures on moist seeds. *Scientific Proceedings of the Royal Dublin Society* 11:1–7.

Adkins, S. W., G. M. Simpson, and J. M. Naylor. 1984. The physiological basis of seed dormancy in *Avena fatua*. IV. Alternative respiration and nitrogenous compounds. *Physiologia Plantarum* 60:234–238.

Al-Ani, A., F. Bruzau, P. Raymond, V. Saint-ges, J. M. Leblanc, and A. Pradet. 1985. Germination, respiration and adenylate energy charge of seeds at various oxygen partial pressures. *Plant Physiology* 79:885–890.

Albaum, H. G., J. Donnelly, and S. Korkes. 1942. The growth and metabolism of oat seedlings after exposure to oxygen. *American Journal of Botany* 29:388–395.

Algarra, P., S. Linder, and F. Thummler. 1993. Biochemical evidence that phytochrome of the moss *Ceratodon purpureus* is a light regulated protein kinase. *FEBS Letters* 315:69–73.

Alpi, A., and H. Beevers. 1983. Effects of oxygen concentration on rice seedlings. *Plant Physiology* 71:30–34.

Amen, R. D. 1968. A model of seed dormancy. *Botanical Review* 34:1–31.

Anslo, R. C. 1962. A quantitative analysis of germination and early seedling growth in perennial ryegrass. *Journal of British Grassland Society* 17:261–266.

Ashworth, E. N., and R. H. Obendorf. 1980. Imbibitional chilling injury in soybean axes: Relationship to stelar lesions and seasonal environments. *Agronomy Journal* 72:923–928.

Association of Official Seed Analysts. 1991. Rules for testing seeds. *Journal of Seed Technology* 12(3):1–109.

Atkin, J. D. 1958. Relative susceptibility of snapbean varieties to mechanical injury of seed. *Proceedings of the American Society for Horticultural Science* 72:370–373.

Bandurski, R. S., A. Schulze, P. Dayanandan, and P. B. Kaufman. 1984. Response to gravity by *Zea mays* seedlings. I. Time course of the response. *Plant Physiology* 74:284–289.

Barton, L. V. 1941. Relation of certain air temperatures and humidities to viability of seeds. *Contributions from Boyce Thompson Institute* 12:85–102.

Bensen, R. J., R. D. Beall, J. E. Mullet, and P. W. Morgan. 1990. Detection of endogenous gibberellins and their relationship to hypocotyl elongation in soybean seedlings. *Plant Physiology* 94:77–85.

Beryln, G. P. 1972. Seed germination and morphogenesis. In: *Seed Biology*, ed. T. T. Kozlowski, pp. 223–228. New York: Academic Press.

Bertani, A., I. Brambilla, and F. Menegus. 1981. Effect of anaerobiosis on carbohydrate content in rice roots. *Biochemic and Physiologie der Planzen* 176:835–840.

Bewley, J. D., and M. Black. 1978. *Physiology and Biochemistry of Seeds in Relation to Germination.* New York: Springer-Verlag.

Black, J. N., and P. F. Wareing. 1955. Growth studies in woody species, VII. Photo-periodic control of germination in *Betula pubescens* Ehr. *Physiologia Plantarum* 8:300–316.

Bliss, R. D., K. A. Platt-Abia, and W. W. Tomson. 1984. Changes in plasmalemma organisation in cowpea radicle during imbibition in water and NaCl solution. *Plant Cell Environment* 7:601–606.

Bond, H. M., and D. J. Bowles. 1983. Characterization of soybean endopeptidase activity using exogenous and endogenous substrates. *Plant Physiology* 72:345–350.

Borthwick, H. A., S. B. Hendricks, E. H. Toole, and V. K. Toole. 1954. Action of light on lettuce seed germination. *Botanical Gazette* 115:205–225.

Bowers, J. L. 1958. Preliminary studies on cucumber seed development as related to viability. *Proceedings of the Association of Southern Agricultural Workers* 55:163–164.

Bramlage, W. J., A. C. Leopold, and J. E. Specht. 1979. Imbibitional chilling sensitivity among soybean cultivars. *Crop Science* 19:811–815.

Briggs, D. E. 1963. Biochemistry of barley germination. Action of gibberellic acid on barley endosperm. *Journal of the Institute of Brewing* 69:13–19.

Brown, H. T., and F. Escombe. 1898. Note on the influence of very low temperatures on the germinative power of seed. *Proceedings of the Royal Society of London* 62:160–168.

Brown, R. 1932. The absorption of the solute from aqueous solutions by the grain of wheat. *Annals of Botany* (London) 46:571–582.

Burrows, W. J. 1975. Mechanisms of action of cytokinins. *Current Advances in Plant Science* 7:837–864.

Burrows, W. J., D. J. Armstrong, M. Kaminek, F. Skoog, R. M. Bock, S. M. Hecht, L. G. Damnan, N. J. Leonard, and J. Occolowitz. 1970. Isolation and identification of four cytokinins from wheat germ transfer ribonucleic acid. *Biochemistry* 9:1867–1872.

Butler, W. L., S. B. Hendricks, and H. W. Siegelman. 1964. Action spectra of phytochrome in vitro. *Photochemical Photobiology* 3:521–528.

Calero, E., S. H. West, and K. Hinson. 1981. Water absorption of soybean seeds and associated causal factors. *Crop Science* 21:926–933.

Cavalieri, A. J., and J. S. Boyer. 1982. Water potentials induced by growth in soybean hypocotyls. *Plant Physiology* 69:492–497.

Chen, S. S. C. 1970. Action of light and gibberellic acid on the growth of excised embryos from *Phacelia tanacetifolia* seeds. *Planta* 95:336–340.

Chen, S. S. C., and J. E. Varner. 1973. Hormones and seed dormancy. *Seed Science and Technology* 1:325–358.

Chippendale, H. G. 1934. The effect of soaking in water on the "seed" of some *Gramineae*. *Annals of Applied Biology* 21:225–232.

Christiansen, M. N. 1968. Induction and prevention of chilling injury to radicle tips of imbibing cottonseed. *Plant Physiology* 43:743–746.

Chung, D. S., and H. B. Pfost. 1967. Adsorption and desorption of water vapor by cereal grains and their products. *Transactions of American Society of Agricultural Engineers* 10:549–555.

Cohen, D. 1958. The mechanism of germination stimulation by alternating temperatures. *Bulletin of the Research Council of Israel Section D Botany* D6:111–117.

Cohn, M. A., and R. L. Obendorf. 1976. Independence of imbibitional chilling injury and energy metabolism in corn. *Crop Science* 16:449–453.

———. 1978. Occurrence of a stelar lesion during imbibitional chilling of *Zea mays* L. *American Journal of Botany* 65:50–56.

Cohn, M. A., R. L. Obendorf, and G. T. Rytko. 1979. Relationship of stelar lesions to radicle growth in corn seedlings. *Agronomy Journal* 71:954–959.

Dassou, S., and E. A. Kueneman. 1984. Screening methodology for resistance to field weathering of soybean seed. *Crop Science* 24:774–779.

Davis, A. C. W. 1964. The relative susceptibility to threshing damage of six varieties of wheat. *Journal of the National Institute of Agricultural Botany* 10:122–128.

Denny, F. E. 1917. Permeability of membranes as related to their composition. *Botanical Gazette* 63:468–485.

Dexter, S. T. 1966. Conditioning dry bean seed (*Phaseolus vulgaris* L.) for better processing quality and seed germination. *Agronomy Journal* 58:629–631.

Dorrell, D. C., and M. W. Adams. 1969. Effect of some seed characteristics on mechanically induced seed coat damage in navy beans (*Phaseolus vulgaris* L.). *Agronomy Journal* 61:672–673.

Dotzenko, A. D., and J. G. Dean. 1959. Germination of six alfalfa varieties at three levels of osmotic pressure. *Agronomy Journal* 51:308–309.

Dudeck, A. E. and C. H. Peacock. 1985. Salinity effects on perennial ryegrass germination. *Horticultural Science* 20:268–269.

Duke, S. H., G. Kakefuda, and T. M. Harvey. 1983. Differential leakage of intracellular substances from imbibing soybean seeds. *Plant Physiology* 72:919–925.

Duke, S. H., G. Kakefuda, C. A. Henson, N. L. Loeffler, and N. M. VanHulle. 1986. Role of the testa epidermis in the leakage of intracellular substances from imbibing soybean seeds and its implications for seedling survival. *Plant Physiology* 60:716–722.

Egley, G. H. 1984. Ethylene, nitrate and nitrite interactions in the promotion of dark germination of common purslane seeds. *Annals of Botany* 53:833–840.

Ellis, R. H., T. D. Hong, and E. H. Roberts. 1986. The response of seeds of *Bromus sterilis* L. and *Bromus mollis* L. to white light of varying photon flux density. *New Phytology* 104:485–496.

Ensor, H. L. 1967. The influence of water content of sand on the germination of dwarf beans (*Phaseolus vulgaris* L.). *Proceedings of the International Seed Testing Association* 32(1):13–30.

Eslick, R. F., and W. Vogel. 1959. Effect of soil moisture tension on the ultimate emergence of grass and legume seed. *Proceedings of Association of Official Seed Analysis* 49(2):154–160.

Evans, M. L., R. Moore, and K. H. Hasenstein. 1986. How roots respond to gravity. *Scientific American* 255:112–119.

Fausey, N. R. and M. B. McDonald, Jr. 1985. Emergence of inbred and hybrid corn following flooding. *Agronomy Journal* 77:51–56.

Feldman, L. J. 1981. Effect of auxin on acropetal auxin transport in roots of corn. *Plant Physiology* 67:278–281.

Flint, L. H. 1934. Light in relation to dormancy and germination in lettuce seed. *Science* 80:38–40.

Flint, L. H., and E. D. McAlister. 1935. Wavelengths of radiation in the visible spectrum inhibiting the germination of light-sensitive lettuce seed. *Smithsonian Miscellaneous Collection* 94:1–11.

———. 1937. Wavelengths of radiation in the visible spectrum promoting the germination of light-sensitive lettuce seed. *Smithsonian Miscellaneous Collection* 96(2):1–8.

Foard, D. E., and A. H. Haber. 1966. Mitosis in thermodormant lettuce seeds with reference to histological location, localized expansion, and seed storage. *Planta* 71:160–170.

Forward, B. F. 1958. Studies on the germination of oats. *Proceedings of the International Seed Testing Association* 23:23.

Fox, J. E., and J. L. Erion. 1975. Cytokinin binding protein from higher plant ribosomes. *Biochemistry Biophysiology Research Communications* 64:694–700.

Fryer, J. R. 1919. Germination of oats exposed to varying degrees of frost at different stages of maturity. *Agricultural Gazette* (Canada) 6:337–339.

Fujii, T., S. Isikawa, and A. Nakagawa. 1960. The effects of gibberellin on the germination of seeds of *Sedum kamtschaticum* Fisch. *Botanical Magazine* (Tokyo) 73:404–411.

Fujii, T., and Y. Yokohama. 1965. Physiology of light-requiring germination in *Eragrostis* seeds. *Plant and Cell Physiology* 6:135–145.

Gadd, I. 1955. Germination of seed of New Zealand Browntop, *Agrostis tenuis* Sibth. *Proceedings of the International Seed Testing Association* 23:41.

Giri, A. 1967. Germination percentage, average height, and girth of seedlings raised from seed-stones extracted from syrupy and firm mango fruits. *Pakistan Journal of Science* 18:79–81.

Grabe, D. F., 1956. Maturity in smooth bromegrass. *Agronomy Journal* 48:253–256.

Haber, A. H., and H. J. Luippold. 1960. Separation of mechanisms initiating cell division and cell expansion in lettuce seed germination. *Plant Physiology* 35:168–173.

Hagon, M. W. 1976. Germination and dormancy of *Themeda australis* Danthonia spp., *Stipa bigeniculata* and *Bothriochloa macra*. *Australian Journal of Botany* 24:319–324.

Harrington, G. T. 1923. Germination of Kentucky bluegrass under alternating versus constant temperature conditions. *Journal of Agricultural Research* 23:298–303.

Harrington, J. F. 1959. Effect of fruit maturity and harvesting methods on germination of muskmelon seed. *Proceedings of the American Society for Horticultural Science* 73:422–430.

Hashimoto, T. 1958. Increase in percentage of gibberellin-induced dark germination of tobacco seeds by N-compounds. *Botanical Magazine* (Tokyo) 71:430–431.

Heydecker, W., P. I. Orphanos, and R. S. Chetram. 1969. The importance of air supply during seed germination. *Proceedings of the International Seed Testing Association* 34(2):297–304.

Hilhorst, H. W. M. 1990. Dose-response analysis of factors involved in germination and secondary dormancy of seeds of *Sisymbrium officinale*. II. Nitrate. *Plant Physiology* 94:1096–1102.

Hill, H. J., S. H. West, and K. Hinson. 1986. Effect of water stress during seed fill on impermeable seed expression in soybean. *Crop Science* 26:807–813.

Hill, M. J., and B. B. Watkin. 1975. Seed production studies on perennial ryegrass, timothy and prairie grass. 2. Changes in physiological components during seed development and time and method of harvesting for maximum yield. *Journal of British Grassland Society* 30:131–140.

Hilton, J. R. 1984. The influence of light and potassium nitrate on the dormancy and germination of *Avena fatua* L. (wild oat) seed and its ecological significance. *New Phytology* 96:31–34.

Hilton, J. R., and J. A. Thomas. 1986. Regulation of pregerminative rates of respiration in seeds of various seed species by potassium nitrate. *Journal of Experimental Botany* 37:1516–1524.

Holm, R. E., and M. R. Miller. 1972. Weed seed germination responses to chemical and physical seed treatments. *Weed Science* 20:150–153.

Holmes, A. D. 1953. Germination of seeds removed from mature and immature butternut squashes after seven months of storage. *Proceedings of the American Society for Horticultural Science* 62:433–436.

Hsiao, A. I., and G. M. Simpson. 1971. Dormancy studies in seed of *Avena fatua*. VII. The effects of light and variation in water regime on germination. *Canadian Journal of Botany* 49:1347–1357.

Hunter, J. R., and A. E. Erickson. 1952. Relation of seed germination to soil moisture tension. *Agronomy Journal* 44(3):107–109.

Ikuma, H., and K. V. Thimann. 1960. Action of gibberellic acid on lettuce seed germination. *Plant Physiology* 35:557–566.

Ingle, J., L. Beevers, and R. H. Hageman. 1964. Metabolic changes associated with the germination of corn. I. Changes in weight and metabolites and their redistribution on the embryo axis, scutellum, and endosperm. *Plant Physiology* 39:735–740.

Isikawa, S. 1962. Light sensitivity against the germination. III. Studies on various partial processes in light sensitive seeds. *Japanese Journal of Botany* 18:105–132.

Isikawa, S., and T. Fujii. 1961. Photocontrol and temperature dependence of germination of *Rumex* seeds. *Plant and Cell Physiology* 2:51–62.

Jacobsen, J. V., and J. E. Varner. 1967. Gibberellic acid-induced synthesis of protease by isolation of aleurone layers of barley. *Plant Physiology* 42:1596–1600.

Johnston, E. H. 1964. Investigations into techniques for the germination of *Paspalum dilatatum*. *Proceedings of the International Seed Testing Association* 29(1):145–148.

Jones, R. L., and J. L. Stoddart. 1977. Gibberellins and seed germination. In: *The Physiology and Biochemistry of Seed Dormancy and Germination,* ed. A. A. Khan, pp. 77–111. New York: North-Holland Publishing Company.

Jones, R. S., and J. E. Armstrong. 1971. Evidence for osmotic regulation of hydrolytic enzyme production in germinating barley seeds. *Plant Physiology* 48:137–142.

Judd, R., D. M. TeKrony, D. B. Egli, and G. M. White. 1982. Effect of freezing temperatures during soybean seed maturation on seed quality. *Agronomy Journal* 74:645–650.

Justice, O. L. 1967. Evaluation of seedlings grown from gamma-irradiated seeds. *Proceedings of the Association of Official Seed Analysts* 57:148–156.

Justice, O. L., and M. M. Kulik. 1970. Some effects of gamma radiation on germination and storage life of seeds of eight crop species. *Proceedings of the International Seed Testing Association* 35:697–712.

Justice, O. L., and M. H. Reece. 1954. A review of literature and investigation on the effects of hydrogen-ion concentration on the germination of seeds. *Association of Official Seed Analysts* 34:144–149.

Kadman-Zahavi, A. 1960. Effects of short and continuous illuminations on the germination of *Amaranthus retroflexus* seeds. *Bulletin of the Research Council of Israel Section D* 9D:1–20.

Karssen, C. M., and H. W. M. Hilhorst. 1992. Effect of chemical environment on seed germination. In: *Seeds: The Ecology of Regeneration in Plant Communities,* ed. M. Fenner, pp. 327–348. Wallingford, U. K.: CAB International.

Kennedy, R. A., T. C. Fox, and J. N. Siedow. 1987. Activities of isolated mitochondrial enzymes from aerobically and anaerobically germinated barnyard grass (*Echinochloa*) seedlings. *Plant Physiology* 85:474–480.

Kidd, F., and C. West. 1918. Physiological predetermination: The influence of the physiological condition of the seed upon the course of subsequent growth and upon the yield. I. The effects of soaking seeds in water. *Annals of Applied Biology* 5:1–10.

Kiesselbach, T. A., and J. S. Ratcliff. 1920. Freezing injury of seed corn. *Nebraska Agricultural Experimental Station Bulletin 16.*

Kincaid, R. R. 1935. Effects of certain environmental factors on germination of Florida cigarwrapper tobacco seeds. *Technical Bulletin 277, University of Florida Experiment Station.*

Kinch, R. C., and D. Termunde. 1957. Germination of perennial sow thistle and Canada thistle at various stages of maturity. *Proceedings of Association of Official Seed Analysts* 47:165–166.

Klein, S. 1955. Aspects of nitrogen metabolism in germinating lettuce seeds with special emphasis on three amino acids (in Hebrew). Ph.D. dissertation, Hebrew University, Jerusalem.

Koller, D. 1971. Analysis of the dual action of white light on germination of *Atriplex dimorphostegia* (*Chenopodiaceae*). *Israel Journal of Botany* 19:499–516.

———. 1972. Environmental control of seed germination. In: *Seed Biology*, ed. T. T. Kozlowski, pp. 1–101. New York: Academic Press.

Koller, D., A. M. Mayer, A. Poljakoff-Mayber, and S. Klein, 1962. Seed germination. *Annual Review of Plant Physiology* 13:437–464.

Leopold, A. C., and M. E. Musgrave. 1979. Respiratory changes with chilling injury of soybeans. *Plant Physiology* 64:702–706.

Leopold, A. C., and C. W. Vertucci. 1986. Physical attributes of desiccated seeds. *In: Membranes, Metabolism and Dry Organs,*ed. A. C. Leopold, Ithaca, N.Y.: Cornell University Press.

Leopold, A. C., and C. W. Vertucci. 1989. Moisture as a regulator of physiological reaction in seeds. *In: Seed Moisture,* eds. P. C. Stanwood and M. B. McDonald, Madison, Wisc.: Crop Science Society of America.

Libbert, E., and H. Lubke. 1957. Physiologische Wirkungen des Scopoletins. I. Mitteilung der Einfluss des Scopoletins auf die Samenkeimung. *Flora* 145:256–263.

Luyet, B. J., and P. M. Gehenio. 1938. The lower limit of vital temperatures, a review. *Biodynamics* 33:1–92.

Martin, B. A., S. F. Cerevick, and L. D. Reding. 1991. Physiological basis for inhibition of maize seed germination by flooding. *Crop Science* 31:1052–1058.

Mayer, A. M., and A. Poljakoff-Mayber. 1989. *The Germination of Seeds*. New York: Pergamon Press.

McDonald, M. B. 1985. Physical seed quality of soybean. *Seed Science and Technology* 13:601–628.

McDonald, M. B. 1994. Seed germination and seedling establishment. *In: Physiology and Determination of Crop Yield,* eds. K. J. Boote and T. R. Sinclair. Madison, Wisc.: Crop Science Society of America.

McDonald, M. B., L. O. Copeland, A. D. Knapp, and D. F. Grabe. 1994a. Seed development, germination and quality. In: *Cool-Season Grass Monograph,* ed. L. Moser. Madison, Wisc.: American Society of Agronomy.

McDonald, M. B., J. Sullivan, and M. J. Lauer. 1994b. The pathway of water uptake in maize seeds. *Seed Science and Technology* 22:79–90.

McDonald, M. B., C. W. Vertucci, and E. E. Roos. 1988a. Seed coat regulation of soybean seed inhibition. Crop Science 28:987–992.

McDonald, M. B., C. W. Vertucci, and E. E. Roos. 1988b. Soybean seed imbibition: Water absorption by seed parts. *Crop Science* 28:993–997.

Mitchell, B., C. Armstrong, M. Black, and J. Chapman. 1980. Physiological aspects of sprouting and spoilage in developing *Triticum aestivum* L. (wheat) grains. In: *Seed Production,* ed. P. D. Hebblethwaite, pp. 339–356. London: Butterworth.

Morinaga, T. 1926. The favorable effect of reduced oxygen supply upon the germination of certain seeds. *American Journal of Botany* 13:159–166.

Mugnisjah, W. Q., I. Shimano, and S. Matsumoto. 1987. Studies of the vigour of soybean seeds. II. Varietal differences in seed coat colour and swelling components of seed during moisture imbibition. *Journal of Faculty of Agriculture of Kyushu University* 31:227–234.

Murdoch, A. J., E. H. Roberts, and C. O. Goedert. 1989. A model for germination responses to alternating temperatures. *Annals of Botany* 63:97–111.

Nagao, A., Y. Esashi, T. Tanaka, T. Kumagai, and S. Fukumoto. 1959. Effects of photoperiod and gibberellin on the germination of seeds of *Begonia evansiana* Andr. *Plant and Cell Physiology* (Tokyo) 1:39–47.

Niffenegger, D., and A. A. Schneiter. 1963. A comparison of methods of germinating green needlegrass seed. *Proceedings of Association of Official Seed Analysts* 53:72–77.

Nordin, P. 1984. Preferential leaching of pinitol from soybeans during imbibition. *Plant Physiology* 76:313–316.

Ogawara, K., and K. Ono. 1961. Interaction of gibberellin, kinetin and potassium nitrate in the germination of light-sensitive tobacco seeds. *Plant and Cell Physiology* (Tokyo) 2:87–98.

Oota, Y., R. Fujii, and S. Osawa. 1953. Changes in chemical constituents during the germination stage of a bean, *Vigna sesquipedalis. Journal of Biochemistry* (Tokyo) 40:649–661.

Paleg, L. G. 1960. Physiological effects of gibberellic acid: I. On carbohydrate metabolism and amylase activity of barley endosperm. *Plant Physiology* 35:293–294.

Parker, K. E., and W. R. Briggs. 1990. Transport of indoleacetic acid in intact corn coleoptiles. *Plant Physiology* 94:417–424.

Pegler, R. A. D. 1976. Harvest ripeness in grass seed crops. *Journal of British Grassland Society* 31:7–13.

Phiney, B. O., and C. A. West. 1960. Gibberellins as native plant growth regulators. *Annual Review of Plant Physiology* 11:411–436.

Pollock, B. M., and H. O. Olney. 1959. Studies of the rest period. I. Growth, translocation, and respiration changes in the embryonic organs of the after-ripening cherry seed. *Plant Physiology* 34:131–142.

Pollock, B. M., and V. K. Toole. 1966. Imbibition period as the critical temperature sensitive stage in germination of lima bean seeds. *Plant Physiology* 41:221–229.

Potts, H. C., J. Duangpatra, W. G. Hairston, and J. C. Delouche. 1978. Some influences of hardseededness on soybean seed quality. *Crop Science* 18:221–224.

Radley, M. 1967. Site of production of gibberellin-like substances in germinating barley embryos. *Planta* 75:164–171.

Resuhr, B. 1941. Uber die Bedeutung konstitutioneller Mangel fur das Aufreten von Keimlingsschaden bei *Soja hispida* Moench. II. *Zeitschrift fur Pflanzenkrank heiten und Pflanzenschutznachrichten* (West Germany) 51:161–192.

Ries, R. E., and L. Hoffman. 1983. Effect of sodium and magnesium sulfate on forage seed germination. *Journal of Range Management* 3677:658–662.

Robbins, W. A., and R. H. Porter. 1946. Germinability of sorghum and soybean seed exposed to low temperatures. *Journal of the American Society of Agronomy* 38:905–913.

Rodger, J. B. B., G. G. Williams, and R. L. Davis. 1957. A rapid method for determining winterhardiness of alfalfa. *Agronomy Journal* 49:88–92.

Rogan, P. G., and E. W. Simon, 1975. Root growth and the onset of mitosis in germinating *Vicia faba. New Phytology* 74:273–275.

Rorabaugh, P. A., and F. B. Salisbury. 1989. Gravitropism in higher plant shoots. VI. Changing sensitivity to auxin in gravistimulated soybean hypocotyls. *Plant Physiology* 79:1329–1339.

Rossman, E. C. 1949a. Freezing injury of inbred and hybrid maize seed. *Agronomy Journal* 41:574–583.

——. 1949b. Freezing injury of maize seed. *Plant Physiology* 24:629–656.

Roundy, B. A. 1985. Germination and seedling growth of tall wheatgrass and Basin wildrye in relation to boron. *Journal of Range Management* 38:270–272.

Rumpho, M. E., A. Pradet, A. Khalik, and R. A. Kennedy. 1984. Energy charge and emergence of the coleoptile and radicle at varying oxygen levels in *Echinocloa crus-galli. Physiologia Plantarum* 62:133–138.

Ryan, C. A. 1973. Proteolytic enzymes and their inhibitors. *Annual Review of Plant Physiology* 24:173–196.

Sahid, and Soemartono. 1974. Pengaruh Penyinaran co60-gamma Dosis Rendah Terhadap Kemungkinan Peningkatan Daya Kecambah Benih Pada. *Kerjasama Badan Tenaga Atom Nasional Lembaga Pendidikan Perkebunan and Himpunanapemulia Tanaman Indonesia,* 28–30.

Saini, H. S., and M. S. Spencer. 1986. Manipulation of seed nitrate content modulates the dormancy breaking effect of ethylene on *Chenopodium album* seed. *Canadian Journal of Botany* 65:876–878.

Schoettle, A. W., and A. C. Leopold. 1984. Solute leakage from artificially aged soybean seeds after imbibition. *Crop Science* 24:835–838.

Seneratna, T., and B. D. McKersie. 1983. Characterization of solute efflux from dehydration injured soybean [*Glycine max* (L.) Merr.] seeds. *Plant Physiology* 72:911–915.

Seyedin, N., J. S. Burris, C. E. LaMotte, and I. C. Anderson. 1982. Temperature-dependent inhibition of hypocotyl elongation in some soybean cultivars. I. Localization of ethylene evolution and role of cotyledons. *Plant Cell Physiology* 23:427–431.

Shakeywich, C. F. 1973. Proposed method of measuring swelling pressure of seeds prior to germination. *Journal of Experimental Botany* 24:1056–1061.

Siegel, S. M., and L. A. Rosen. 1962. Effects of reduced oxygen tension on germination and seedling growth. *Physiologia Plantarum* 15:437–444.

Simpson, G. M. 1990. *Seed Dormancy in Grasses.* Cambridge, U.K.: Cambridge University Press.

Snyder, F. W. 1975. Personal communication, based on unpublised data.

Stanwood, P. C. 1986. Dehydration problems associated with the preservation of seed and plant germplasm. In: *Membranes, Metabolism, and Dry Organisms,* ed. A. C. Leopold, pp. 327–340. Ithaca, N.Y.: Cornell University Press.

Takahashi, N. 1985. Inhibitory effects of oxygen on the germination of *Oryza sativa* seeds. *Annals of Botany* 55:597–600.

Taylorson, R. B., and S. B. Hendricks. 1976. Aspects of dormancy in higher plants. *Bioscience* 26(2):95–101.

Thill, D. C., K. G. Beck, and R. H. Callihan. 1984. The biology of down brome (*Bromus tectorum*). *Weed Science* 32(Supplement 1):7–12.

Thiselton-Dyer, Sir William. 1899. On the influence of temperature of liquid hydrogen on the germination power of seeds. *Royal Society of London* 65:361–368.

Thomas, T. H. 1977. Cytokinins, cytokinin-active compounds and seed germination. In: *The Physiology and Biochemistry of Seed Dormancy and Germination,* ed. A. A. Khan, pp. 111–145. New York: North-Holland Publishing Company.

Thomson, C. J., and H. Greenway. 1991. Metabolic evidence of stelar anoxia in maize roots exposed to low O_2 concentrations. *Plant Physiology* 96:1294–1302.

Thornton, N. C. 1936. Carbon dioxide storage. IX. Germination of lettuce seeds at high temperatures in both light and darkness. *Contributions from Boyce Thompson Institute* 8:25–40.

Tilford, P., C. F. Able, and R. P. Hibbard. 1924. An injurious factor affecting the seeds of *Phaseolus vulgaris* soaked in water. *Papers of the Michigan Academy of Science* 4:345–356.

Tilsner, H. R. and M. K. Upadhyaya. 1985. Induction and release of secondary dormancy in genetically pure lines of *Avena fatua. Physiologia Plantarum* 64:377–382.

Toole, E. H., H. A. Borthwick, S. B. Hendricks, and V. K. Toole. 1953. Physiological studies of the effects of light and temperature on seed germination. *Proceedings of International Seed Testing Association* 18:270–275.

Toole, E. H., S. B. Hendricks, H. A. Borthwick, and V. K. Toole. 1956. Physiology of seed germination. *Annual Review of Plant Physiology* 7:299–324.

———. 1955a. Interaction of temperature and light in germination of seeds. *Plant Physiology* 30:473–478.

———. 1955b. Photocontrol of *Lepidium* seed germination. *Plant Physiology* 30:15–21.

Toole, E. H., V. K. Toole, H. A. Borthwick, and S. B. Hendricks. 1957. Changing sensitivity of seeds to light. *Plant Physiology* 32(suppl.):xi.

Toole, E. H., V. K. Toole, H. A. Borthwick, S. B. Hendricks, and R. J. Downs. 1958a. Action of light on germination of seeds of *Paulownia tomentosa*. *Plant Physiology* 33(abstract):xxiii.

Toole, V. K. 1938. Germination requirements of the seed of some introduced and native range grasses. *Proceedings of the Association of Official Seed Analysts* 28:227–243.

Toole, V. K., H. A. Borthwick, E. H. Toole, and A. G. Snow, Jr. 1958b. The germination response of seeds of *Pinus taeda* to light. *Plant Physiology* 33(suppl.):xxiii.

Toole, V. K., and E. J. Koch. 1977. Light and temperature control of dormancy and germination in bentgrass seeds. *Crop Science* 17:806–811.

Toole, V. K., E. H. Toole, H. A. Borthwick, and A. G. Snow, Jr. 1962. Responses of *Pinus taeda* and *Pinus strobus* to light. *Plant Physiology* 37:228–233.

Toole, V. K., E. H. Toole, S. B. Hendricks, H. A. Borthwick, and A. G. Snow, Jr. 1961. Responses of seed of *Pinus virginiana* to light. *Plant Physiology* 368:285–290.

Torrent, M., M. I. Geli, and M. D. Ludevid. 1989. Storage protein hydrolysis and protein body breakdown in germinated *Zea mays* L. seeds. *Planta* 180:90–95.

Tully, R. E., M. E. Musgrave, and A. C. Leopold. 1981. The seed coat as a control of imbibitional chilling injury. *Crop Science* 21:312–317.

Ungar, I. A. 1987. Halophyte seed germination. *Botany Review* 44:233–264.

Van Toai, T., N. R. Fausey, and M. B. McDonald, Jr. 1985. Alcohol dehydrogenase and pyruvate decarboxylase activities in flood-tolerant and susceptible corn seeds during flooding. *Agronomy Journal* 77:753–758.

Van Toai, T., N. Fausey, and M. B. McDonald, Jr. 1988. Oxygen requirements for germination and growth of flood-susceptible and flood-tolerant corn lines. *Crop Science* 28:79–83.

Vegis, A. 1963. Control of germination, bud break, and dormancy. In: *Environmental Control of Plant Growth*, ed. L. T. Evans, pp. 269–270, New York: Academic Press.

Vertucci, C. W., and A. C. Leopold. 1987. Water binding in legume seeds. *Plant Physiology* 85:224–231.

Viestra, R. D. 1993. Illuminating phytochrome functions. There is light at the end of the tunnel. *Plant Physiology* 103:679–684.

Villiers, T. A., and P. F. Wareing. 1960. Interaction of a growth inhibitor and a natural germination stimulator in the dormancy of *Fraxinus excelsior* L. *Nature* (London) 185:112–114.

Weatherby, J. H. 1943. Concerning the mechanism of membrane semipermeability. *Journal of Cellular and Comparative Physiology* 21:1–17.

Williams, S. 1972. The effects of harvest date on the yield and quality of seed of tetraploid hybrid ryegrass. *Journal of British Grassland Society* 27:221–227.

Willing, R. P., and A. C. Leopold. 1983. Cellular expansion at low temperatures as a cause of membrane lesions. *Plant Physiology* 71:118–122.

Wilson, R. F. 1987. Seed metabolism. p. 643–686. In: *Soybeans: Improvement, Production and Uses*, ed. J. R. Wilcox, Madison, Wisc.: American Society of Agronomy, 2nd ed.

Wolf, M. J., C. L. Buzan, M. M. MacMasters, and C. E. Rist. 1952. Structure of the mature corn kernel. II. Microscopic structure of pericarp, seed coat, and hilar layer of dent corn. *Cereal Chemistry* 29:334–348.

Yaklich, R. W., E. L. Vigil, and W. P. Wergin. 1986. Pore development and seed coat permeability in soybean. *Crop Science* 26:616–624.

Young, J. A., and R. A. Evans. 1981. Germination of Great Basin wildrye seeds collected from native stands. *Agronomy Journal* 73:917–920.

Young, L. M., M. L. Evans, and R. Hertel. 1990. Correlations between gravitropic curvature and auxin movement across gravistimulated roots of *Zea mays*. *Plant Physiology* 92:792–797.

5
Seed Viability Testing

Although the concept of seed viability is well known, there is considerable disagreement and confusion as to its precise meaning. To most seed technologists and commercial dealers, viability means that a seed is capable of germinating and producing a "normal" seedling. Therefore, it is used synonymously with germination capacity. In this sense, a given seed is either viable or nonviable, depending on its ability to germinate and produce a normal seedling; thus, only seed lots representing populations of seeds may exhibit levels of viability.

In another sense, viability denotes the degree to which a seed is alive, metabolically active, and possesses enzymes capable of catalyzing metabolic reactions needed for germination and seedling growth. In this context, a given seed may contain both live and dead tissues, and may or may not be capable of germination. This meaning deals with tissue viability as well as viability of the entire seed.

In either context, seed viability is probably highest at the time of physiological maturity, though environmental conditions on the parent plant may not permit germination. After physiological maturity, the viability of seeds gradually declines. Their longevity depends on the environmental conditions to which they are exposed.

Numerous tests exist for determining seed viability; these are discussed on the following pages.

GERMINATION TEST

The germination test is most commonly used to determine seed viability. It has become so universally accepted that seed germination and viability are probably considered one and the same by most people. Regardless of its acceptance, the germination test is merely an estimate and has certain limitations as a universal estimate of seed quality. However, if these limitations are recognized, the germination test is a useful viability index.

Germination Terminology

Why test seeds for germination? The intuitive answer to this question is "Because we need to know how the seeds will perform in the field." However, there are more reasons than this. For example, we know that a seed lot consists of a population of seeds, each with its own distinct capability to produce a plant. A germination test is an analytical procedure to evaluate seed germination under standardized, favorable conditions that are seldom, if ever, encountered in the field. Consequently, germination tests results often overestimate field performance of a seed lot. It is now increasingly recognized that seed vigor tests more accurately reflect the ability of a seed to perform in the field (see Chapter 7). However, the use of favorable conditions leads to a maximum germination percentage and this result can be consistently reproduced. This permits different analysts from different laboratories to obtain similar germination percentages for the same seed lot. This allows verification of germination results on seed labels, which permits interstate and global shipment of seed lots of known value to move throughout commerce. A germination test allows a seed producer to determine and compare the quality of a seed lot before it is planted. This information can be used to determine the planting value of a seed lot and its storage potential to satisfy labeling laws and provide for standardized marketing of seed.

To the seed analyst, germination has a definite and standardized meaning, which must lead to uniform interpretations made by analysts in different laboratories. In the Rules for Testing Seed of the Association of Official Seed Analysts, seed germination is described as "the emergence and development from the seed embryo of those essential structures which, for the kind of seed in question, are indicative of the ability to produce a normal plant under favorable conditions." (AOSA 1978) The seed analyst looks for the emergence of an embryonic plant that must consist of a complete root and shoot axis that has the capacity of normal growth under favorable conditions.

Seed testing associations (Association of Official Seed Analysts and International Seed Testing Association—see Chapter 13) have developed sets of standardized conditions for germinating seeds. These conditions are published in their official rules, which specify the optimum germination conditions for hundreds of kinds of agricultural, vegetable, flower, and tree seeds. The rules provide optimum germination conditions for each kind of seed. The rules also provide additional measures that may aid in germination, particularly of dormant seeds. An excerpt of the rules of the Association of Official Seed Analysts (AOSA) is given in Table 5.1.

Rules for Germination Testing Procedures

In the AOSA rules for testing seeds, seeds are divided into four groupings: agricultural, vegetable and herb, flower, and tree and shrub seeds. Each kind of seed is listed according to its recognized scientific nomenclature as well as common name. Following the name of the seed is the substrate (germination media) recommended (e.g., petri dishes (P), blotters (B), sand or soil (S), paper towels (T), or kimpac (C)). Substrata should (1) be nontoxic to germinating seedlings, (2) be free of molds, microorganisms, and their spores, and (3) provide adequate aeration and moisture for germination.

Table 5.1. Some Methods for Testing for Germination

Kind of seed	Substrata	Temperature °C	First count days	Final count days	Specific requirements and photo numbers	Fresh and dormant seed
Eragrostis trichodes sand love grass	P	20–30	5	14	Light; KNO$_3$	Prechill at 5° or 10°C for 6 weeks
Erodium cicutarium alfileria	B,T	20–30	3	14	Clip seeds	
Fagopyrum esculentum buckwheat	B,T	20–30	3	6		
Festuca arundinacea tall fescue	P	15–25; 20–30	5	14	Light and KNO$_3$ optional	Prechill at 5° or 10°C for 5 days and extend test to 21 days
Festuca capillata hair fescue	P	10–25	10	28	KNO$_3$	
Festuca elatior meadow fescue	P	15–25; 20–30	5	14	Light and KNO$_3$ optional	
Festuca ovina sheep fescue alternate method	P	15–25	7	21		
	P	20–30	7	28	Light	
Festuca ovina var duriuscula hard fescue	P	15–25	7	21	Light and KNO$_3$ optional	

Data from AOSA (1978).

Timothy seed should be germinated on any substratum suspected of being toxic to seedlings since its roots will not develop normally in the presence of toxic substances.

Recommended temperatures (in degrees Celsius) for conducting germination tests are followed by the first and final evaluation times (counts). Temperature recommendations are based on research that has shown optimum temperatures for each species of seed. The Association of Official Seed Analysts permits only a 1° C variation from the recommended temperature. One number in the temperature column indicates a constant temperature throughout the testing period. Two numbers separated by a dash indicate an alternating temperature during a 24-hour period—the first temperature maintained for 16 hours followed by the second temperature for eight hours. Two temperature recommendations separated by a semicolon are considered alternate recommendations. In such cases, both temperatures are permissible and a laboratory may use the temperature that is most convenient. The time of the first count is considered to be approximate and a deviation of one to three days is permitted. At this time, the analyst counts and discards germinated seedlings and inspects the test for potential problems such as lack of moisture or pathogen infestation. The final count must be on the day specified since it provides sufficient time for even weak seedlings to germinate. Thus, the test should be terminated by the final count unless dormancy is suspected.

Seeds of certain species require special treatment for maximum germination or breaking dormancy. Such requirements are listed in the two columns, "Special requirements" (Table 5.1) and include germination-promoting environments such as light and the addition of KNO_3. Light is required for the germination of most grasses, many tree and shrub seeds, and some vegetable seeds. The Association of Official Seed Analysts recommends that light be evenly distributed with an intensity range of 810 to 1620 lux (75 to 150 foot-candles). For most seed germination tests, seeds should be subjected to light for only a part of the test period. Eight hours of light per day is usually sufficient. Potassium nitrate solution (0.2%) is used to promote germination of some dormant seeds.

Special treatments for overcoming seed dormancy are listed under "Fresh and dormant seed." In most instances, directions for prechilling are provided (i.e., placing the seed on a moist substratum at low temperatures—generally 10°C—for a specified period.) Then the seed is transferred to optimum temperature conditions for the duration of the test. Although prechilling treatment (stratification) prolongs a germination test substantially, it is essential to break the dormancy of many species.

Although moisture and aeration are not specified in the germination methods, enough moisture should be provided so that the seeds can imbibe the water before it is evaporated from the substratum. Too much water creates an anaerobic environment where essential oxygen is not available to seeds. A formula determining optimum moisture for seeds germinating in sand is provided in the "Rules for Testing Seeds" (AOSA 1978). Regardless of the kind of seed being tested, all germination tests should be examined daily to ensure that the moisture content of the substratum is optimal. The Association of Official Seed Analysts further recommends that germination tests be conducted in germinators or germinator boxes that maintain a relative humidity of 95% or higher to minimize moisture loss from the germination substrata.

The rules for germination testing specify only environmental conditions and dormancy-breaking procedures that are of proven effectiveness in promoting germination and lead to standardized interpretations in routine seed testing. Too frequently, they do not reflect new knowledge that is available through scientific research. For example, it is widely known that many growth regulators are effective in breaking dormancy, yielding a more accurate estimate of viability; yet only one inorganic chemical (potassium nitrate) is recognized by the AOSA "Rules for Testing Seed" (AOSA 1994) as an aid in performing germination tests. Of the plant hormones, only ethylene is recognized as an approved aid in breaking of dormancy in peanut seeds.

TETRAZOLIUM TEST

The tetrazolium test is widely recognized as an accurate means of estimating seed viability. This method was developed in Germany in the early 1940s by Professor Georg Lakon (1928) who had been trying to distinguish between live and dead seeds by exposing them to selenium salts. He then tried tetrazolium salts and found them more effective. Today the test is used throughout the world as a highly regarded method of estimating seed viability and is a routine test in many seed testing laboratories. It is often referred to as a "quick test," since it can be completed in only a few hours (as compared to regular germination tests that require as long as two months for some species.) Tetrazolium test results can be extremely valuable for providing labeling information for immediate shipment of seed lots without waiting for completion of germination tests. It is also a valuable research technique for estimating seed viability and determining reasons for poor germination.

Principle

The tetrazolium test distinguishes between viable and dead tissues of the embryo on the basis of their relative respiration rate in the hydrated state. Although many enzymes are active during respiration, the test utilizes the activity of *dehydrogenase* enzymes as an index to the respiration rate and seed viability. Dehydrogenase enzymes react with substrates and release hydrogen ions to the oxidized, colorless, tetrazolium salt solution, which is changed into red *formazan* as it is reduced by hydrogen ions (see Figure 5.1). Seed viability is interpreted according to the topographical staining pattern of the embryo and the intensity of the coloration.

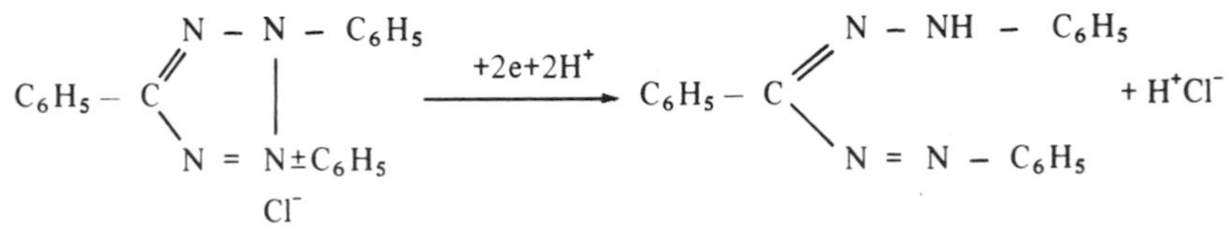

$$C_6H_5 - C \underset{\underset{Cl^-}{N = N \pm C_6H_5}}{\overset{N - N - C_6H_5}{\Big|}} \quad \xrightarrow{+2e+2H^+} \quad C_6H_5 - C \underset{N = N - C_6H_5}{\overset{N - NH - C_6H_5}{}} \quad + H^+Cl^-$$

2,3,5-triphenyl tetrazolium chloride   formazan (red)

Figure 5.1. *The chemical reaction that changes the colorless tetrazolium solution into formazan.*

Procedure

Seeds are first imbibed on a wet substratum to allow complete hydration of all tissues. For many species, the tetrazolium solution can be added to the intact seed. Other seeds must be prepared by cutting and puncturing in various ways (see Figure 5.2) to permit access of the tetrazolium solution to all parts of the seed. After hydration, the seeds are placed in a tetrazolium salt solution and held in a warming oven at about 35°C for complete coloration. Two hours is usually adequate for seeds that are bisected through the embryo, but others require longer periods of staining. If seeds are held too long in contact with the tetrazolium solution, they tend to become overstained, making interpretation difficult. Handbooks of instructions for performing tetrazolium tests and

Preparation of Seed

Method 1 Section laterally; discard apex potion of seed.

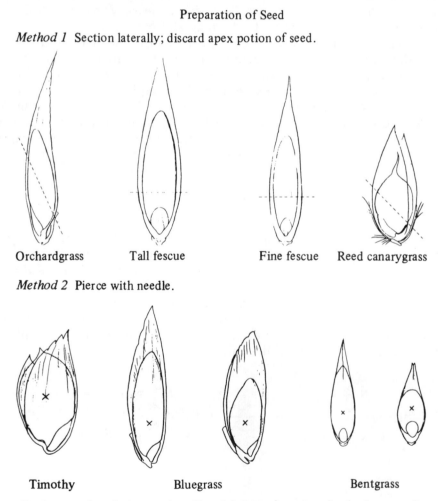

Method 2 Pierce with needle.

Figure 5.2. *Cutting procedure for large and small seeded grasses in preparation for the tetrazolium test. (From AOSA (1970).)*

interpretation instructions has been published by the Association of Official Seed Analysts (AOSA 1970) and International Seed Testing Association (1985).

Evaluation

Although the living tissues of seeds stain red, interpreting these results to estimate viability requires considerable skill and experience. Sound embryo tissues absorb tetrazolium slowly and tend to develop a lighter color than embryos that are bruised, aged, frozen, or disturbed in other ways. The experienced analyst learns to distinguish between seeds with the capacity to produce normal seedlings and those that stain abnormally (see Figure 5.3).

The tetrazolium test is often called the topographical tetrazolium test because the pattern, or topography, of staining is an important aspect of its interpretation. Many seeds are neither completely dead nor completely alive. The staining pattern reveals the live and dead areas of the embryo and enables the analyst to determine if seeds have the capacity to produce normal seedlings. The cell division areas of the embryo are most critical during germination, and if they are unstained, or abnormally stained, a seed's germination potential is weakened. The analyst must be familiar with crucial cell division areas of the embryo and learn to interpret their staining pattern in terms of seed germinability.

Other factors must be carefully observed when interpreting a tetrazolium test. Among these are flaccid tissues (lack of adequate turgor) and critically located fractures, bruises, and insect cavities. Any of these factors, when present in a vital position, may cause an otherwise sound seed to be nongerminable.

Advantages and Disadvantages

The rapidity of the test is its most obvious advantage and may justify its use when speed is important. Another advantage is its usefulness for dormant as well as for nondormant seeds, although its nondetection of dormancy (other than hard seed) is sometimes cited as a disadvantage. Actually, when used in combination with a germination test it can be useful for testing dormant seed. The germination test tells the percentage of immediate germination; the tetrazolium test tells the percentage of live seed; and the difference between the tetrazolium and germination tests represents the percentage of dormant seed.

Perhaps the greatest disadvantage of the tetrazolium test is its difficulty and the experience required to interpret it correctly. Another disadvantage has been the lack of acceptance of tetrazolium test results by seed law enforcement agencies, the seed trade, and certification agencies; however, this prejudice is gradually disappearing, as tetrazolium testing procedures become standardized and more analysts are trained to use them.

OTHER BIOCHEMICAL TESTS FOR SEED VIABILITY

The tetrazolium method has largely supplanted other biochemical viability tests for routine testing of seed quality. An excellent review of rapid biochemical viability

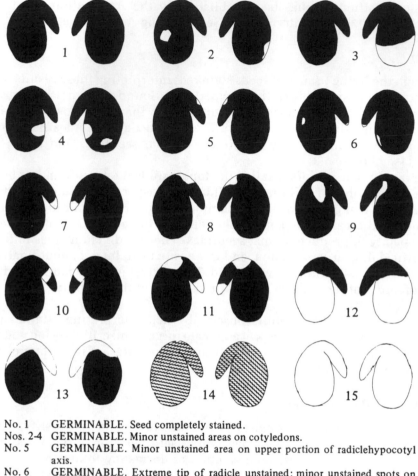

No. 1   GERMINABLE. Seed completely stained.
Nos. 2-4 GERMINABLE. Minor unstained areas on cotyledons.
No. 5   GERMINABLE. Minor unstained area on upper portion of radiclehypocotyl axis.
No. 6   GERMINABLE. Extreme tip of radicle unstained; minor unstained spots on cotyledons.
No. 7   NON-GERMINABLE. More than extreme tip of radicle unstained.
No. 8   NON-GERMINABLE. Unstained area at juncture of cotyledons and radicle-hypocotyl axis.
No. 9   NON-GERMINABLE. Unstained area near point of attachment of cotyledons and radicle-hypocotyl axis over location where plumule develops.
No. 10   NON-GERMINABLE. Radicle-hypocotyl axis bisected by unstained area.
No. 11   NON-GERMINABLE. Unstained areas on radicle-hypocotyl axis and at point of attachment of cotyledons to axis.
No. 12   NON-GERMINABLE. More than one-half of cotyledonary tissue unstained.
No. 13   NON-GERMINABLE. Radicle-hypocotyl axis unstained.
No. 14   NON-GERMINABLE. Seed stained off color, grayish-red, orange-red or glassy or transparent red color.
No. 15   NON-GERMINABLE. Seed completely unstained.

Figure 5.3. *Tetrazolium staining patterns and their interpretation for clover seeds. Black areas represent stained, living tissue; white areas represent unstained, dead tissue. Pairs represent both sides of the seed. (From AOSA (1970).)*

tests that have been attempted (with varying success) has been prepared by Gadd (1950) of the Swedish seed testing station. The following information was taken largely from his review.

Vital Coloring Methods

The principle of the vital coloring methods is the differential coloration of live versus dead tissues when exposed to certain dyes. An early method used sulfuric acid. More recently, indigo carmine and other aniline dyes have been used. This dye stains dead tissue blue, but it is incapable of penetrating live tissue, which remains unstained. Gadd states that it is particularly useful for determining viability of forest tree seeds.

Enzyme Activity Methods

These methods measure enzyme activity of imbibed seeds as an indication of their viability. Some enzymes that have been measured are the hydrolyzing enzymes, which are capable of splitting high-molecular organic compounds of proteins, starch, and fats into less complicated soluble substances. Examples of such enzymes are lipase, diastase and amylase. Another group of enzymes are the desmolases, which may be divided into two groups, the oxidases (catalase and peroxidase), or oxidizing agents, and the dehydrogenases. They are directly involved in the respiration process and, thus as a group, are closely correlated with seed viability.

Oxidase Method 1-Catalase. Many workers have used the catalase content of seeds as an estimate of seed viability (Leggatt 1929), although others have doubted its usefulness because of its lack of absolute correlation with seed viability. Catalase activity varies over the course of germination, and the catalase content of immature grains has been reported to be higher than that of ripe grains. Although some workers have reported good correlations between viability and catalase content, others have not. The probability of errors in the detailed and complex method of measuring catalase activity makes it an unreliable measure.

Oxidase Method 2-Peroxidase. The peroxidase content seems to be more closely correlated with viability than is the catalase content. McHargue (1920) used a technique involving guaiacol which, in the presence of H_2O_2, is transformed to blue tetraguaiacoquinone. By pretreating the seed sample with H_2O_2 and then grinding the seed, followed by colorimetric determinations, he achieved results that closely agreed with germination tests. Brucher (1948) used a similar method without the disadvantage of grinding the seed. He soaked the seeds for 12 hours in a mixture of guaiacol and benzidine in 10% dilutions of saturated alcohol solutions and then treated them with the reagent. Unlike McHargue's technique, this method permitted evaluation of each individual seed, and close correlation was found between its results and those of regular germination tests. Brucher considered embryos that were slightly stained or well stained to be capable of germinating. The disadvantage of these methods is that the color disappears rapidly.

Other Oxidases. Gadd reported that other oxidases generally disappeared gradually from mature seeds and were unsuitable for viability tests. Phenolase has been tested but with unreliable results.

Dehydrogenase Activity Tests. These chemical tests are based on the color changes of certain substances, depending on whether they are in an oxidized or a reduced state. The tetrazolium test, described earlier, is the best example of a successful dehydrogenase activity test. Another method is based on the change of methylene blue to colorless methylene white; however, the seed must be ground and the determination made in a vacuum, since exposure to air reoxidizes the methylene blue. This process is also complicated by the presence of microorganisms that cause a similar color change. Another test utilizes the dehydrogenase reduction of dinitrobenzene to a red nitrophenylhydroxylamine compound in the presence of ammonia. Gadd reported that the color reaction was rapid, and at 40°C takes place in one to two hours; however, the color quickly disappears and the substance itself is poisonous.

The Selenite Method. The selenite test was a biochemical test of the 1930s based on the reduction of colorless selenium salts to red elementary selenium by dehydrogenase activity of living cells. Gadd states that the selenite principle was used as early as 1900 with bacteria cultures; its potential in seed viability testing was discussed by Hasegawa (1935). It was popularized by Eidmann (1938) in the 1930s. Eidmann worked primarily with forest tree seeds, which he prepared by halving the embryo, piercing the seed coat or excising the embryo. The usual method, according to Gadd, is to soak the seeds for about 24 to 48 hours at 20 to 30°C, at which time they are washed in water and the color reaction determined. As in the tetrazolium test, both the distribution and intensity of coloration are important in interpreting seed viability. Gadd suggested the following classification of selenite results: (1) completely and intensively red-colored embryos (germinable), (2) slightly colored to intensively colored embryos with pale spots (germinable), (3) very slightly colored to entirely uncolored embryos and those with larger uncolored spots amounting to more than one-third of the surface (nongerminable). Gadd and Kjaer (1940) reported results of testing cereals with a combination indigo-carmine-biselenite method in which viability estimates were improved. A distinct disadvantage of the selenite method is that a poisonous gas slowly develops, making it dangerous for large-scale testing.

Free Fatty Acid Tests

The breakdown of fats during seed germination was discussed in Chapter 4. Similar degradative reactions may occur in seeds as deterioration progresses, especially under high moisture levels, high temperatures, and microorganism infestation, resulting in the formation of free fatty acids. Consequently, under such conditions, the level of free fatty acids has been suggested as an index of viability. At best, the test provides only a broad quantitative estimate of the general level of viability of a seed lot. It has never attained status as a recognized viability test and is not applicable to the modern seed industry.

OTHER VIABILITY TESTS

Hydrogen Peroxide (H_2O_2) Test

In early seed germination tests, particularly those with tree seeds, hydrogen peroxide was used as a seed treatment to minimize the effects of fungi. It was soon recognized, however, that the compound stimulated germination. The reason(s) for this stimulation remain unknown, although it has been suggested that the hydrogen peroxide degrades to water and one-half molecule of oxygen, which enhances the immediate oxygen environment surrounding the seeds and thus stimulates germination. The test is conducted by cutting the seed coat at the radicle end allowing a 1% solution of hydrogen peroxide to permeate the interior of the seed. This treatment results in more rapid root protrusion compared to the standard germination test.

There are a number of other seed viability tests that focus on the integrity of the seed coat which can have an influence on imbibition damage, seed leakage, and susceptibility to invasion by pathogens. These include the ferric chloride, indoxyl acetate, fast green, and sodium hypochlorite tests.

Ferric Chloride Test

Mechanically injured legume seeds turn black when placed in a solution of ferric chloride (Hardin 1980). This is a very fast and useful test that provides the farmer or seed producer a rapid estimate of the percentage of abnormal seedlings expected from a crop. The seed is placed in a 20% solution of $FeCl_3$ for 15 minutes, at which time all black staining seeds are separated. Because of the rapidity of this test, a seed producer can examine seeds immediately after conditioning and make adjustments to equipment needed to reduce mechanical damage.

Indoxyl Acetate Test

Any damage to seed coats, particularly in soybeans and other large-seeded legumes, is an indication of mechanical abuse during harvesting and conditioning and serves as an entry site for pathogen infestation. The indoxyl acetate test (French et al. 1962) is a rapid laboratory test that reveals seed coat damage in soybeans and other light-colored legume seeds. Seeds are soaked for 10 seconds in a 0.1% solution of indoxyl acetate prepared in 95% ammonia for 10 seconds and allowed to air dry. Lesions in the seed coats that are difficult to detect become visible because they turn purplish-green against the yellow or white seed cotyledon background.

Fast Green Test

The fast green test reveals physical fractures in the seed coat of light-colored seeds such as corn. Seeds are soaked in a 0.1% fast green solution for only 15–30 seconds. During this period, the fast green penetrates any area of the seed coat which has been fractured and stains the endosperm green. After the soak period, the seeds are washed and the fractures then become apparent in the seed coat.

Sodium Hypochlorite Test

The sodium hypochlorite test is used to identify soybean seeds with seed coat damage. Seeds are immersed in a dilute solution of sodium hypochlorite for 10 minutes. Seeds with seed coat cracks rapidly absorb the sodium hypochlorite and swell two to three times their original size. This enables identification of seeds with seed coat cracks from those without seed coat deformations.

Other seed viability tests focus on seed membrane and embryo integrity such as the conductivity, excised embryo, and x-ray tests.

Conductivity Test

Seed producers have long dreamed of a simple method for determining seed viability by merely subjecting them to an electrical current and noting the different responses of live and dead seeds. According to Gadd this concept has been used with varying degrees of success beginning with Waller in 1901, who showed that live and dead seeds gave different "blaze currents" that could be measured with a galvanometer. A more reliable electrical method was suggested by Fick and Hibbard (1925) who found that after soaking a seed sample in water for a few hours under controlled temperature conditions, the conductivity of the solution reflected the general level of viability of the seed sample.

Conductivity tests are based on the premise that as seed deterioration progresses, the cell membranes become less rigid and more water-permeable, allowing the cell contents to escape into solution with the water and increasing its electrical conductivity. In the past, the conductivity test suffered from the disadvantage of being a bulk test where 50–100 seeds were soaked simultaneously and a mean conductivity value obtained for all seeds. Recently, a commercial instrument (Figure 5.4), has been developed and possesses the capability of monitoring individual seed conductivity values. The instrument provides a rapid indication of seed viability for seed lots (McDonald and Wilson 1979) and brings a sophistication to the concept of conductivity tests by improving the value and reliability of test results. However, doubt remains about the reliability of test results in the medium quality range.

Excised Embryo Tests

The excised embryo test (Figure 5.5) provides a unique way of assessing seed viability that can greatly reduce the time required for viability estimates of dormant seed. It is particularly useful for seed of woody species for which the time for viability estimates can be dramatically reduced. Most of the original research on this technique was performed by Flemion (1936, 1938, 1941, 1948) at the Boyce Thompson Institute. She observed that if such embryos of dormant seeds were carefully removed without injury and placed on a moist blotter or filter paper under favorable conditions they would readily grow and turn green.

When embryo dormancy occurred, growth was slow, although much more rapid than in the intact seed. Where extreme embryo dormancy existed, no growth occurred. Even in the latter case, dormancy was more easily overcome when the embryos were excised than when in the intact seed.

The excised embryo method is especially valuable with seeds of trees and shrubs where dormancy is a major problem in evaluating seed viability, and germination sometimes takes as long as six months. Laboratories that test considerable quantities

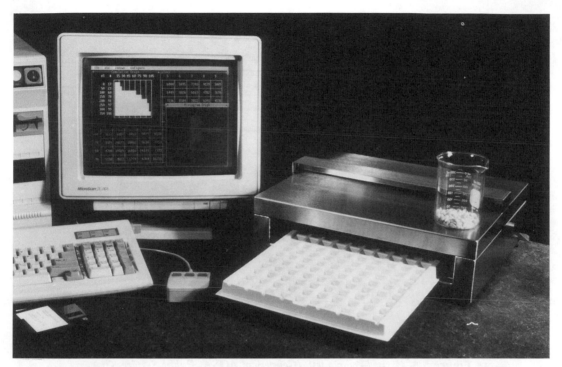

Figure 5.4. *This machine has the capability of measuring the conductivity of leachate from 100 seeds sequentially (Courtesy of Wayfront, Inc., Ann Arbor, MI).*

of tree and shrub seeds routinely conduct excised embryo tests; however, a standard germination test may be performed for comparative results.

One disadvantage of the excised embryo test is the high degree of skill and time required to prepare the embryos for the test. Considerable caution must be taken to avoid injuring the embryo during removal. Another disadvantage is that the test does not reveal damage to the embryo other than in the root-shoot axis that might prevent normal germination of the intact seed.

X-Ray Tests

Although the x-ray test is not a viability test, it does provide information that can help assess viability. It can reveal morphological deficiencies that indicate the structural potential for viability. However, by using different metallic salts with different absorption capacity into live and dead tissue, it can be used to distinguish viable from nonviable seeds.

The most beneficial use of x-rays in seed testing is to obtain a quick indication of abnormal morphology or mechanical damage that might impair germination capacity. Probably most x-ray work has been done on seeds of sugar beets and forest tree seeds. In both cases, it is particularly useful, since it reveals the inner seed structure within the hard seed coat and shows developmental deficiencies (Figure 5.6) as well as mechanical

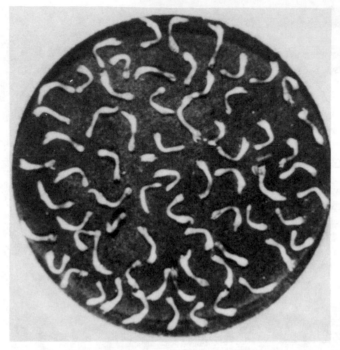

Figure 5.5. Excised embryo test. (Courtesy of AOSA, Public Service Kit.)

fractures. The Association of Official Seed Analysts has recently developed a handbook for x-ray testing of seeds (AOSA 1979).

Questions

1. What do you consider to be the most useful seed viability test and why?
2. Do you believe the germination test will ever be replaced as the most common seed viability test?
3. Does the seed analyst's concept of germination differ from that of the layperson? How?
4. What is the principle of the tetrazolium test?
5. Do any of the other biochemical viability tests have use in routine seed analysis?
6. List the advantages of other viability tests, including those for free fatty acid, indoxyl acetate, conductivity, excised embryo, and x-ray tests.
7. List the substrate, temperature, and final count days recommended for germination of tall fescue seeds according to the AOSA "Rules for Testing Seeds."

General References

Association of Official Seed Analysts. 1970. Tetrazolium testing handbook for agricultural seeds. *Association of Official Seed Analysts Handbook No. 29*, ed. D. F. Grabe.

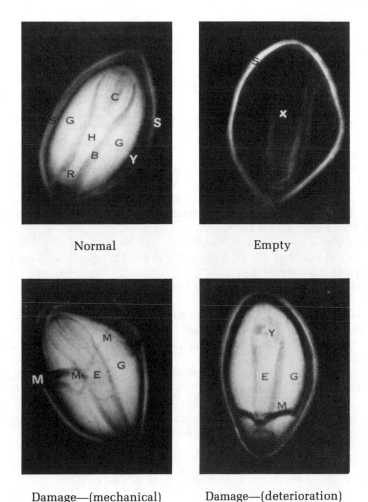

Normal	Empty
Damage—(mechanical)	Damage—(deterioration)

Figure 5.6. *Radiograph of longleaf pine seed showing normal embryo development, empty seed, damaged (mechanical) and damaged (deterioration). (From "Radiographic Analysis of Agricultural and Forest Tree Seed," Handbook No. 31)*

Association of Official Seed Analysts. 1978. Rules for Testing Seeds. *Proceedings of the Association of Official Seed Analysts* 60(2):39.

Association of Official Seed Analysts. 1979. *Radiographic Analysis of Agricultural and Forest Tree Seed: Handbook No. 31*, by E. Belcher and J. Bozzo.

Brücher, H. 1948. Eine Schnellme node zur Bestimmung der Keimfahigkeit von Samen. *Physiologia Plantarum* 1:343–358.

Eidmann, F. E. 1938. Eine neue biochemische Methode zur Erkennung des Aussaatwertes von Samen, *Proceedings of the International Seed Testing Association* 10:203–211.

Fick, G. L., and R. P. Hibbard. 1925. A method for determining seed viability by electrical conductivity measurements. *Michigan Academy of Science, Arts, and Letters* 5:95–103.

Flemion, F. 1936. A rapid method for determining the germinative power of peach seeds. *Contributions from Boyce Thompson Institute* 8:2154–293.

————. 1938. A rapid method for determining the viability of dormant seeds. *Contributions from Boyce Thompson Institute* 9:339–351.

————. 1941. Further studies on the rapid determination of the germinative capacity of seeds. *Contributions from Boyce Thompson Institute* 11:455–464.

————. 1948. Reliability of the excised embryo method as a rapid test for determining the germinative capacity of dormant seeds. *Contributions from Boyce Thompson Institute* 15: 229–241.

French, R. C., J. A. Thompson, and C. H. Kingsolver. 1962. Indoxyl acetate as an indicator of cracked seed coats of white beans and other light colored legume seeds. *Proceedings of the American Society for Horticultural Science* 80:377–386.

Gadd, I. 1950. Biochemical tests for seed germination. *Proceedings of the International Seed Testing Association* 16:235–253.

Gadd, I., and A. Kjaer. 1940. Uber die Yerwendbarkeit der Selenund Indigokarmin-methoden bei der Prufung von frost-und fusariumgeschadigten Getreide. *Proceedings of the International Seed Testing Association* 12:140–149.

Hardin, E. E. 1980. Personal communication.

Hasegawa, K. 1935. On the determination of vitality in seed by reagents. *Proceedings of the International Seed Testing Association* 7:148–153.

International Seed Testing Association. 1985. Handbook on tetrazolium testing. Zurich: ISTA. 72 pp.

Lakon, G. 1928. Ist die Bestimmung der Keimfahigkeit der Samen ohne Keimversuch moglich. *Angewandte Botanik (Zeitschrift der Vereinigung fur angewandte Botanik)* 10:470.

Leggatt, C. W. 1929–1930. Catalase activity as a measure of seed viability. *Scientific Agriculture* (Ottawa) 10:73–110.

McDonald, M. B., Jr., and D. O. Wilson. 1979. An assessment of the standardization and ability of the ASA-610 to rapidly predict potential soybean germination. *Journal of Seed Technology* 4(2):1–12.

McHargue, J. S. 1920. The significance of the peroxidase reaction with reference to the viability of seeds. *Journal of the American Chemical Society* 42:612–615.

6

Seed
Dormancy

The ability of seeds to delay their germination until the time and place are right is an important survival mechanism in plants. Seed dormancy may be a complex and puzzling challenge to the seed analyst and the seed researcher, but it is the method through which plants are able to survive and adapt to their environment.

Seed dormancy is a genetically inherited trait (Naylor 1983) whose intensity is modified by the environment during seed development. Plants with a long history of domestication generally show less seed dormancy than wild or recently domesticated species. When domesticated species exhibit dormancy, they become a problem to seed producers, their customers, and seed analysts. However, a degree of dormancy in certain crops (e.g., winter cereals) is desirable since it prevents preharvest sprouting and helps maintain seed quality. However, dormancy may cause seeds of numerous species to remain ungerminated in the soil for many years. This explains the presence of unwanted crop plants or weeds in fields that are cultivated regularly. Considerable attention is now given to the composition of species in soil seed banks (Fenner 1992; Leck et al. 1989).

A common misconception of seed dormancy is that it is merely a resting state in the absence of suitable germination conditions. This state is often called *quiescence*. However, true dormancy is defined as a state in which seeds are prevented from germinating even under environmental conditions normally favorable for germination. Several physical and physiological mechanisms of dormancy, including both primary and secondary dormancy, occur in seeds.

PRIMARY DORMANCY

Primary dormancy is the most common form of dormancy and takes two forms: exogenous and endogenous dormancy.

Exogenous Dormancy

Exogenous dormancy is a condition in which the essential germination components (e.g., water, light, and temperature) are not available to the seed and thus it fails to germinate. This type of dormancy is generally related to physical properties of the seed coat. However, proper light conditions and other environmental stimuli favorable for germination may also be absent. Thus, this form of dormancy is under *exogenous* control.

Factors Responsible for Exogenous Dormancy

There are three factors responsible for exogenous dormancy: water, gases, and mechanical restriction.

Water. The impermeability of seed coats to water is typical of many species in a number of families (e.g., *Fabaceae, Malvaceae, Chenopodiaceae, Liliaceae*). Seeds that exhibit water impermeability are known as *hard seeds*. Water impermeability is caused by both genetic and environmental factors.

The effect of genotype (variety), or inheritance, on seed coat impermeability is variable. Dexter (1955) reported only small differences in seed impermeability of different alfalfa varieties. However, most workers have found this trait to be highly heritable. One study showed that 80% of 388 bean varieties studied had impermeable seed coats with a range of 1–79% (Gloyer 1932). Genetic control of seed impermeability was demonstrated in another study in which crimson clover breeding lines were selected for water impermeability for nine generations during which the impermeable seed content was increased from 1% to 63% (Bennett 1959). The result of this breeding program was the crimson clover variety "Chief" with superior reseeding characteristics. Two genes control hardseededness in cotton seeds (Lee 1975). Similarly, two or more genes control the semihard condition in garden bean seeds (Dickson and Boettger 1982) and a single recessive gene modifies the expression of hard seeds in lentil (Ladizinsky 1985).

The environment also influences seed impermeability, although little is known about the nature of this control. Weather and soil conditions during the final stages of seed maturation are especially influential. Lee (1975) showed that the two genes which control hardseededness in cotton were expressed more in dry conditions during seed maturation. Aitken (1939) reported that subterranean clover seeds produced on plants under moisture stress were more prone to hardseededness than seeds produced on plants without a moisture stress. Hill et al. (1986) concluded that high soil moisture availability during soybean seed fill reduced seed coat impermeability. Thicker and more impermeable seed coats occurred on soybean seeds developing under mineral nutrient-deficient conditions (Nooden et al. 1985).

While genetics and the environment modify the expression of hardseededness, the cause of this trait is a subject of considerable debate. What is known is that impermeability continues to increase with decreases in seed moisture content (Hyde 1954; Quinlivan 1971; Standifer et al. 1989) and seeds do not become impermeable until their moisture content decreases to about 14 %. Hyde (1954) demonstrated that the dry-down process in hardseeded white clover seeds was regulated by the hilum which acted as a one-way valve, closing when relative humidities were high and opening when relative humidities were low. This results in a continual loss of moisture content in hard seeds (Figure 6.1). A similar mechanism has been proposed for species in the *Caesalpinaceae* and *Mimosaceae* (Werker 1980/1981).

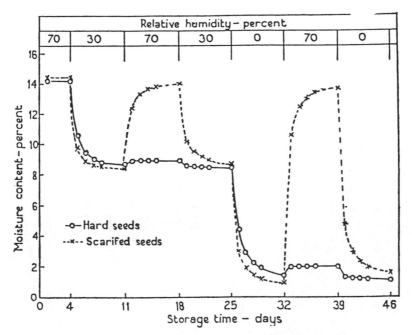

Figure 6.1. *Changes in moisture content in white clover (Trifolium repens) hard or scarified seeds transferred once a week successively to chambers of different relative humidity (70%, 30%, 0%). (From Hyde (1954).)*

The actual cause of hardseededness has been attributed to both physical and chemical attributes of the seed coat (Tran and Cavanagh 1984; Egley 1989). The impermeability to water may be due to the presence of a cuticle and a well-developed layer of palisade cells or both. Heavy deposits of suberin, lignin, or cutin are common in the integuments of many legume seeds (Corner 1976) as well as those of other hard-coated species. The cuticle of bean seeds can effectively limit water penetration, leaving only the micropyle for imbibition (Pammel 1899). It has been demonstrated that Great Northern bean seeds with the micropyle sealed off gain only 0.25% of their weight while control seeds gain 79%. According to Kyle (1955), the raphe of Great Northern bean seed contributes more to water entry during the first 24 hours than either the micropyle or seed coat. Cutin deposits have been reported in the nucellar layers of watermelon seeds (Thornton 1968).

In other seeds, the impermeability to water may be related to the fine structure of the hilum (Gutterman 1993). In some hard seeds, a mucilaginous layer beneath the cuticle which is hard and thick has been observed in legume seeds (Werker et al. 1979). In cotton (Figure 6.2), the palisade layer becomes discontinuous at the chalazal region and the cap is blocked by a plug of closely packed parenchyma and mesophyll cells that are high in tannin content. In those cotton seeds that are hard, this palisade layer is tightly adherent to the chalazal cap and its cells are heavily lignified (Werker 1980/1981). In other seeds, the strophiolar cleft, which is a small opening in the seed coat near the hilum, is thought to control water uptake. The strophiolar cleft is filled with a corklike substance of suberin called the strophiolar plug (Hagon and Ballard 1970;

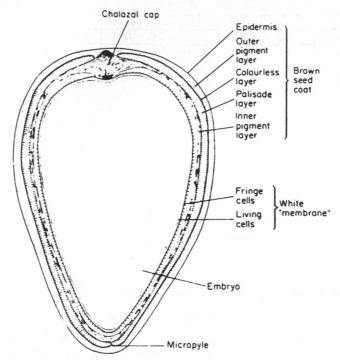

Figure 6.2. *A schematic drawing of cotton seed coat. (From Marchaim et al. (1974).)*

Dell 1980; Kelly and Van Staden 1987). This plug must be removed or loosened before water uptake can occur. Vigorous shaking of seeds possessing this mechanism can either loosen or remove the strophiolar plug and permit water entry. This process is called *impaction* and may be used to break the dormancy of sweetclover (*Melilotus alba*). Other water-impermeable seeds do not possess a strophiolar plug and the seed coat must be abraded or punctured to eliminate dormancy. Thus, different parts of the seed are important in the control of water entry during the initial stages of imbibition.

   Gases. The impermeability of gases through the seed coat has been described as a mechanism of dormancy governed by the seed coat. This is somewhat surprising since water and oxygen represent small molecules of similar molecular weight, yet the seed coat is able to be selectively permeable to one but not the other. In addition, the mechanism by which gases are restricted from the embryo is difficult to determine because of the volatile nature of the gases and the small seed tissues through which the gases flow. Still, the nucellar membrane of cucumbers (Brown 1940), the pericarp of *Fraxinus excelsior* (Villiers and Wareing 1964), and the endocarp of coffee "seeds" (Huxley 1965) are known to restrict the entry of oxygen. The general mechanism of action is thought to be related to seed coat permeability to oxygen: Dormant seeds are less permeable to oxygen, thus retarding the aerobic metabolism required for germination. This appears to be the case in a number of grasses (Major and Roberts 1968; Landgraff and Juntilla 1979; Probert et al. 1985).

   The causes of gas impermeability in seeds have been attributed to physical and biochemical barriers. From a physical perspective, it may be that the mere process of imbibition replaces pore spaces in the seed coat with water that hinders the ready movement of gases to the embryo. Many cruciferous seeds also have epidermal cells in the seed coat that are mucilaginous and swell when wetted (Vaughan and Whitehouse 1971). This makes the path of oxygen diffusion longer before it actually reaches the embryo. From a biochemical perspective, chemical compounds in the seed coat may consume oxygen as it permeates the testa, thus reducing the amount of oxygen to the embryo. For example, apple seeds have restricted oxygen permeability at 20°C compared to 4°C (Come 1968) indicating a temperature-oxygen permeability interaction. Later studies suggested that the high level of phenolic compounds found in the seed coats were oxidized to quinones by enzymatic reactions which were more rapid at the higher temperatures, thus reducing the rate of oxygen diffusion to the embryo (Come and Tissaoui 1973). Peroxidases found in seed coats also have been reported to reduce the availability of oxygen to the embryo (Renard and Capelle 1976). Other biotic factors also influence seed coat permeability to gases. The decomposition of soil plant materials produces saponins which alter oxygen permeability to alfalfa seeds (Marchaim et al. 1972) and possibly other seeds as well.

Other seeds have been shown to be differentially permeable to oxygen and carbon dioxide. The nucellar membrane of cucumber seeds is one example in which the inner membrane is more permeable to CO_2 (15.5 ml/cm^2/hr) than to oxygen (4.3 ml/cm^2/hr) (Brown 1940). Perhaps the best-known example of seed coat impermeability to oxygen is *Xanthium* or cocklebur. For many years, we have known that the dimorphic seeds present in the bur differ in their germination capacity. The upper, smaller seed requires pure oxygen for 100% germination, while the lower seed needs only 6% oxygen for complete germination. The germination of these dimorphic seeds under varying oxygen levels was used as the classical example to illustrate the selective gas permeability in some seeds. However, other data now indicate that the small, upper seed of *Xanthium* contains a much greater quantity of inhibitor than the larger, lower seed (Wareing and Foda 1957). It has been suggested that the small seed requires more oxygen to oxidize and inactivate the inhibitor before 100% germination is achieved (Porter and Wareing 1974). Thus, this classical example of *Xanthium* demonstrating seed coat impermeability to oxygen documents the complexity of seed coat effects on impermeability of gases as a cause of seed dormancy.

Mechanical Restriction. Dormancy has also been attributed to the physical restraint by the seed coats on an enlarging embryo. This assumes that the thrust developed during imbibition and growth is inadequate to rupture the seed coat and permit germination. Most evidence to support this contention is based on the observation that removal of a thick-walled seed coat produces germination. According to Koller (1955), this type of dormancy has been described in seeds of water plantain (*Alisma plantago*), pigweed (*Amaranthus retroflexus*), raspberry (*Rubus idaeus*), peach (*Prunus persica*), and cherry (*Prunus carasus*). However, it should be noted that seed coats are often the source of inhibiting substances which are also eliminated during seed coat removal. Additionally, seed coats may interfere with the leaching of inhibitors or restriction of water flow, accounting for a myriad of changes that occur with the removal of seed coats. To date, no experimental evidence unequivocably demonstrates that seed coats act as a mechanical obstruction to the germinating embryo. More studies are needed to measure the mechanical resistance of the seed coats *as well as* the force produced by a germinating embryo. Only one study has attempted to provide this information. Esashi and Leopold (1968) concluded that dormant seeds of *Xanthium* required slightly more force to rupture the seed coats than nondormant seeds. More importantly, they showed that the nondormant embryo developed twice the thrust of a dormant embryo, clearly indicating that the physical restraint of the seed coat was not the major dormancy-imposing mechanism for these seeds (Table 6.1). Further studies of this nature are needed to establish whether seed coat restriction is a causal factor in seed dormancy.

Methods of Breaking Exogenous Dormancy

Under natural conditions, exogenous dormancy is overcome by the freezing-thawing of the soil, ingestion by animals, microorganism activity, forest fires, natural soil acidity, and other factors. All of these factors affect the integrity of the seed coat in some way. For example, temperature fluctuations cause a gradual expansion and contraction of cells that eventually disrupt the testa at weak points such as the strophiole (Quinlivan 1966; Taylor 1981; Egley 1989) and leads to germination (Figure 6.3). Such processes may take many years to complete and thus serve to expand the range of time over which seeds germinate. While this may represent an advantage to some species, in many cases we need to ensure that crop seeds will germinate more rapidly and uniformly. Overcoming this dormancy is accomplished by the mechanical and chemical removal of the seed coat, a process called *scarification*.

Mechanical Scarification. Grinding seeds with abrasives or sand or shaking them (impaction) are techniques that are often used to scarify seed coats. Other techniques such as heating, chilling, drastic temperature shifts, brief immersion in boiling water, piercing the seed coat with a needle (Table 6.2), or exposure to certain radio frequencies alter seed coat integrity, permitting penetration of both water and gases. However, the duration of these treatments is critical, since prolonged treatment may result in seed damage while brief treatments may not be sufficient to break dormancy.

Chemical Scarification. Seeds may also be treated with chemicals to cause degradation of the seed coat. Sulfuric acid has been used most widely and is effective in industrial and concentrated forms. Other compounds such as sodium hypochlorite and hydrogen peroxide have also been reported to scarify seeds (Hsiao and Quick 1984). However, chemical scarification has not been commercially popular because the materi-

Table 6.1. The Amount of Embryonic
Thrust Developed and Force Required to Rupture
the Seed Coats of Nondormant and Dormant Cocklebur Seeds

Physiological State	Thrust Developed	Force Required to Rupture
Nondormant	84	67
Dormant	41	56

From Esashi and Leopold (1968).

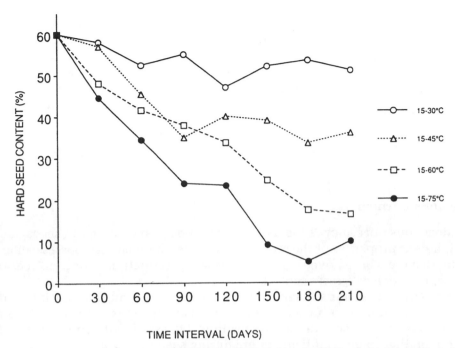

Figure 6.3. *The rates of softening of Geraldton subterranean clover (Trifolium subterraneum) seeds, during 210 days of dry storage with increasing amplitudes of the temperature fluctuations. (From Quinlivan (1966).)*

Table 6.2. The Effects of Piercing and Hull
Removal on the Germination Percentage of Dormant Oat Seeds

Test No.	Tested at 20° C for 10 days	Hulls pierced with needle, then tested at 20° C for 10 days	Hulls removed, then tested at 20° C for 10 days
1	43	94	98
2	36	90	79
3	47	95	91
4	10	84	87
5	77	96	96
6	74	90	92
7	75	91	84
8	71	88	93
9	75	88	93
10	89	95	97
Average	59.7	91.1	91

From Forward (1958).

als are hazardous to handle, the seeds must be thoroughly washed and dried after treatment, and reduction of germination may occur from even slight overscarification.

Recent techniques include the use of selective seed coat enzymes such as cellulase and pectinase to degrade seed coats (Brant et al. 1971; Lester 1985). Since many seed coats contain water-insoluble compounds that retard water entry into the seed, organic solvents such as alcohol and acetone have been used to dissolve and remove these insoluble constituents and permit imbibition (Rolston 1978). It should be emphasized that while scarification enhances germination, it invariably leads to seed injury due to the disruption of essential cells. This enhances fungal invasion and mechanical injury.

Endogenous Dormancy

Endogenous dormancy is the most prevalent dormancy found in seeds and is due to the inherent properties of the seed. For example, the seed may possess an excess of inhibitor that must be removed or reduced prior to germination. This form of dormancy, therefore, is under *endogenous* control.

Causes of Endogenous Dormancy. Environmental conditions during seed development and maturation influence the duration of endogenous dormancy. Among these factors are day length, moisture status, position of the seed in the fruit or inflorescence, age of the mother plant, and temperature during seed maturation.

The day length experienced by the mother plant influences the dormancy of developing seeds. This effect is greatest in the final stages of seed maturation (Gutterman 1978). Plants differ in their response to photoperiod and their expression of seed dormancy (Gutterman 1993). For example, longer days promote the germination of *Polypogon monspeliensis* (Table 6.3), *Carrichtera annua*, and *Cucumis prophetarum* seeds. Short days enhance the germination in *Chenopodium album*, *Portulaca oleracea*, and *Lycopersicon esculentum*. The reasons for the day-length effect on seed dormancy are still not known. In some cases, long days result in increased seed coat thickness (Dorne 1981; Pourrat

Table 6.3. The Influence of Different Day Lengths, Under Greenhouse and Outdoor Conditions, on *Polypogon Monspeliensis* During Growth and Seed Maturation and on Seed Germinability

Day length during growth of mother plant and seed maturation (h)	Seeds from greenhouse plants Germination (%)		Seeds from outdoor plants Germination (%)	
	3 days	7 days	3 days	7 days
9.0	0.0	0.0	1.5	10.0
11.0	0.0	0.0	9.5	17.0
12.0	17.5	19.0	64.0	86.5
13.5	38.0	44.5	98.5	98.5
15.0	60.5	66.0	97.0	98.5
18.0	90.5	91.0	98.0	99.0
Control natural daylength	93.0	96.0	91.5	93.0

From Gutterman (1982).

and Jacques 1975). In other cases, fruits matured under long days evolve more ethylene gas (Gutterman 1978).

The moisture status of the mother plant or developing seed also influences the degree of seed dormancy. Water deficits increase barley seed dormancy when they occur close to flowering but decrease dormancy at the final stages of seed maturation (Aspinall 1965). In *Avena fatua*, water stress on the parent plant during seed maturation reduces the level of dormancy (Simpson 1990). Desiccation of the developing seed decreases dormancy in wheat and barley (Nicholls 1986) and exposure of developing cereal seeds to high relative humidities causes preharvest sprouting (Black et al. 1987). It has been proposed that water stress during seed maturation causes the seed to switch from a developmental to germinative system (Kermode et al. 1986).

Position of the seed on the mother plant also influences dormancy. These effects are primarily attributed to differences in seed maturation which are often expressed by seed weight. The classic example is with members of the *Apiaceae* (carrot) family where the inflorescence is an umbel which is produced in sequential order from primary to secondary, etc. Table 6.4 shows that seeds from primary umbels are heavier, more mature, and more dormant than those produced elsewhere on the plant (Thomas et al. 1979). These differences in dormancy serve as an effective mechanism for maintaining viable but dormant seeds in the soil seed bank (Gutterman 1992).

Age of the mother plant at the time of flower induction can also affect seed dormancy. For example, dormancy of seeds from *Amaranthus retroflexus* (Kigel et al. 1979) and *Oldenlandia corymbosa* (DoCao et al. 1978) plants increased as the age of the plant increased. Temperature during seed maturation also influences the expression of dormancy. In general, higher temperatures produce less dormant seeds. For example, Datta et al. (1972) showed that *Aegilops ovata* plants exposed to 28/22°C produced more germinable seeds than when exposed to 15/10°C.

Table 6.4. Seed Position of Umbel, Weight (mg) and Germination (%) After 21 Days at 18°C in Light, in Three Celery (*Apium graveolens*) Cultivars. LSD at 5% in Parenthesis

Cultivars	Umbel position	Mean seed weight (mg)	Germination (%)
Green snap	Primary	0.590	51
	Secondary	0.440	85
	Tertiary	0.386	94
	Quaternary	0.382	80
		(0.069)	(9.8)
Lathom Blanching	Primary	0.474	50
	Secondary	0.438	72
	Tertiary	0.380	94
	Quaternary	0.348	82
		(0.069)	(9.2)
Ely White	Primary	0.590	59
	Secondary	0.468	62
	Tertiary	0.490	80
	Quaternary	0.520	87
		(0.086)	(7.3)

From Thomas et al. (1979).

Environmental conditions during seed development and maturation also influence the duration of endogenous dormancy. The level of dormancy in seeds of touch-me-not (*Impatiens balsamina*) depends on water supply and mineral nutrition, especially nitrogen, during seed formation and development. Seeds from plants that had been adequately watered and well supplied with nitrogen had less dormancy than those from plants deficient in these factors (Junges and Ludwig 1963). Lettuce (*Lactuca sativa*) seed germination requirements have been shown to differ when grown under different environments (Koller 1962). Stage of maturity has little effect on the duration of endogenous dormancy in Kentucky bluegrass seeds following harvest, but the level of dormancy was directly associated with the seed moisture content at the time of harvest: The higher the moisture content, the greater the degree of dormancy (Delouche 1958). Table 6.5 shows the effect of maturity on endogenous seed dormancy in barley.

Unlike exogenous dormancy, which requires a physical alteration of the seed coat, only physiological changes such as rudimentary embryo maturation, response to growth regulators, changes in temperature, exposure to light, and endogenous rhythms are able to relieve endogenous dormancy in seeds.

Rudimentary Embryo Dormancy

Seeds of some species are shed before they are morphologically mature. This results in dormancy because the immature embryo is unable to germinate. Rudimentary embryo dormancy occurs in *Ranunculus*, *Plantago*, *Fraxinus*, *Viburnum*, *Ilex*, and *Pinus*. Further embryo maturation occurs following seed dispersal and may take a few days or several months. The embryo of holly (*Ilex opaca*) is an undifferentiated mass of cells when the seed is shed, but during subsequent maturation the cells become a well-defined structure (Ives 1923). Cherry seed embryos increase in size, weight, length of leaf primor-

Table 6.5. Timetable of Barley Seed Dormancy During Seed Development (Germination Tests at 15°C)

1–4 weeks after anthesis	4–6 weeks	6–8 weeks	Rain in 9th week	Condition of ripe seed sample
No germination	Rise of germination percentage	Germination percentage over 90%	Reduction of germinability of 14%	Majority nondormant
Moisture content over 80%	Moisture content decreasing	Moisture content ca. 16%	Rise of moisture content to ca. 30%	Some still in primary dormancy
Isolated embryo germinated				Some still in secondary dormancy
Covering layers green	Covering layers turn yellow	Collapse of cells of covering layers, disorganization of their contents		
Primary dormancy	Dormancy disappearing	No dormancy	Secondary dormancy	

From Evenari (1965).

dium, and oxygen uptake of the embryonic axis after being shed from the tree (Pollock and Olney 1959). Similar changes occur in other species (Zagaja 1962; Scott and Waugh 1941; Zagaja and Czapski 1962). In *Isopyrum biternatum*, seed dispersal occurs in late May but embryo growth is delayed by high temperatures until October (Baskin and Baskin 1986). In contrast, *Apium graviolens* seeds require light before embryo growth continues (Jacobsen and Pressman 1979). The changes that occur in dry, dormant seeds, after dispersal is known as *after-ripening*. This process will be more fully discussed later in this chapter.

Physiological Dormancy

Seed dormancy in higher plants is generally believed to be regulated by a balance of endogenous growth inhibitors and promotors (Amen 1968). Thus, dormancy may be considered a result of the presence of growth inhibitors, the absence of growth promotors, or a combination of both. The levels of these endogenous compounds are controlled by certain environmental stimuli such as light and temperature. Many substances that are present in seeds help determine whether a seed will be dormant. The goal of research workers is to identify these componds and determine their metabolic function in order to better understand their role in the regulation of seed dormancy. A large number of compounds have been isolated that can induce dormancy through their influence on metabolic and osmotic inhibition.

Metabolic Inhibition. Certain compounds that are present in seeds inhibit specific metabolic pathways. An example of this type of inhibitor is cyanide, which is found in apple or peach seeds. Such compounds act by suppressing germination through their effect on respiration. However, at very low concentrations cyanide stimulates germination (Taylorson 1988; Tilsner and Upadhyaya 1987) and this effect is pH-dependent (Cohn and Hughes 1987).

Phenolic compounds also inhibit germination, and because of their widespread occurrence, have been regarded as natural germination inhibitors. Compounds such as substituted phenols and cresols have been shown to inhibit germination, although they are not true dormancy-inducing compounds (Picman and Picman 1984; Rice 1983). The first dormancy-inducing inhibitor found was *coumarin*, which is characterized by an aromatic ring and represents an unsaturated lactone structure. Coumarin is widely distributed and rapidly metabolized in seeds and is considered a natural germination inhibitor. Additionally, coumarin derivatives such as the glycosides of the lactone or substituted coumarins have also been found in many fruits, supporting the role of coumarin as a natural seed inhibitor. The exact mechanism of coumarin inhibition has yet to be elucidated, although it is suspected that it interferes with respiration and oxidative phosphorylation and indirectly with the availability of energy by its effect on phosphorous metabolism.

Another extremely active inhibitor of seed germination was discovered in 1966 (Cornforth et al. 1966). Because of its dormancy-inducing properties, the compound was initially named dormin. Later, chemical characterization revealed that dormin had the identical chemical structure of abscisic acid—a compound that was being investigated for its promotion of the abscission of young cotton fruit and its ability to induce bud dormancy in trees. By general agreement, the name of *abscisic acid* (ABA) was given to dormin in 1967.

Since these initial reports, many studies have documented that ABA is present in a wide range of seeds and is a naturally occurring endogenous hormone that is active at very low concentrations. Further, reports have shown that the levels of endogenous ABA are reduced following stratification of dormant ash, rose, and other seeds (Sondheimer et al. 1968; Williams et al. 1973; Webb et al. 1973). ABA has also been shown to be localized in the testa of apple seeds (Lewak and Rudnicki 1977) and is translocated to the embryo during imbibition to induce dormancy. The exact mechanism of ABA inhibition of germination has not been clearly defined. It is known to inhibit the synthesis of enzymes that are important in the early stages of germination (Varty et al. 1983) perhaps by inhibiting their translation from mRNA (Ho and Varner 1976).

Due to the ubiquitous occurrence and reported dormancy-inducing capability of ABA, attempts have been made to determine whether promotive hormones such as gibberellins could interact with ABA in the recognized inhibitor-promotor hypothesis. Many studies have, in fact, shown that ABA can completely or partially reverse the promotive action of either the gibberellins or cytokinins (Rudnicki et al. 1972). Additionally, it has been demonstrated that as ABA levels decrease, gibberellin and cytokinin levels increase during stratification in seeds such as sugar maple (Van Staden et al. 1972; Webb et al. 1973). Thus, an inverse relationship between inhibitors and promotors occurs during the dormancy-breaking (after-ripening) process. However, it should be emphasized that these reports are still preliminary and many were conducted using bioassay techniques. Correlations between inhibitor levels and dormancy in seeds do not demonstrate direct physiological roles until these are determined. More refined analyses are necessary before a general role in the induction of dormancy can be attributed to ABA. We should also emphasize that ABA is not the only natural germination inhibitor in seeds. Other compounds such as coumarin may serve a similar function.

Osmotic Inhibition.   Many substances possessing high osmotic pressures can inhibit the germination of seeds. Compounds such as sugars or salts in sufficient concentration may compete so successfully for water that the seed never becomes fully imbibed and thus remains ungerminated (quiescent). However, such seeds readily germinate when removed from the osmotic-inhibiting environment. Many such substances are located in the fruits or fruit walls that surround the seeds. One study concluded that the osmotic pressure exerted by inorganic substances in the fruit ball of sugar beet is responsible for the inhibition of germination (Duym et al. 1947). In another study, the concentration of electrolytes from sugar beet fruit ball extracts retarded the growth of wheat seedlings (Snyder et al. 1965). Demineralized extracts of sugar beet fruits are also known to inhibit germination of peppergrass and other species (Froschell 1957); the cause is attributed to the presence of a specific organic substance or substances rather than osmotic inhibition. Ferulic and caffeic acids occur in tomato fruit (Ackerman and Veldstra 1949); parasorbic acid in fruits of European mountain ash (*Sorbus aucuparia*) (Kuhn et al. 1943); and a mixture of organic acids exists in fruits of lemons, strawberries, and apricots (Varga 1957). These substances are known to be capable of inhibiting germination. Tomato juice completely inhibits the germination of garden cress (*Lepidium sativum*) seeds even when diluted in a proportion of 1:25 (Meyer et al.

1960). It is well known that seed germination is delayed in most fleshy fruits while the seed is still embedded in the fruit itself. Thus, the high osmotic pressure of such fruit juices undoubtedly contributes to seed dormancy.

Physiological studies of imbibition have frequently utilized the concept of osmotic inhibition to allow partial intake of water into the seed while preventing it from germinating. Initially, many of these investigations were conducted with salts until it was revealed that the seeds often experience deleterious effects due to ionic toxicity. Subsequently, such studies began to employ mannitol as the osmotic-inhibiting compound. However, it was again demonstrated that mannitol was imbibed into the seed and altered later metabolic functions. More recent studies have utilized polyethylene glycol, which is a long-chain polymer incapable of being taken into the intact seed but still water soluble and, therefore, osmotically active.

Methods of Breaking Physiological Dormancy. There are various methods of alleviating physiological dormancy. Seeds that are dormant due to osmotic inhibition can be germinated after removing the seed from the influence of the inhibitor or diluting the inhibitor from around the seed, a process called *leaching*. Leaching typically requires exposing the seeds to an excess of water that dilutes or removes the inhibitor from the seed. An example of seeds that are typically leached to remove an inhibitor from the fruit wall is sugar beet seeds, compared to tomato seed that can be germinated only after removing from the juice which otherwise causes osmotic inhibition.

Seeds that possess metabolic inhibitors located in the seed coat can be relieved of dormancy by removal of the seed coat by mechanical or chemical *scarification*. Conventional scarification to break down otherwise hard, impermeable seed coats to allow water penetration is usually accomplished by various procedures including the use of abrasives or acids, or by simply piercing the seed coat. However, scarification may also be effective in degradation of the seed coat, alleviating physiological dormancy and permitting germination by one of two processes. First, seed coats have often been shown to be the source of inhibitors, and when they are removed, the inhibitor is also removed. Second, seed coats can function as differentially permeable membranes, permitting the entry of water but retarding the loss of inhibitors. Removing the seed coats eliminates the barrier to inhibitors, resulting in leaching of the inhibitors from their site of action, and thus permitting germination. An example of this process is provided by Glory cabbage seed which has an inhibitor in the seed coats (Cox et al. 1945). Soaking these seeds in dilute sulfuric acid for several minutes results in rapid germination.

Temperature Requirement

Seeds with a specific temperature requirement for germination often contain both inhibitors and promotors. Evidence now exists to support the view that dormancy of this type is controlled by an inhibitor-promotor balance that is altered by exposing seeds to low temperatures under imbibed conditions (stratification) or under higher temperatures while unimbibed.

Stratification. It is now well documented that both physical and physiological changes may occur in imbibed seeds exposed to low temperatures. Embryo growth occurs in stratified apple embryos (Table 6.6). Similarly, the embryonic axis of stratified cherry seeds increases in cell number, dry weight, and total length (Olney and Pollock 1960).

Table 6.6. Development of Apple Embryos Isolated After Different Times of Stratification and Subsequently Cultured at 25°C for 9 Days

Weeks of stratification	Germination (%)	Embryos with both greening cotyledons (%)	Embryos with only one greening cotyledon (%)	Mean length of main root (mm)
0	9	7	42	1.9
2	12	22	27	2.2
4	24	73	5	3.7
6	39	97	2	9.1
8	51	100	0	9.9
10	59	100	0	10.9
12	65	100	0	11.9
14	79	100	0	21.4

From Lewak and Rudnicki (1977).

On a cellular basis, increased oxygen uptake and energy supply to the embryonic axis of stratified seeds has been observed. Increases in catalase, phosphatases, alkaline lipase, and peroxidase enzymes have also been detected (Zarska-Maciejewska and Lewak 1976). Thus, it appears that many phases of embryonic development and metabolism are influenced by stratification. All of these may be considered as after-ripening processes.

Shifts in hormonal levels also occur in stratified seeds. For example, the level of ABA drops during stratification of apple, ash, walnut, and hazelnut seeds. However, the addition of exogenous gibberellins can substitute for the stratification requirement in some seeds (Pinfield 1968), implicating this hormone as a promotive agent. Increases in the levels of endogenous gibberellins have been observed during stratification, supporting this premise (Frankland and Wareing 1966). Using gibberellin-deficient mutants, Karssen et al. (1989) showed that low-temperature exposure of imbibed seeds increased their sensitivity to gibberellins. Thus, beyond the observed increases in growth and metabolic activity of stratified seeds, shifts in the levels of inhibitors and promotors have also been detected that further contribute after-ripening and the release from dormancy.

The Rules for Testing Seeds of both the Association of Official Seed Analysts and the International Seed Testing Association (see Chapter 13) list stratification as an aid for germinating seeds of many species. Since the analyst is seldom aware of the history of the seed lot being tested, stratification is ordinarily performed as a routine part of the seed testing procedure. The effect of prechilling (stratification) is shown in Table 6.7.

The germination rate of many seeds may also be increased if exposed to daily alternating temperature cycles. Germination of Chinese red pine (*Pinus densiflora*) and Japanese black pine (*P. thunbergii*) seeds was improved under such conditions. Seeds of tigertail spruce (*Picea polita*), which needed light to germinate at a constant temperature, germinated in the dark under alternating temperatures (Asakawa 1959). The need for alternating temperatures is apparently associated with endogenous dormancy and characterizes many agricultural species.

Stratified seeds are usually preconditioned at temperatures between 3° and 10°C, although the specific temperatures and duration of exposure may vary. For some species, low-temperature stratification is an absolute requirement for germination; for

Table 6.7. The Effect of Prechilling (vs. Control) on the Germination of Oat Seed

Test No.	Tested at 20° C for 10 days	Prechilled at 10° C for 5 days, then tested at 20° C for 5 days
1	43	99
2	36	85
3	47	97
4	10	91
5	77	97
6	74	97
7	75	98
8	71	96
9	75	98
10	89	100
Average	59.7	95.8

From Forward (1958).

others, it may only hasten germination and increase the speed of growth. For some species, stratification may decrease the sensitivity to external conditions, thus increasing the range of temperatures under which germination can proceed. Such is the case for sugar pine (*Pinus lambertiana*) seeds that germinated only at temperatures above 25°C, but after stratification were able to germinate at lower temperatures (Stone 1957). The length of time required for stratification varies depending on the species. Seeds of wild rose (*Rosa multiflora*) require a two-month stratification period (Crocker and Barton 1931), however swamp persicaria (*Polygonum coccineum*) and mild water pepper (*P. hydropiperoides*) may require an 8½-month stratification period (Justice 1944). The seeds of more than 60 species, many of them woody plants, have been listed as requiring low-temperature stratification for germination (Crocker and Barton 1957).

The stratification requirement of a particular seed lot also depends on the intensity of the temperature exposure and seed age. Meyer et al. (1990) showed that the depth of dormancy for a range of ecotypes of *Artemisia tridentata* correlated with the mean January temperature: colder temperatures required longer stratification periods. Freshly harvested seeds of Chinese maple (*Acer truncatum*), which require a two-month stratification, germinated well without pretreatment after a year of storage (Ackerman 1957). Decrease in the persistence of dormancy with increasing age of seeds is a universal characteristic of endogenous dormancy, but the speed at which dormancy is lost varies among species.

After-Ripening. The disappearance of dormancy in dry seeds over time at room temperature is widespread among many seeds and almost universal in cereals. For most cereals, storage for 1 to 2 months at 15 to 20°C suffices to allow maximum germination. In *Phleum arenarium* seeds, longer periods of dry seed after-ripening not only result in increased germination but also increased germination over a broader range of temperatures (Figure 6.4).

Most seeds that undergo after-ripening during dry storage are also responsive to stratification in breaking dormancy. Furthermore, the treatment of non-after-ripened

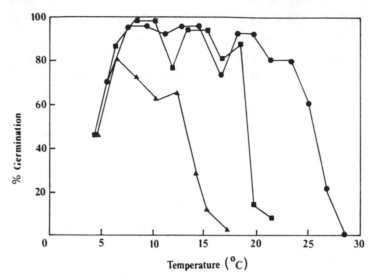

Figure 6.4. *Dry after-ripening in seeds of* Phleum arenarium. *Changes in the germination response to temperature during storage at 15°C and 15% r.h. Germination was tested on a thermogradient bar after 1, (▲); 6 (■); and 13 months (●) storage. The seeds were incubated on wetted filter paper and received an 8-h photoperiod/day. (From Probert (1992).)*

dry seeds with gibberellins can remove the dormancy block present in most cereals and many grasses (Simpson 1990). Wild oat (*Avena fatua*), for example, exhibits this behavior. Although these seeds are dry, there is measurable metabolic activity. In wild oat, significant CO_2 fixation and protein synthesis have been demonstrated (Naylor and Simpson 1961). Other studies have shown that the rate of respiration is fivefold greater in fully imbibed seeds compared to those that experience after-ripening in dry seeds. Because of this reduced rate of respiration, the breaking of dormancy in non-after-ripened seeds is slow. Since gibberellins and stratification help accelerate after-ripening, this process appears essential to allow the embryos to synthesize gibberellins, again suggesting that a shift from inhibitor to promoter must occur before germination can occur.

Light Requirement

The light quantity and quality perceived by a seed depend on its position in the soil, the vegetative covering, and the light absorption characteristics of the seed coat or comparable structure(s). Measurable quantities of light do not penetrate the soil to greater depths than a few millimeters or centimeters depending on soil type (Tester and Morris 1987). Sandy soil has the greatest light transmission which is reduced with increasing levels of humus. In addition, the quality of the light is also modified by soils. Shorter wavelengths of light are absorbed more by soils than longer wavelengths (Bliss and Smith 1985). The presence of a leafed canopy over the soil also reduces the photosynthetically active (400–700 nm) portion of the light spectrum. Thus, canopy shade is richer in far-red and poorer in red portions of the spectrum. The light-absorbing phytochrome pigment which controls germination in seeds is believed to be located in the embryo. As a result, seed coats that are thick and/or intensely pigmented can alter both the quantity and quality of the light perceived by the embryo (Widell and Vogelmann 1988).

Light intensity, wavelength, and photoperiod are all known to affect the germination of seeds that have physiological dormancy. Three well-known species whose dormancy is broken by exposure to red light (670 nm) are lettuce, birch, and Virginia pine.

Duration of light exposure (photoperiod) can also affect seed germination. The seeds of some species respond to short days, others to long days, while still others are unaffected by daylength.

Continuous light may inhibit germination of some seeds, and thus impose a different kind of dormancy. Continuous light is reported to inhibit both onion and leek seeds (Lovato and Amaducci 1965). Although light may initially promote the radicle growth of Douglas fir, an extended period of light may become inhibitory, especially for seeds with large radicle size (Villiers 1961).

For a more complete discussion of the influence of light on seed germination, refer to Chapter 4.

Circadian Rhythms

Plants have long been observed to follow an orderly sequence of growth and developmental processes. There is evidence that they are somehow able to measure time independently of the outside environment, and thus they are able to regulate certain growth and developmental processes (Cummings and Wagner 1968). This orderly sequence of growth and development is referred to as *circadian* rhythms.

Circadian rhythms also seem to influence the pattern of seed germination. They have been classified as those of a single yearly cycle and those with more than one cycle per year (Kummerow 1965). Several hundred species have been reported to show yearly oscillations (Bünning 1965). A periodicity of pigweed (*Amaranthus retroflexus*) seed germination has been reported when the seed was held moist at 20°C for 78 months (Crocker and Barton 1957); two maxima occurred, one at 8–10 months and the other at 20–22 months. Maguire (1969) reported circadian-controlled rhythm patterns in bluegrass seed germination (Figure 6.5). In a study of over 300 species, Baskin and Baskin (1988) showed that seasonal temperature changes influenced the breaking of dormancy in a number of species.

Interaction of Primary Dormancy Mechanisms

The types of dormancy described are by no means mutually exclusive, and more than one mechanism for the imposition of dormancy may be possessed by seeds. Such an example is provided by Indian rice grass (*Oryzopsis hymenoides*) seeds which possess both seed coat (exogenous) and physiological (endogenous) dormancy (McDonald and Khan 1977). In this species, seed coat dormancy can be effectively removed by scarification with sulfuric acid. However, addition of exogenous GA_3 further enhances germination to 70% for freshly harvested seeds (Table 6.8). Dry seeds that have been after-

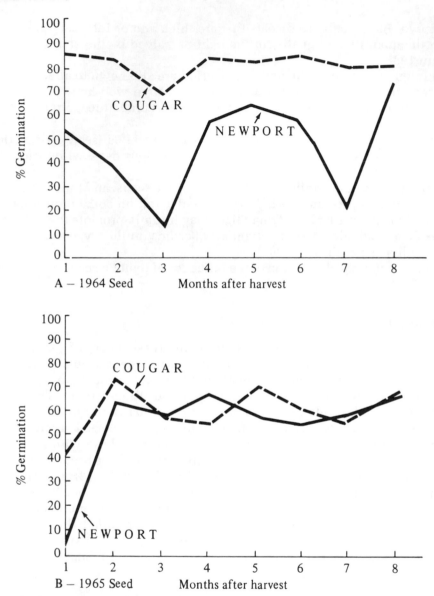

Figure 6.5. *Endogenous germination rhythms observed in Cougar and Newport varieties of Kentucky bluegrass seed (stored at 5°C). (From Maguire (1969).)*

ripened by natural storage are not as responsive to exogenous GA_3, presumably because the seed has had adequate time to synthesize this compound.

Embryo Excision

The complexity of dormancy and its exact cause can often be shown by removing the embryo and growing it independently of the seed. Flemion (1936) was the first to

Table 6.8. Influence of Various Exogenous Hormonal Treatments on the Germination of 1- and 2-Year-Old Intact and Scarified Indian Rice Grass Seeds After 72 Hours Soaking

Treatment	1 year		2 years	
	Intact	Scarified	Intact	Scarified
H_2O	0	31	5	67
10 µM GA_3	0	65	5	76
10 µM Kinetin	0	54	6	72
10 µM ABA	0	1	2	2

From McDonald and Khan (1977).

report that the embryos of dormant seed could be germinated if they were excised and grown separately. Other studies have shown that birch seeds, which normally require light, can be germinated in the dark if excised (Redmond and Robinson 1954). Excised embryos of Canadian hemlock (*Tsuga canadensis*) can be germinated in the dark, whereas intact seeds or those with only the integuments removed required light for germination (Stearns and Olson 1958). The light requirement of Grand Rapids lettuce seed can be eliminated by removing or puncturing the endosperm surrounding the embryo.

It is possible to force germination of peach seeds normally requiring stratification by embryo excision or by seed coat removal, but such forced seeds usually produce dwarfed or deformed seedlings (Mayer and Poljakoff-Mayber 1989).

The excised embryo test for seed viability was discussed in Chapter 5.

SECONDARY DORMANCY

Sometimes nondormant seeds encounter conditions that subsequently cause them to become dormant. This may be caused by exposure of the seed to conditions that favor germination in all respects except one. A study of spring wheat and winter barley reported that secondary dormancy could be imposed by: (a) exposure of dry barley seed to temperatures between 50 and 90°C, (b) seven-day storage of winter barley at high moisture contents at 20°C, (c) one-day storage of spring wheat at high moisture contents in airtight containers at 50°C, and (d) placement of the seed under water and in darkness for one to three days at 20°C (Frischnicht et al. 1961). Exposures of dry seeds to 50°C required four days for imposition of secondary dormancy; to 70°C required four hours; and to 90°C only one hour. Induction of secondary dormancy was possible one and one-half months after seeds reached physiological maturity, although the persistence of dormancy obtained in this way decreased almost continuously as the time between physiological maturity and treatments increased. For example, compared to primary dormancy, which was partly broken, secondary dormancy in spring wheat could not be broken by two weeks of storage at 40°C. However, it was completely broken by treatments of 0.1% gibberellic acid, 0.5 to 1.0% ethanol, low-temperature stratification, removal of the pericarp above the embryo, and storage at 20°C.

We have discussed secondary dormancy as being either thermo- (temperature), photo- (light), or skoto- (darkness) imposed (Evenari 1965), though other causes such as imposition by excess or adverse amounts of water, chemicals, and gases might also be involved. Two suggestions have been made to explain the mechanism of secondary dormancy: (1) the imposition of a block at crucial points in the metabolic sequence leading to germination, or (2) an unfavorable balance of growth-promoting versus growth-inhibiting substances.

Questions

1. Is it correct to describe dormancy as merely a resting state? Explain.
2. What is the difference between hard seed and firm seed?
3. What is reseeding?
4. How can preharvest factors influence hard-seed coat dormancy? Many students confuse scarification with stratification. What is the difference?
5. What is after-ripening?
6. Name at least two species whose seeds have each type of dormancy.
7. What is the argument, if any, against prechilling or stratifying seeds during laboratory germination testing?
8. Do you think that more than one inhibitor is responsible for chemical inhibition of germination in most seeds?
9. Name several chemicals with germination-inhibiting ability.
10. How can secondary dormancy be induced in nondormant seeds?
11. Do you believe there is adequate evidence for endogenous rhythms in seed germination?
12. Distinguish between primary and secondary dormancy.
13. Name at least three ways that seed coats can regulate seed dormancy and cite two procedures used to overcome this dormancy.
14. Describe the inhibitor-promoter concept for the regulation of seed dormancy and suggest potential hormones that might be involved in this interaction.
15. Describe endogenous dormancy and provide at least four factors that can relieve this condition.
16. Differentiate between stratification and after-ripening.

General References

Ackerman, A. M., and H. Veldstra. 1949. The chemical nature of Kockerman's blastocholine from *Lycopersicon esculentum* Mill. *Recueil des Travaux Chimiques des Pays-bas et de la Belgique* 66:411–412 (*Biological Abstracts*, 1949. 23:12727).

Ackerman, W. L. 1957. After-ripening requirements for germination of *Acer truncatum* Bunge. *Proceedings of the American Society for Horticultural Science* 69:570–573.

Aitken, Y. 1939. The problem of hard seeds in subterranean clover. *Proceedings of the Royal Society of Victoria* 51:187–192.

Amen, R. D. 1968. A model of seed dormancy. *Botany Review* 34:1–31.

Asakawa S. 1959. Germination behavior of several coniferous seeds. *Journal of the Japanese Forestry Society* 41:430–435.

Aspinall, D. J. 1965. Effects of soil moisture stress on the growth of barley. III. Germination of grain from plants subjected to water stress. *Journal of Institute of Brewing*. 72:174–176.

Baskin, C. C., and J. M. Baskin. 1988. Germination ecophysiology of herbaceous plant species in a temperate region. *American Journal of Botany* 75:286–305.

Baskin, J. M., and C. C. Baskin. 1986. Germination ecophysiology of the mesic deciduous forest herb *Isopyrum biternaturm*. *Botany Gazette* 147:152–155.

Bennett, H. W. 1959. The effectiveness of selection for the hard-seeded character in crimson clover. *Agronomy Journal* 51:15–16.

Black, M., J. Butler, and M. Hughes. 1987. Control and development of dormancy in cereal. In: *4th International Symposium, Preharvest Sprouting in Cereals*, ed. D. J. Mares, pp. 379–392. Boulder, Colo.: Westview Press.

Bliss, D., and H. Smith. 1985. Penetration of light into soil and its role in the control of seed germination. *Plant Cell Environment* 8:475–483.

Brant, R. E., G. W. McKee, and R. W. Cleveland. 1971. Effect of chemical and physical treatment on hard seed of Penngift crownvetch. *Crop Science* 11:1–5.

Brown, R. 1940. An experimental study of the permeability to gases of the seed-coat membranes of *Curcurbita pepo*. *Annals of Botany* (London) 4:379–395.

Bünning, E. 1965. Endogenous rhythmic activities (in German). *Encyclopedia of Plant Physiology* 15:879–907.

Cohn, M. A., and J. A. Hughes. 1987. Seed dormancy in red rice. V. Response to azide, cyanide, and hydroxylamine. *Plant Physiology* 80:531–536.

Come, D. 1968. Relations entre l'oxygene et les phenomenes de dormance embryonaire et d'inhibition tegumentaire. *Bulletin de la Societe Francaise de Physiologie Vegetale* 14:31–45.

Come, D., and T. Tissaoui. 1973. Interrelated effects of imbibition, temperature and oxygen on seed germination. In: *Seed Ecology*, ed. W. Heydecker. London: Butterworths.

Corner, E. J. H. 1976. *The Seeds of Dicotyledons*. Cambridge: Cambridge University Press.

Cornforth, J. W., B. V. Milforrow, and G. Ryback. 1966. Identification and estimation of (–) – abscisin II ("dormin") in plant extracts by spectropolorimetry. *Nature* (London) 210:627–628.

Cox, L. G., H. N. Munger, and E. A. Smith. 1945. A germination inhibitor in the seed coats of certain varieties of cabbage. *Plant Physiology* 20:289–294.

Crocker, W., and L. V. Barton. 1931. After-ripening, germination, and storage of certain *Rosaceous* seeds. *Contributions from Boyce Thompson Institute* 3:385–404.

———. 1957. *Physiology of Seeds: An Introduction to the Experimental Study of Seed and Germination Problems*. Waltham, Mass.: Chronica Botany Company.

Cumming, B. C., and E. Wagner. 1968. Rhythmic processes in plants. *Annual Review of Plant Physiology* 19:381–416.

Datta, S. C., Y. Gutterman, and M. Evenari. 1972. The influence of the origin of the mother plant on yield and germination of their caryopses in *Aegilops ovata* L. *Planta* 105:155–164.

Dell, B. 1980. Structure and function of the strophiolar plug in seeds of *Albizia lophantha*. *American Journal of Botany* 67:556–561.

Delouche, J. C. 1958. Germination of Kentucky Bluegrass harvested at different stages of maturity. *Proceedings of the Association of Official Seed Analysts* 48:81–84.

Dexter, S. T. 1955. Alfalfa seedling emergence from seed lots varying in origin and hard seed content. *Agronomy Journal* 47:357–361.

Dickson, M. H., and M. A. Boettger. 1982. Heritability of semihard seed induced by low seed moisture in beans (*Phaseolus vulgaris* L.). *Journal of American Society of Horticultural Science* 107:69–74.

DoCao, T., Y. Attims, F. Corbineau, and D. Come. 1978. Germination des grains produits par les plantes de deux lignees d'*Oldenlandia corymbosa* L. (*Rubiaceae*) cultivees dans des conditions controllees. *Physiologia Vegetale* 16:521–531.

Dorne, C. J. 1981. Variation in seed germination inhibition of *Chenopodium bonus-henricus* in relation to altitude of plant growth. *Canadian Journal of Botany*. 59:1893–1898.

Duym, C. P. A., J. G. Komen, A. J. Ultee, and B. M. van der Weide. 1947. The inhibition of germination, caused by extracts of seed balls of sugarbeet (*Beta vulgaris*). *Proceedings, Koninklijke Nederlandsche Akamedie Van Wetenschappen* 50:527–585 (First published in *Biological Abstracts*, 1949, 23:30204).

Egley, G. H. 1989. Water-impermeable seed coverings as barriers to germination. In: *Recent Advances in the Development and Germination of Seeds*, ed. R. B. Taylorson. New York: Plenum Press.

Esashi, Y., and A. C. Leopold. 1968. Physical forces in dormancy and germination of *Xanthium* seeds, *Plant Physiology* 43:871–876.

Evenari, M. 1965. Light and seed dormancy. *Encyclopedia of Plant Physiology* 15:805–847.

Fenner, M., ed. 1992. *Seeds: The Ecology of Regeneration in Plant Communities*. Wallingford, U.K.: CAB International.

Flemion, F. 1936. A rapid method for determining the germination power of peach seeds. *Contributions from Boyce Thompson Institute* 8:298–293.

Forward, B. F. 1958. Studies of germination in oats. *Proceedings of the International Seed Testing Association* 23:20.

Frankland, B., and P. F. Wareing. 1966. Hormonal regulation of seed dormancy in hazel (*Corylus avellana* L.) and beech (*Fagus sylvatica* L.). *Journal of Experimental Botany* 17:596–611.

Frischknicht, O., M. Thielebein, and A. Grahl. 1961. Sekundare Keimruhe bei Getreide. *Proceedings of the International Seed Testing Association* 26:89–114.

Froschell, P. 1957. Over de remstoffen in de zaden van *Beta saccharifera* (On the inhibiting substances in the seedballs of *Beta saccharifera*). *Natuurwetenschappelijk Tijdschrift* 37(5/6):97–100 (First published in *Biological Abstracts*. 1957, 31:9077).

Gloyer, W. O. 1932. Percentage of hardshell in pea and bean varieties. *New York Agricultural Experiment Station Technical Bulletin* 195.

Gutterman, Y. 1978. Germination of seeds as a function of the maternal environments. *Acta Hortic.* 83:49–55.

Gutterman, Y. 1982. Phenotypic maternal effect of photoperiod on seed germination. In: *The Physiology and Biochemistry of Seed Development, Dormancy, and Germination*, ed. A. A. Khan, pp. 67–79. Amsterdam: Elsevier.

Gutterman, Y. 1992. Maternal effects on seeds during development. In: *Seeds: The Ecology of Regeneration in Plant Communities*, ed. M. Fenner, pp. 27–59. Wallingford, U.K.: CAB International.

Gutterman, Y. 1993. *Seed Germination in Desert Plants*. Berlin: Springer-Verlag.

Hagon, M. W., and L. A. T. Ballard. 1970. Reversibility of strophiolar permeability to water in seeds of subterranean clover (*Trigonium subterraneum* L.). *Australian Journal of Biological Science* 23:519–528.

Hill, H. J., S. H. West, and K. Hinson. 1986. Effect of water stress during seed fill on impermeable seed expression in soybean. *Crop Science* 26:807–813.

Ho, D. T., and J. E. Varner. 1976. Response of barley aleurone layers of abscisic acid. *Plant Physiology* 57:175–178.

Hsiao, A. I., and A. W. Quick. 1984. Actions of sodium hypochlorite and hydrogen peroxide on seed dormancy and germination of wild oats, *Avena fatua* L. *Weed Research.* 24:411–419.

Huxley, P. A. 1965. Coffee germination test recommendations and defective seed types. *Proceedings of the International Seed Testing Association* 30:705–714.

Hyde, E. O. C. 1954. The function of the hilum in some Papilionaceae in relation to the ripening of the seed and the permeability of the testa. *Annals of Botany* 18:251–256.

Ives, S. A. 1923. Maturation and germination of seeds of *Ilex opaca*. *Botanical Gazette* 76:60–77.

Jacobsen, J. V., and E. Pressman. 1979. A structural study of germination in celery (*Apium graveolens* L.) seed with emphasis on endosperm breakdown. *Planta* 144:241–248.

Junges, W., and H. Ludwig. 1963. Einfluss von Wasserversorgung, Dungung und Licht wahrend der Samenbildung auf die Samenruhe bei *Impatiens balsamina* L. *Proceedings of the International Seed Testing Association* 28:71–96.

Justice, O. L. 1944. Viability and dormancy in seeds of *Polygonum amphibium* L., P. *coccineum* Muhl., and P. *hydropiperoides* Michx. *American Journal of Botany* 31:369–377.

Karssen, C. M., S. Zagorski, J. Kepczynski, and S. P. C. Groot. 1989. Key role for endogenous gibberellins in the control of seed germination. *Annals of Botany* 63:71–80.

Kelly, K. M., and J. Van Staden. 1987. The lens as the site of permeability in the Papilionoid seed *Aspalathus linearis*. *Journal of Plant Physiology* 128:395–400.

Kermode, A. R., J. D. Bewley, J. Dasgupta, and S. Misra. 1986. The transition from seed development to germination: A key role for desiccation? *Horticultural Science* 21:113–118.

Kigel, J., A. Gibly, and M. Negbi. 1979. Seed germination in *Amaranthus retroflexus* L. as affected by the photoperiod and age during flower induction of the parent plants. *Journal of Experimental Botany* 30:997–1002.

Koller, D. 1955. The regulation of germination in seeds (review). *Bulletin of the Research Council of Israel* 5D:85–108.

———. 1962. Preconditioning of germination in lettuce at time of fruit ripening. *American Journal of Botany* 49:841–844.

Kuhn, R., D. Jerchel, F. Moewus, and E. F. Moller. 1943. Uber die chimische Natur der Blastokoline und ihre Einwirkung auf kiemende, Samen, Pollenkorner, Hefen, Bakterien, Epithelgewebe und Fibroblasten. *Naturwissenschaften* 3:468.

Kummerow, J. 1965. Endogenous rhythmic oscillations in the germination of seeds (in German). *Encyclopedia of Plant Physiology* 15:721–726.

Kyle, J. H. 1955. A study of the relationship of the micropyle opening to hard seeds in the Great Northern bean. Master's thesis, University of Idaho.

Ladizinsky, G. 1985. The genetics of hard seed coat in the genus *Lens*. *Euphytica* 34:539–544.

Landgraff, A., and O. Juntilla. 1979. Germination and dormancy of reed canary grass seeds (*Phalaris arundinacea*). *Physiology Plantarum* 45:96–102.

Leck, M. A., V. T. Parker, and R. L. Simpson, eds. 1989. *Ecology of Soil Seed Banks*. New York: Academic Press.

Lee, J. A. 1975. Inheritance of hard seed in cotton. *Crop Science* 15:149–154.

Lester, R. N. 1985. Seed germination stimulated by enzyme etching. *Biochemie und Physiologie der Pflanzen*. 180:709–714.

Lewark, S., and R. M. Rudnicki, 1977. After-ripening in cold-requiring seeds. In: *The Physiology and Biochemistry of Seed Dormancy and Germination*, ed. A. A. Khan, pp. 193–217. New York: North-Holland Publishing Company.

Lovato, A., and M. T. Amaducci. 1965. Examination of the problem of whether dormancy exists in seeds of onion (*Allium cepa* L.) and leek (*Allium porrum* L.). II. Effects of temperature, prechilling, and light on germination. *Proceedings of the International Seed Testing Association* 30:803–820.

Maguire, J. D. 1969. Endogenous germination rhythms in seeds. *Proceedings of the Association of Official Seed Analysts* 59:95–100.

Major, W., and E. H. Roberts. 1968. Dormancy in cereal seeds. I. The effects of oxygen and respiratory inhibitors. *Journal of Experimental Botany* 19:77–89.

Marchaim, U., Y. Birk, A. Dovrat, and T. Berman. 1972. Lucerne saponins as inhibitors of cotton seed germination. Their effect on diffusion of oxygen through the seed coats. *Journal of Experimental Botany* 23:302–309.

Mayer, A. M., and A. Poljakoff-Mayber, 1989. *The Germination of Seeds*. New York: Pergamon Press.

McDonald, M. B., Jr. and A. A. Khan. 1977. Factors determining germination in Indian Ricegrass seeds. *Agronomy Journal* 69:558–563.

Meyer, B. S., D. B. Anderson, and R. H. Bohning. 1960. *Introduction to Plant Physiology*. New York: D. Van Nostrand Company.

Meyer, S. E., S. B. Monsen, and E. Durrant McArthur. 1990. Germination response of *Artemisia tridentata* (*Asteraceae*) to light and chill: Patterns of between-population variation. *Botany Gazette* 151:176–183.

Naylor, J. M. 1983. Studies on the genetic control of some physiological processes in seeds. *Canadian Journal of Botany* 61:3561–3567.

Naylor, J. M., and G. M. Simpson. 1961. Dormancy studies in seed of *Avena fatua*. 2. A gibberellin sensitive inhibitory mechanism in the embryo. *Canadian Journal of Botany* 39:281–295.

Nicholls, P. B. 1986. Induction of sensitivity to gibberellic acid in wheat and barley caryopses: Effect of dehydration, temperature and the role of the embryo during caryopsis maturation. *Australian Journal of Plant Physiology* 13:785–794.

Nooden, L. D., K. A. Blakley, and J. M. Grzybowski. 1985. Control of seed coat thickness and permeability in soybean—A possible adaptation to stress. *Plant Physiology* 79:543–548.

Olney, H. O., and B. M. Pollock. 1960. Studies of rest period. II. Nitrogen and phosphorus changes in embryonic organs of after-ripening cherry seed. *Plant Physiology* 35:970–975.

Pammel, L. H. 1899. Anatomical characters of the seeds of *Leguminosae*, chiefly genera of Gray's Manual. *Transactions of the St. Louis Academy of Sciences* 9:91–263.

Picman, J., and A. K. Picman. 1984. Autotoxicity in *Parthenium hysterophorus* and its possible role in control of germination. *Biochemical Systems and Ecology* 12:287–292.

Pinfield, N. J. 1968. The promotion of isocitrate lyase activity in hazel cotyledons of exogenous gibberellin. *Planta* 82:337–341.

Pollock, B. M., and H. O. Olney. 1959. Studies of the rest period. Growth, translocation, and respiration changes in the embryonic organs of the after-ripening cherry seed. *Plant Physiology* 34:131–142.

Porter, N. G., and P. F. Wareing. 1974. The role of the oxygen permeability of the seed coat in the dormancy of seed of *Xanthium pennsylvanicum* Wallr. *Journal of Experimental Botany* 25:583–594.

Pourrat, Y., and R. Jacques. 1975. The influence of photoperiodic conditions received by the mother plant on morphological and physiological characteristics of *Chenopodium polyspermum* L. seeds. *Plant Science Letters* 4:273–279.

Probert, R. J. 1992. The role of temperature in germination ecophysiology. In: *Seeds: The Ecology of Regeneration in Plant Communities*, ed. M. Fenner. Wallingford, U.K.: CAB International.

Probert, R. J., R. D. Smith, and P. Birch. 1985. Germination responses to light and alternating temperatures in European populations of *Dactylis glomerata* L. III. The role of the outer covering structures. *New Phytology* 100:447–455.

Quinlivan, B. J. 1966. The relationship between temperature fluctuations and the softening of hard seeds of some legume species. *Australian Journal of Agricultural Research* 17:625–631.

Quinlivan, B. J. 1971. Seed coat impermeability in legumes. *Journal of Australian Institute of Agricultural Science* 37:283–295.

Redmond, D. R., and R. C. Robinson. 1954. Viability and germination of yellow birch. *Forestry Chronicle* 30:79–87.

Renard, C., and P. Capelle. 1976. Seed germination in ruzizi grass (*Brachiaria ruziziensis* (Germain and Evrard)). *Annuals Physiologie Vegetale* 2:99–107.

Rice, E. L. 1983. *Allelopathy*, 2nd ed. New York: Academic Press.

Rolston, M. P. 1978. Water impermeable seed dormancy. *Botany Review* 44:365–385.

Rudnicki, R. M., I. Sinska, and S. Lewak. 1972. The influence of abscisic acid on the gibberellin content in apple seeds during stratification. *Biologia Plantarum* (Praha) 14:325–329.

Scott, D. H., and J. G. Waugh. 1941. Treatment of peach seed as affecting germination and growth of seedlings in the greenhouse. *Proceedings of the American Society for Horticultural Science* 38:291–298.

Simpson, G. M. 1990. *Seed Dormancy in Grasses*. Cambridge, U.K.: Cambridge University Press.

Snyder, F. W., J. M. Sebeson, and J. L. Fairley. 1965. Relation of water soluble substances in fruits of sugar beet to speed of germination of sugar beet seeds. *Journal of the American Society of Sugar Beet Technology* 13(5):379–388.

Sondheimer, E., D. S. Tzou, and E. C. Galson. 1968. Abscisic acid levels and seed dormancy. *Plant Physiology* 43:1443–1447.

Standifer, L. C., P. W. Wilson, and A. Drummond. 1989. The effects of seed moisture content on hard-seededness and germination of four cultivars of okra (*Abelmoschus esculentus* (L.) Moench). *Plant Variety and Seeds* 2:149–154.

Stearns, F., and J. Olson. 1958. Interactions of photoperiod and temperature affecting seed germination in *Tsuga canadensis*. *American Journal of Botany* 45:53–59.

Stone, E. C. 1957. Embryo dormancy and embryo vigor of sugar pine as affected by length of storage and storage temperatures. *Forest Science* 3:357–371.

Taylor, G. B. 1981. Effect of constant temperature treatments followed by fluctuating temperatures on the softening of hard seeds of *Trifolium subterraneum* L. *Australian Journal of Plant Physiology* 35:201–210.

Taylorson, R. B. 1988. Anaesthetic enhancement of *Echinochloa crus-galli* (L.) Beauv. seed germination: Possible membrane involvement. *Journal of Experimental Botany* 39:50–55.

Tester, M., and C. Morris. 1987. The penetration of light through soil. *Plant Cell Environment* 10:281–286.

Thomas, T. H., N. L. Biddington, and D. F. O'Toole. 1979. Relationship between position on the parent plant and dormancy characteristics of seed of three cultivars of celery (*Apium graveolens*). *Physiologia Plantarum* 45:492–496.

Thornton, M. L. 1968. Seed dormancy in watermelon *Citrullis vulgaris* Shrad. *Proceedings of the Association of Official Seed Analysts* 58:80–84.

Tilsner, H. R., and M. K. Upadhyaya. 1987. Action of respiratory inhibitors on seed germination and oxygen uptake in *Avena fatua* L. *Annuals of Botany* 59:477–482.

Tran, V. N., and A. K. Cavanagh. 1984. Structural aspects of seed dormancy. In: *Seed Physiology*, Vol. 2, ed. D. R. Murray. New York: Academic Press.

Van Staden, J., D. P. Webb, and P. F. Wareing. 1972. The effect of stratification and endogenous cytokinin levels in seeds of *Acer saccharum*. *Planta* 104:110–114.

Varty, K., L. B. Arreguin, T. M. Gomez, T. P. Lopez, and L. M. A. Gomez. 1983. Effects of abscisic acid and ethylene on the gibberellic acid-induced synthesis of α-amylase by isolated wheat aleurone layers. *Plant Physiology* 73:692–697.

Varga M. 1957. Examination of growth-inhibiting substances separated by paper chromatography in fleshy fruits. II. Identification of the substance of growth-inhibiting zones on the chromatograms. *Acta Biologica Szegediensis* 3:212–223.

Vaughan, J. G., and J. M. Whitehouse. 1971. Seed structure and the taxonomy of the Cruciferae. *Botanical Journal of the Linnean Society* 64:383–409.

Villiers, T. A. 1961. Dormancy in tree seeds. A brief review of recent work. *Proceedings of the International Seed Testing Association* 26:516–536.

Villiers, T. A., and P. F. Wareing. 1964. Dormancy in fruits of *Fraxinus excelsior* L. *Journal of Experimental Botany* 15(44):359–367.

Wareing, P. F., and H. A. Foda. 1957. Growth inhibitors and dormancy in *Xanthium* seeds. *Physiologia Plantanum* 10:266–280.

Webb, D. P., J. Van Staden, and P. F. Wareing. 1973. Seed dormancy in *Acer*. Changes in endogenous cytokinins, gibberellins and germination inhibitors during the breaking of dormancy in *Acer saccharum* Marsh. *Journal of Experimental Botany* 24:105–116.

Werker, E. 1980/1981. Seed dormancy as explained by the anatomy of embryo envelopes. *Israel Journal of Botany* 29:22–44.

Werker, E., I. Marbach, and A. M. Mayer. 1979. Relation between the anatomy of the testa, water permeability and the presence of phenolics in the genus *Pisum*. *Annuals of Botany* 43:765–771.

Widell, K. O., and T. C. Vogelmann. 1988. Fibre optics studies of light gradients and spectral regime within *Lactuca sativa* achenes. *Physiologia Plantarum* 72:706–712.

Williams, P. M. and J. D. Ross, and J. W. Bradbeer. 1973. Studies in seed dormancy: VII. The abscisic acid content of the seeds and fruits of *Corylus avellana* L. *Planta* 110:303–310.

Zagaja, S. W. 1962. After-ripening requirements of immature fruit tree embryos. *Horticultural Research* 2(1):19–34.

Zagaja, S. W., and J. Czapski. 1962. Some effects of low temperature treatments of phosphorus uptake by immature apple embryos. *Horticultural Research* 2:13–18.

Zarska-Maciejewska, B., and S. Lewak. 1976. The role of lipases in the removal of dormancy in apple seeds. *Planta* 132:177–181.

7

Seed Vigor and Vigor Tests

Seeds, as reproductive units, are expected to produce plants in the field. However, farmers and seed producers have long recognized that the labeled percent germination often overestimates the actual field emergence of seed lots. This observation is attributed to the objective of a standard germination test which states that germination is the emergence and development from the seed embryo of those *essential structures* which, for the kind of seed in question, are indicative of the ability to produce a normal plant under *favorable* conditions. (AOSA, 1991) As a result, the standard germination test fails to provide accurate information concerning a seed lot's field performance potential for at least four reasons. These include the following.

1. The definition of seed germination emphasizes that the seed analyst must focus on *essential structures* which lead to the production of a normal plant. But, this emphasis on seedling morphology may have little relationship with rapidity of growth; a prime criterion of the potential for successful stand establishment.

2. Methodology for the conduct of a germination test is standardized so that test results are reproducible within and among seed-testing laboratories. This process means that *favorable* conditions are utilized as described in the definition to ensure greater uniformity in test results. Tests must be conducted on artificial, standardized, essentially sterile media in humidified, temperature-controlled chambers; conditions that are so synthetic that they seldom relate to field conditions that seeds likely encounter. In essence, because the standard germination test is conducted under *favorable* conditions, it basically establishes the maximum plant-producing ability of the seed lot. When field conditions are optimum, the standard germination test may correctly predict field performance of the seed lot. For the most part, however, standard germination values overestimate actual field emergence. We know, for example, that when the standard germination test result is 80%, actual emergence under field conditions seldom reaches 80%. In most instances, the field emergence is considerably less.

3. The standard germination test is designed to provide for a first and final count. The first count has a purpose of basically removing most of the strong seedlings that have already germinated. The final count is designed to provide a sufficiently long period that even weak seeds are coaxed or provided every opportunity to be considered germinable. The germination percentage, therefore, is the sum of strong and weak seedlings. The difficulty with such a process is that weak seedlings seldom perform adequately when provided environmental stresses associated with field emergence.

4. By definition, germination is scaleless. A seed is considered either germinable or it is not. There are no distinctions provided for strong or weak seedlings. Those considered germinable may vary from weak to semilame to robust in field performance. This inability to document the quality of the seed fails to take into account the progressive nature of seed deterioration, which has a major impact on stand establishment.

These deficiencies have led to a continuously disquieting murmur for years that not all facets of seed quality were being properly identified by the standard germination test. As a result, it is useful to review the history and development of seed vigor testing.

HISTORY OF SEED VIGOR TESTING

In 1876, Fredrich Nobbe first distinguished the concept of seed vigor from that of germination. He introduced the term *triebkraft,* which means driving force or shooting strength to convey the idea that, in addition to germination, speed and uniformity of emergence were important parameters of seed quality. However, it was not until the 1950 International Seed Testing Congress that renewed interest was focused on seed vigor. At that time, European and American laboratories were expressing concern that germination test results were not standardized. In an attempt to explain these disparities, Franck (1950) pointed out the differing concepts of germination testing between European and American laboratories. He noted that in Europe, germination tests were made under optimum, reproducible conditions, to assure that seed lots could be sold across national boundaries, while *special tests,* such as the brick grit and soil tests, were developed to evaluate "seedling vigor." The American concept of germination, which was based on soil test results, was to determine the plant-producing ability of a seed lot. Franck (1950) contended that both groups needed to come to grips with these differing philosophies. To start the debate, he proposed that germination testing should be conducted under favorable conditions in order that uniform test results be obtained. The plant-producing ability in the field of a seed lot was to be defined by a new term: *vigor*.

Definition of Seed Vigor

The development of a satisfactory definition of seed vigor has been a central theme in the development of vigor tests. Without a definition, the ability to measure or test this undefined entity becomes difficult, if not impossible. Fortunately, many definitions have been proposed and a study of their evolution portrays the initially confusing and changing status in the expectations for seed vigor. As an example, in 1957, Isely defined seed vigor as "the sum total of all seed attributes which favor stand establishment under favorable conditions." Building on this definition, Delouche and Caldwell (1960) stated that "seed vigor is the sum of all seed attributes which favor rapid and uniform stand establishment." Note the subtle differences from Isely's definition. Delouche and Caldwell clarified stand establishment to emphasize rapid and uniform performance and they also deleted the reference to favorable conditions. It was clear at this point that rapid and uniform field performance were acceptable parameters of seed vigor. However, the reference to " . . . sum total of all seed attributes . . ." still left unresolved what the factors were that determined seed vigor. To address this issue, Woodstock (1965) proposed that seed vigor was "that condition of good health and natural robustness in seed, which, upon planting, permits germination to proceed rapidly and to completion under a wide range of environmental conditions." Perry (1973) identified seed vigor as the "physiological property determined by the genotype and modified by the environment which governs the ability of a seed to produce a seedling rapidly in soil and the extent to which the seed tolerates a range of environmental factors." He clearly emphasized that seed vigor was determined by both genetic and environmental components. By this time, consensus was rapidly emerging on a definition of seed vigor. In 1977, the International Seed Testing Association (ISTA) defined vigor as "the sum total of those properties of the seed which determines the potential level of activity and performance of the seed or seed lot during germination and seedling emergence" (Perry 1978). Among the aspects of performance are: (1) biochemical processes and reactions during germination such as enzyme reactions and respiration activity, (2) rate and uniformity of seed germination and seedling growth, (3) rate and uniformity of seedling emergence and growth in the field, and (4) emergence ability of seedlings under unfavorable environmental conditions. Factors that influence the level of seed vigor include the genetic constitution of the seed; environment and nutrition of the mother plant; stage of maturity at harvest; seed size, weight, and specific gravity; mechanical integrity; deterioration and aging; and pathogens. This definition is considered an *academic* definition because it discusses, identifies, and describes seed vigor (i.e., it attempts to relay what seed vigor *is*).

In 1979, the Association of Official Seed Analyst's Vigor Committee defined seed vigor as "those seed properties which determine the potential for rapid uniform emergence and development of normal seedlings under a wide range of field conditions" (McDonald 1980b). This definition quantifies vigor in terms of rapid uniform emergence and development of normal seedlings. Thus, it focuses on what seed vigor *does* and is considered to be an *operational* definition.

Progress in Seed Vigor Testing

Initially, progress in seed vigor testing was slow in the United States. While the 1950 ISTA Congress and the perceptive comments of Franck on the need for vigor testing were emphasized, the United States still considered this issue primarily a European concern. It wasn't until the publication of two articles on seed vigor testing (Isely 1957; Delouche and Caldwell 1960) that the key stimulus was provided in the

United States to refocus and redirect the development of the concept of seed vigor. In 1961, the first AOSA vigor test committee was formed and was chaired by Dr. R. P. Moore. The principal objectives of that committee were to bring into focus the advantages and disadvantages of direct vs. indirect vigor tests as well as outlining various concepts of seed vigor. It seemed at this point that the challenge of vigor testing was straightforward and the solutions imminent. But a review of the history associated with vigor test development (McDonald 1994) demonstrates how naive this notion was. By 1966, Moore noted that "Since quite *diverse* points of interest are involved, the progress of the Committee could no doubt be promoted by restriction of the assignment to measurement of vigor which commonly conveys rate and magnitude of growth." Clearly, the more the topic was studied, the more challenging it became.

The next major advance occurred in 1974 when the increasing attention given to seed vigor and its potential for advancing the ability to better estimate field performance resulted in a clamor for standardized vigor tests. As a result, the Association of American Seed Control Officials formally resolved that AOSA develop standardized seed vigor test procedures. This resolution prompted the AOSA vigor test committee to even greater activity. Then Chairman Lowell Woodstock wrote "there has been more real movement towards consensus in seed vigor testing and more real progress by the AOSA Vigor Testing Committee in meeting its responsibilities for developing, evaluating, codifying, and standardizing vigor testing procedures during the past nine months than during any recent period." Under Dr. Woodstock's leadership, the "Seed Vigor Testing Progress Report" was published as a special edition of the AOSA Newsletter in 1976. This document was a significant milestone because it provided specific guidelines for the conduct of eight proposed vigor tests. These tests could be evaluated using a referee format.

The AOSA vigor test committee set out to determine the standardization capability of each of these tests. Corn and soybeans served as the crops receiving the greatest emphasis. Concurrently, the committee was deriving a satisfactory definition of seed vigor. By 1980, a seed vigor definition had been approved by every major organization involved with the commerce and testing of seeds. As referee and research results were studied, the committee became convinced that seven useful vigor tests were available. The procedures for these tests were published in the AOSA Vigor Testing Handbook in 1983. The tests were placed into a suggested vigor test section because of the remaining concern that the tests needed to stand the scrutiny of routine use before they could be declared standardized. This handbook also contained a historical perspective of vigor test development, the types of vigor tests available, and the applications of vigor testing information. By 1987, the accelerated aging test was significantly revised and improved and became the first vigor test for soybeans to be moved from the suggested to recommended vigor test section. With increasing reliability and standardization of the tests, the committee set out to evaluate acceptable tolerances for vigor test results. This first evaluation was completed in 1991.

Clearly, the AOSA vigor testing committee in conjunction with its ISTA counterpart have provided effective leadership in this important seed quality testing area. Its success can be followed by monitoring the number of laboratories routinely using seed vigor tests. In 1978, 52% of seed-testing laboratories were conducting vigor tests (Mc-

Donald 1994). In 1983, the number was 60% and by 1990, 75% of the seed-testing laboratories were using one or more vigor tests. These data indicate that these committees not only provided useful vigor testing protocols but have also educated the users so that the value and limitations of vigor testing are fully understood. Useful reviews on the many facets of seed vigor have been provided by Heydecker (1972, 1977), Pollock and Roos (1972), McDonald (1975), Cantliffe (1981), Halmer and Bewley (1984), and TeKrony and Egli (1991). To better understand the importance of seed vigor, we must first identify the factors that influence this important seed quality attribute.

FACTORS INFLUENCING SEED VIGOR

The development of a seed encompasses a series of important ontogenetic stages from fertilization, to accumulation of nutrients, to seed dry down, to dormancy. Each of these stages represents a change in morphological and physiological ontogeny that can alter seed performance potential. The point at which the seed achieves its maximum dry weight is called *physiological maturity*. At this point, it has its greatest potential for maximum germination and vigor (Delouche 1974). However, since seeds generally achieve physiological maturity at high moisture levels unsafe for storage, seed is typically not harvested until it attains *harvest maturity*, which is low enough for safe storage, but high enough to minimize mechanical injury. Between physiological maturity and harvest maturity, the seed is essentially stored on the plant where it may be exposed to severe environmental conditions that adversely affect seed quality.

Among the factors that influence seed vigor are genetic constitution, environment during seed development, and seed storage environment.

Genetic Constitution

For many years, plant breeders have inadvertently selected for increased seed vigor. In their attempts to increase yields, they have also improved such seed characteristics as mechanical integrity (hard-seededness), resistance to disease, protein content, and seed size. These factors lead to better field emergence and often result in enhanced yields. In addition to such physical manifestations of seed vigor, plant breeders have introduced hybrid vigor. Thus, factors that are under genetic control such as hybrid vigor, hard-seededness, susceptibility to seed damage, and the seed chemical composition influence the expression of seed quality.

Hybrid Vigor. Hybrid vigor is a component of heterosis and represents the measurable superiority of the hybrid progeny over its inbred parents. The superiority of the hybrid is often greater under conditions of stress than under optimal conditions. For example, seeds of hybrid corn and barley germinate faster and grow more rapidly than their inbred parents (Whaley 1950; McDaniel 1969). This increased growth potential has been attributed to more efficient mitochondria and extra enzyme systems for carbon assimilation (McDaniel and Sarkissian 1968). The production of hybrid corn seed is discussed in more detail in Chapter 9.

Hard Seed. For the most part, hard-seededness is an undesirable genetic trait because it results in extreme variations in germination and hence, stand establishment. However, recent efforts have been made to reintroduce hard-seededness into some cultivars to help protect the seed from aging and to protect against leakage of nutrients during imbibition (Potts et al. 1978). Hard-seededness can be eliminated from cultivars relatively easily. Lebedeff (1947) showed that, in crosses between soft lines and lines with varying degrees of hard-seededness, the F_1 was intermediate or approached the soft-seeded parent. The F_2 seed showed all possible degrees of permeability between the parental extremes. The data indicated that hard-seededness was controlled by several genes.

Kyle and Randall (1963) examined the nature of hard seed in the great northern and red Mexican dry bean classes and reported that the micropyle was the site of water imbibition in great northern, while in red Mexican it was the raphe and hilum. The impermeability of the hilum and raphe was due to a simple recessive gene and was closely associated with the p gene for white seed. Gloyer (1932), however, showed in a cross of red and white kidney beans that is was easy to select for soft seed in white-seeded segregants.

Susceptibility to Mechanical Damage. Susceptibility to mechanical damage, whether induced by harvesting or conditioning equipment, has been shown to be under genetic control. For example, Barriga (1961) demonstrated that 41 strains of navy beans possessed differing tolerances to mechanical abuse. Atkin (1958) and Wester (1970) reported that colored snap bean cultivars were more resistant to mechanical damage than white-seeded cultivars (Figure 7.1). Kannenberg and Allard (1964) showed that seed thickness in lima beans directly affected the extent of seed coat cracking. Green and Pinnell (1968) studied visible seed defects in soybeans and found them inherited quantitatively. Along with Walters and Caviness (1973), they were the first to suggest the need for enhanced seed quality through improved breeding lines.

Seed Chemical Composition. Breeding for improved nutritional quality such as high-lysine corn has often resulted in increased seed quality problems. For example, this process often produces small, shrunken, low-vigor seeds. Plant breeders are now attempting to find gene systems that control nutritional quality but do not result in reduced seed vigor. Nass and Crane (1970) found that various genes for endosperm expression influenced seed germination at 15°, 20°, and 25°C. Seeds with the A_1 gene produced more vigorous seeds than those without this gene. Ullrich and Eslick (1978) also compared the effect of different shrunken endosperm genes on kernel weight for barley. They found that mutant kernel weights ranged from 38 to 95% of normal and suggested that seed quality could be improved by selecting those segregations with the greatest kernel weight.

Environment During Seed Development

The concentration of seed production (Chapter 9) for some crops in specific areas is persuasive testimony to the environmental influence on seed development and quality (Delouche 1980). For example, a major portion of the seed of temperate climate forage and lawn or turf grasses is produced in the Pacific Northwest, especially in Oregon where the climate is favorable for high-quality seed production. In these locales, the seeds complete maturation, dry down, and are harvested with little risk from rain and adverse humidity. Other examples are the arid irrigated regions of California, Idaho,

Figure 7.1. *(Top) Faster germination and more vigorous seedlings from green (nonbleached) than from white (bleached) bean seed. (Bottom) Field view showing much better stand from green-sorted (nonbleached) than from white (bleached) seed, especially when the green-sorted seed is sound. (From Wester (1970).)*

and Arizona where high-quality cotton, vegetable, flower, and forage legume seeds are produced because of the low humidity, minimal rainfall, and favorable temperatures that occur during seed maturation. These environmental conditions reduce the spread of seed-borne diseases as well as the risks associated with inclement weather during late harvesting periods. However, most of the seed used in crop production is produced in the same area where the commercial crop is grown. Consequently, environmental conditions may range from bad to good for high-quality seed production. Among the factors that affect seed quality are soil moisture and fertility, seed maturity, and the postmaturation preharvest environment (Delouche 1980).

Soil Moisture and Fertility. Moisture stress during seed development often results in light, shriveled seed which, in turn results in poor-vigor seed. However, it is generally agreed that when soil fertility becomes limiting, plants respond by producing less seed. Thus, the fewer seeds produced under marginal fertility conditions are usually as viable and vigorous as are the greater number of seeds produced under favorable conditions. Exceptions do occur, however. Legatt (1948) demonstrated that boron-deficient pea seeds produced abnormal seedlings and such abnormalities could be corrected only with the addition of borax. In contrast, soybean seeds produced in areas with high molybdenum soil concentrations, possessed such high levels of molybdenum that they did not require sodium molybdate seed treatment for planting in molybdenum-deficient soils (Harris et al. 1965). Peanuts are particularly susceptible to soil mineral deficiencies. Soils low in boron and calcium produce peanut seeds that exhibit a discoloration of the cotyledons associated with boron deficiency, and a watery hypocotyl and physiological root breakdown associated with calcium deficiency (Cox and Reid 1964; Sullivan 1973).

In other instances, a direct relationship between soil fertility and seed vigor has been reported. For example, nitrogen fertilization of wheat is known to cause increased protein levels in seed (Fernandez and Laird 1959; McNeal et al. 1971). Other studies have shown that high-protein wheat seed results in increased germination (Fox and Albrecht 1957), seed vigor (Lowe et al. 1972; Lowe and Ries 1973), and subsequent crop yield (Ries et al. 1970).

Seed Maturity. Abundant information demonstrates an association between parameters of maturity such as seed size (and weight) and seed vigor (Austin 1972; Heydecker 1972; McDonald 1975). Consequently, the environment during seed maturation has an indirect effect on its potential vigor. Large soybean seeds have been shown to be superior to small seeds in germination and vigor, as well as crop yield potential (Burris et al. 1971, 1973; Fontes and Ohlrogge 1972). Other studies have reported no difference in performance among soybean seeds of varying size (Singh et al. 1972; Johnson and Leudders 1974), while others (Aguiar 1974) have found that medium-sized soybean seeds were superior in vigor to both large and small seeds. In hybrid corn, the influence of seed size is reported to have little or no effect on corn performance in terms of emergence rate (Hunter and Kannenberg 1972), date of tasseling (Hicks et al. 1976), final leaf number (Hawkins and Cooper 1979), and grain yield (Hicks et al. 1976; Hawkins and Cooper 1979). Similarly, Shieh and McDonald (1982) reported that seed size of two corn inbreds had no effect on seed quality, although flat seeds were shown to be superior to round seeds.

Postmaturation-Preharvest Environment. Deterioration of seed during the postmaturation-preharvest environment is a serious seed production problem in the eastern half of the United States because of high humidity, frequent rainfalls, and warm temperatures—conditions that produce a rapid loss in seed viability and vigor. For example, Simpson and Stone (1935) reported a 20 to 30% loss in cotton seed viability after only one week's exposure to rainy conditions. Caldwell (1972) and Woodruff et al. (1967) showed that cottonseed from lower bolls that opened first and were exposed longer to the field environment before harvest were consistently lower in seed vigor than seed from bolls in the upper half of the plant.

In soybeans, delayed harvest of seed because of inclement weather results in loss of viability (TeKrony et al. 1980b) and an increase in mechanical damage during harvest (Green et al. 1966). In a study on the effect of planting and maturity dates on soybean seed quality, Green et al. (1965) found that soybeans from early planting dates that matured during hot, dry weather produced lower-quality seeds compared to those from later plantings. It has also been shown that early-maturing soybean cultivars are more susceptible to seed deterioration, not because they are inherently predisposed to deterioration, but because the seed matures in the warmer temperatures of late September and early October compared to cultivars maturing in late October (Delouche 1980). Preharvest loss of seed quality is known to be increased by fungal invasion, which increases during warm and humid conditions (Nicholson et al. 1972; Roncadori et al. 1972; TeKrony et al. 1980a).

Seed Storage

Seldom are seeds harvested and immediately planted without undergoing at least a brief storage period. Consequently, the time of storage, type of seed stored, and storage environment (temperature, relative humidity, and oxygen levels) influence seed vigor. A more detailed discussion of these factors is provided in Chapter 8.

SEED VIGOR TESTS

The challenge of vigor testing has been to identify one or more quantifiable parameters that are common to seed deterioration. Although not all changes that occur during seed deterioration are understood, we can speculate on the probable sequence of events. A hypothetical model (Figure 7.2) has been developed by Delouche and Baskin (1973) that outlines some of the major parameters used in measuring seed vigor.

Because a vigor test is a more sensitive index of seed quality than the standard germination test, any of the events that precede loss of germination could serve as a basis for vigor tests. The earlier the parameter can be measured during the loss of germination, the more sensitive the index of seed vigor. Thus, since the onset of membrane degradation precedes loss of germination (Figure 7.2), the most sensitive vigor test should be one that monitors membrane integrity.

Considerable experimental evidence supports this contention. Membranes are essential for many metabolic events occurring in the seed, including respiration (cristae in mitochondria), which provides the seed with the energy required for subsequent growth. The endoplasmic reticulum is also a membraneous organelle on which many

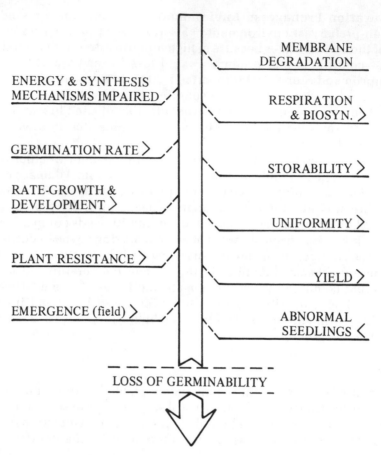

ENERGY & SYNTHESIS
MECHANISMS IMPAIRED

GERMINATION RATE >

RATE-GROWTH &
DEVELOPMENT >

PLANT RESISTANCE >

EMERGENCE (field) >

MEMBRANE
DEGRADATION

RESPIRATION
& BIOSYN. >

STORABILITY >

UNIFORMITY >

YIELD >

ABNORMAL
SEEDLINGS <

LOSS OF GERMINABILITY

Figure 7.2. *Probable sequence of changes in seed during deterioration. (From Delouche and Baskin (1973).)*

enzymes are formed as ribonucleic acid is translated. Thus, any impairment of membrane function can decrease the amount of ATP formed as an energy source, as well as retard the synthesis of specific enzymes essential for growth. Koostra and Harrington (1969) showed that membrane degradation had occurred in deteriorated seeds, and this observation has been supported by others (Streeter 1965; Villiers 1973; Spencer et al. 1973; Simon 1974; McDonald 1975; Harman and Mattick 1976; Stewart and Bewley 1980), though at least one report challenges this concept (Priestley and Leopold 1979). Beyond these studies, ATP (Ching 1973), RNA (Van Onckelen et al. 1973), and respiration (Woodstock and Grabe 1967) levels have been shown to decrease following storage in accelerated or natural aging conditions. Subsequent to the loss in respiration and biosynthetic capacity, the rate of germination declines, culminating in a loss of seed lot uniformity. Other associated events that occur during deterioration are loss in storability and ability to resist disease infection. Deteriorated seeds that are subjected to biological and environmental stresses also exhibit reduced field emergence which may (Johnson and Wax 1978) or may not (TeKrony and Egli 1977; Yaklich and Kulik

1979; Burris 1976) be related to final yields of certain crops. Eventually, these subtle manifestations of loss in seed quality are expressed by an increasing incidence of abnormal seedings—a component of the germination test. The final parameter of seed deterioration (and the one most often employed) is seed germination, underscoring the need for seed vigor tests to supplement routine standard laboratory germination tests.

As a result of the biochemical and physiological changes known to occur during seed deterioration, most vigor tests have focused on measuring one or more of these parameters. These vigor tests may be separated into various categories based on the parameters monitored. For example, Isley (1957) divided vigor tests into *direct* and *indirect* tests. Direct tests imitate the field environment in some way and measure the ability of seeds to emerge under simulated field stress conditions. The cold test is an example of a direct test because it subjects seeds to adverse conditions by placing them in cold, wet soil under direct stress from temperature, moisture, and microorganisms. However, direct tests have been criticized because they may not detect differences in quality when seeds are exposed to favorable soil conditions.

Indirect tests measure specific physiological components of seeds. For example, the conductivity test is an indirect test that monitors cell leakage. However, indirect tests fail to evaluate all the physical and physiological factors that determine field establishment. In the case of the conductivity test, improved performance due to seed treatments, the degree of morphological damage, the influence of soil microorganism attack, and other factors are often not adequately assessed.

Vigor tests may also be classified on the basis of the component of seed vigor measured. For example, Woodstock (1973) separated vigor tests into *physiological* and *biochemical* tests. Physiological tests measure some aspect of germination or seedling growth while biochemical tests evaluate a specific chemical reaction or reactions (e.g., enzymatic activity or respiration) related to the expression of germination (and hence, vigor). McDonald (1975) added one additional grouping, a *physical* category that included seed size, shape, and density, factors long associated with seed vigor because of their relation to seed maturity.

Other investigators have classified vigor tests into *stress* and *quick* test categories (Pollock and Roos 1972). Stress tests consist of subjecting tests to one or more of the environmental stresses that might be encountered under soil conditions. These tests usually involve measurement of some aspect of germination (e.g., hypocotyl length) while only the conditions of stress are varied. Stress conditions may include high temperatures and relative humidity as in the accelerated aging test, low temperatures with or without soil as in the cold test or cool germination test, or osmotic stress imposed by using solutions such as polyethylene glycol. Quick tests are tests in which some chemical reaction associated with seed vigor is monitored; these usually require much less time than stress tests. Examples of quick tests include the tetrazolium test, conductivity test, and various tests associated with enzymatic activity.

Characteristics of a Seed Vigor Test

A vigor test should possess certain essential characteristics that can make it useful to the seed producer and consumer. These characteristics have been described by McDonald (1980a) as follows:

Inexpensive. Due to limited budgets for seed testing, it is important that a vigor test be reasonably priced and require a minimum investment in labor, equipment, and supplies.

Rapid. Every seed laboratory has periods of peak activity, thus it is important that the vigor test be conducted rapidly to minimize analyst time and germinator space. Furthermore, seed producers desire a quick turnaround time for samples submitted for vigor tests since such quick information on seed quality can provide them with a competitive marketing advantage.

Uncomplicated. Where possible, vigor test procedures should be simple so that they can be performed in seed laboratories without requiring additional staff with special backgrounds and training.

Objective. For a vigor test to be easily standardized, a quantitative or numerical index of quality that avoids subjective interpretations by analysts should be utilized.

Reproducible. The success of any test depends on its reproducibility. If these results cannot be repeated because of intricate procedures or subjectivity of interpretation, then comparison of results among laboratories becomes meaningless.

Correlated with Field Performance. Most definitions of seed vigor emphasize the relationship between seed vigor and field performance, and many studies have demonstrated this association exists. Consequently, the ultimate value of any vigor test may be its ability to predict field performance.

Types of Seed Vigor Tests

The standard germination test is conducted under optimum conditions for seed germination. Consequently, when field conditions at planting are near optimum, the results usually correlate well with field emergence (Perry 1977; Egli and TeKrony 1979; Luedders and Burris 1979). However, under suboptimal field conditions, standard germination results usually overestimate field emergence (Sherf 1953; Delouche and Caldwell 1960; TeKrony and Egli 1977; Johnson and Wax 1978; Tao 1978a; Yaklich and Kulik 1979). Therefore, additional tests are needed to better predict seedling emergence under a wide range of field conditions. Many vigor tests have been suggested; however, only a few have attained acceptance by seed analysts and seed testing organizations (AOSA 1983; Perry 1981). These are discussed below.

Cold Test. The cold test is one of the oldest methods of stressing seeds and is most often employed for evaluating seed vigor in corn and soybeans. Seeds are placed in soil or paper towels lined with soil and exposed to cold for a specified period, during which stress from imbibition, temperature, and microorganisms occurs (Figure 7.3). Following the cold treatment, the seeds are placed under favorable growth conditions and allowed to germinate.

The greatest difficulty with the cold test is the lack of uniformity in field soil. Soils differ in moisture, pH, particle composition, and pathogen levels, all of which contribute to divergent results. Use of vermiculite, a more uniform medium, has been suggested as a possible solution to the variability of soil conditions. However, it is widely believed that a cold test requires field soil to be successful. Regardless of these difficulties, cold-test vigor rankings of seed lots tend to remain consistent within laboratories, which support this test as a most useful in-house vigor test.

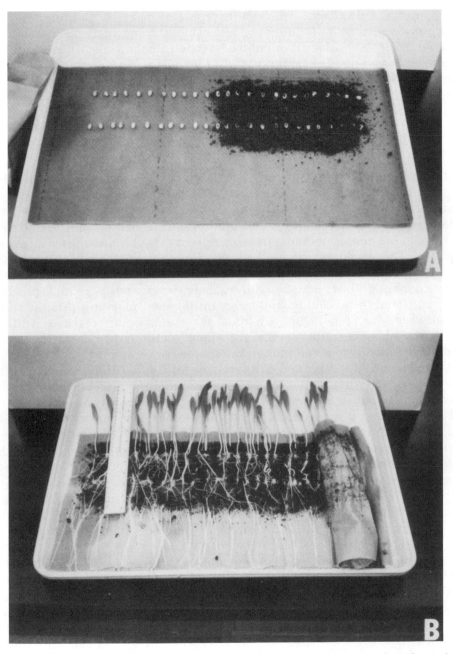

Figure 7.3. *The rolled towel cold test procedure (A) Planting seeds on cold wet towels and covering with a sand-soil mixture (B) Emergence of corn seedlings following the rolled towel cold test procedure.*

Accelerated Aging Test. This test incorporates many of the important traits desired in a vigor test. Initially proposed as a method to evaluate seed storability, the accelerated aging test subjects unimbibed seeds to conditions of high temperature (41°C) and relative humidity (around 100%) for short periods (3 to 4 days). The seeds are then removed from the stress conditions and placed under optimum germination conditions.

The accelerated aging test is rapid, inexpensive, simple and useful for all species; it can be used for individual seed evaluation and requires no additional training for correct evaluation. Modifications of the aging chamber have proven beneficial (Baskin 1977; McDonald and Phaneendranath 1978) and resulted in this vigor test being the first to be standardized. Another study has shown that differences in initial seed moisture should be considered when interpreting this test (McDonald 1977b).

Conductivity Test. Low-vigor seeds have been shown to possess decreased membrane integrity as a result of storage deterioration and mechanical injury. During imbibition, seeds having poor membrane structure release cytoplasmic solutes into the imbibing medium. These solutes with electrolytic properties carry an electrical charge that can be detected by a conductivity meter.

Measurement of the conductivity of leachates from seeds is a rapid, precise, inexpensive, and simple procedure. However, initial seed moisture (Simon and Wiebe 1975) and seed size (Tao 1978b) can affect the rate of solute leakage. Additionally, treatment of seeds with antibiotics may influence conductivity measurements, necessitating their removal before determinations are made.

One limitation of the present conductivity test is that it expresses results as an average conductivity evaluation for 25 seeds. Such an expression presumes that all seeds are equally deteriorated and will provide the same quantity of electrolyte leakage. A seed lot, however, is composed of a population of individuals—each with its own unique potential to perform in the field. Conductivity test results, therefore, would better reflect the vigor capability of a seed lot if they were presented on an individual seed basis. A commercial instrument is now available to monitor the electrolyte leakage of individual seeds. Recent studies suggest that this instrument provides a more accurate appraisal of seed vigor in soybeans (McDonald and Wilson 1979, 1980; Miles and Copeland 1980), cotton (Hopper and Hinton 1980), cowpea (Beighley and Hopper 1981), navy bean (Suryatmana et al. 1980) and corn (Joo et al. 1980) than does the present conductivity method.

Cool Germination Test. Unlike the cold test, the cool germination test is conducted under standard laboratory conditions at low temperatures (18°C) and does not rely on the activity of microorganisms to stress the germinated seeds. It has been demonstrated that low-vigor seeds from warm-season crops, such as cotton, have decreased growth rate and lower germination under these conditions. The major advantage of this test is that it is similar to the standard germination test and the same criteria for interpretation of normal seedlings are employed. Its principal limitation is that it is currently limited to use in cotton.

Seedling Growth Rate Test. Vigorous seeds are able to efficiently synthesize new materials and rapidly transfer these new products to the emerging embryonic axis, resulting in increased dry weight accumulation. The seedling growth rate test is based on this concept and vigor results are expressed as mg dry weight/germinable seedlings. This test is generally conducted according to the standards for the routine germination test. After evaluations are made, the growing segments of the embryos from normal seedlings are excised from the storage organs (cotyledons or endosperm), dried in beakers at 80°C for 24 hours, and weighed to determine their increase in dry weight. Since seedling growth rate is correlated with vegetative development in the field (Burris 1976; Pinthus and Kimel 1979), this test offers substantial promise. However, certain standardization problems still need to be addressed. For example, small differences in moisture and light intensity can have significant effects on the rate of seedling growth. Also, the test may require standardization for specific cultivars since rate of seedling growth can be genetically controlled (Burris 1975).

Seedling Vigor Classification Test. This vigor test is an expansion of the routine germination test, requiring the seed analyst to further classify "normal" seedlings into "strong" and "weak" categories (Figure 7.4). The test requires no additional equipment and employs concepts and terms familiar to seed analysts; thus it is particularly attractive to seed analysts. Despite its advantages, it has one serious difficulty. To further

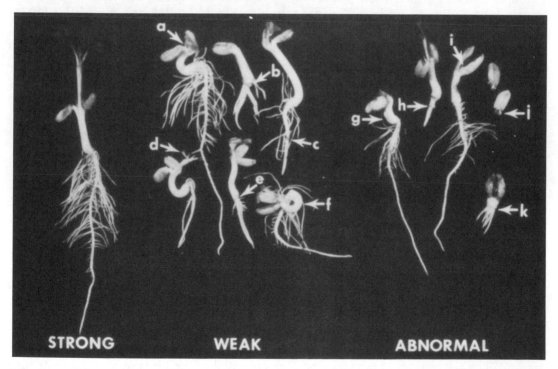

Figure 7.4. *Classification of peanut seedlings at the final count in the seedling vigor classification test. Groups of seedlings from left to right are normal strong, normal weak, and abnormal. Deficiencies of normal weak seedlings are:* **(a)** *partial decay of epicotyl,* **(b)** *no primary root,* **(c)** *primary root cracked,* **(d)** *one primary leaf missing,* **(e)** *split primary root,* **(f)** *curled hypocotyl. Deficiencies of abnormal seedlings are:* **(g)** *shortened hypocotyl,* **(h)** *stubby primary root with no secondary roots,* **(i)** *no epicotyl,* **(j)** *hypocotyl and roots missing,* **(k)** *poorly developed hypocotyl and roots.*

separate "normal" seedlings into two additional categories is a subtle task and represents additional sources of variability. For this reason, seed testing organizations maintain active referee programs to help laboratories identify their interpretation difficulties.

Tetrazolium (TZ) Test. The TZ test is one of the most valuable techniques for analyzing seed quality. It relies on the action of the TZ molecule to react with hydrogen atoms released as a result of dehydrogenase enzyme activity in living tissue. This results in the formation of a water-insoluble red pigment called formazan which a trained seed analyst evaluates for staining pattern and color intensity. The analyst then subjectively places the seeds into vigor categories ranging from strong to weak (Figure 7.5). Though this test correlates well with seed vigor when interpreted by a qualified analyst, it is still subject to certain standardization difficulties: first, the ability of the analyst to ascertain whether a seed is vigorous; second, the failure of the test to detect seed treatment phytotoxicity and reveal seed dormancy.

Speed of Germination. Speed of germination is one of the oldest seed vigor concepts. Seed lots with similar total germination often vary in their rate of germination and growth. Many methods for determining germination rate have been employed (Nichols and Heydecker 1968; Tucker and Wright 1965; Timson 1965). The number of days a lot requires to reach 90% germination was used by Belcher and Miller (1974) as an index of vigorous seed. For lower-quality lots, another percentage value (e.g., 50%) could be used. A different approach was proposed by Maguire (1962) who suggested the following formula:

$$X = \frac{\text{number of normal seedlings}}{\text{days of first count}} + \ldots + \frac{\text{number of normal seedlings}}{\text{days of final count}}$$

Similar germination indexes were suggested by Czabator (1962) and Djavanshir and Pourbeik (1976) for tree seeds based on the following formula:

$$\text{Germination value} = \text{peak value of } \frac{\text{cumulative number of normal seedlings}}{\text{days of germination counts}}$$
$$\times \frac{\text{total number of normal seedlings}}{\text{days of final count}}$$

Normal seedlings ranging from those with radicles emerged and small hypocotyls visible to those with fully developed seedlings with seed coats completely shed, were classified by Wang (1973) into six different classes depending on the stage of seedling growth. He tested red pine seeds from seven sources and found that the combined percentages of the top three classes were significantly correlated with the nursery emergence.

In the standard germination test, a preliminary and final count are routinely performed. The percentage of normal seedlings recorded in the first count represents fast-germinating seeds and can be used as an index of vigor. (This index is similar to that from the seedling vigor classification test except that the strong normal seedlings in the second count are excluded.) In soybean studies, the first count (four days) was found to provide a good estimate of seedling vigor (Burris et al. 1969) and has been used as a vigor index component (TeKrony and Egli 1977). More recently, it was

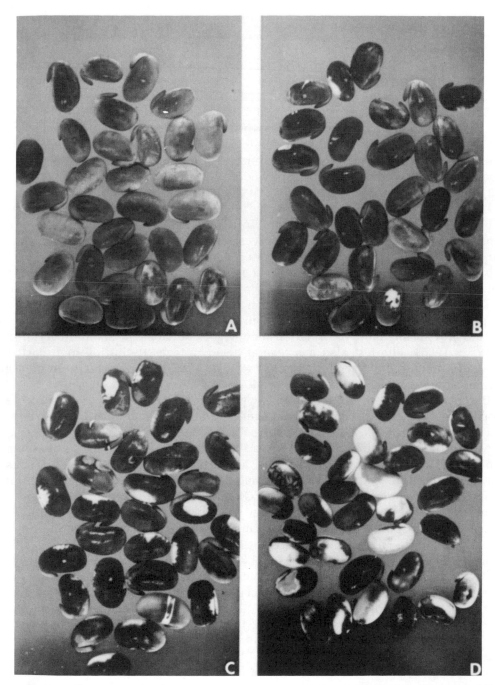

Figure 7.5. *Examples of (A) high, (B) medium, (C) low vigor and (D) ungerminable soybean seeds following staining with tetrazolium chloride.*

reported that foliage development, dry matter accumulation, and crop yield from rapidly germinating soybean seeds exceeded those from slowly germinating seeds (Pinthus and Kimel 1979).

The advantage of the speed of germination test is that little additional work is required compared to the standard germination test. However, variations in temperature or moisture in the test chamber and substrate may affect test results.

Brick Grit Test. The brick grit test is also known as the Hiltner test. It was originally developed by Hiltner and Ihssen (1911) for detecting seed-borne *Fusarium* infection in cereals. Results of further studies indicated that the test also detected seed weaknesses other than those caused by fungi. For example, it revealed cereal injury caused by frost, preharvest sprouting (Schoorel 1960), and hot-water treatment (Tempe 1963) which makes it useful as a vigor test (Schoorel 1960; Tempe 1963).

In the Hiltner test, seeds are planted on damp brick grit or in a container of sand and covered with 3 cm of damp brick grit, then germinated in darkness at room temperature for a specific time. Seeds weakened by pathogenic fungi, mechanical injury, or storage deterioration are unable to penetrate the brick grit layers. The percentage of normal seedlings from this test is considered to be an indication of the vigor level.

The Hiltner test has not been popular in the United States. Comparative trials between germination in sand and ground brick (similar to brick grit) have shown that it fails to provide any more information about vigor than does the standard germination test (Fritz 1965). The test also has several disadvantages including high cost, large space requirement, and variability in test results, as well as difficulties in obtaining, washing, and drying of brick grit (Perry 1978).

Osmotic Stress. When seeds are sown in the field, they are often subjected to drought stress which results in poor emergence. Such drought conditions can be simulated in a laboratory test by use of soil, soil solution, and other solution systems (Parmar and Moore 1968; McWilliams and Phillips 1970; Sharma 1973). Since standardization of soil conditions is difficult to achieve, a solution system is preferred. Seeds are germinated in solutions with specific osmotic potentials such as sodium chloride, glycerol, sucrose, polyethylene glycol (PEG), and mannitol (Parmar and Moore 1968; Sharma 1973). There is evidence, however, that some low molecular weight osmotic substances (sucrose, sodium chloride, glycerol, and mannitol) enter germinating seeds and cause toxicity. High molecular weight PEG (4000 or more) is a satisfactory compound for simulating true drought (Manohar 1966; Parmar and Moore 1968) without causing toxic side effects. The osmotic potentials of PEG 6000 solutions at various concentrations and temperatures have been determined (Michel and Kaufmann 1973). The rate of germination under such conditions is markedly reduced, and emergence of the plumule is generally more affected than that of the radicle (El-Sharkawi and Springuel 1977). Since vigorous seeds can tolerate greater osmotic stress, this method has been suggested as a vigor test (Hadas 1977).

The advantage of the osmotic test is that no special equipment or training is required. However, small corn seeds reportedly germinate better than large seeds under such conditions because of their lower water requirement (Muchena and Grogan 1977).

A significant interaction of osmotic stress and temperature of germination has been reported (El-Sharkawi and Springuel 1977).

Respiration. Seed germination and seedling growth require the use of metabolic energy acquired from respiration. Thus, a decrease in the rate of respiration of germinating seeds has been shown to precede a decline in the rate of seedling growth (Woodstock 1968). Respiration rate, measured during the first 18 hours of germination, can be used to detect injury from gamma radiation in corn, sorghum, wheat, and radish (Woodstock and Combs 1965; Woodstock 1968) and chilling injury in lima bean (Woodstock and Pollock 1965) and cacao (Woodstock et al. 1967). Positive correlations have been reported between rate of oxygen uptake during imbibition and seedling growth (Woodstock and Grabe 1967). However, this relationship has not been confirmed by other studies (Abdul-Baki 1969; Anderson 1970; Byrd and Delouche 1971; Bonner 1974).

Respiration tests are rapid and quantitative, but require a respirometer and trained personnel. Furthermore, mechanical injury (which lowers seed vigor) may increase respiration rates (Woodstock 1969), thus producing confusing results.

Standardization of Vigor Tests

For any seed quality test to be useful, it must provide reproducible tests. To evaluate reproducibility of various vigor test results, both the AOSA and ISTA have conducted extensive referees (McDonald 1977a, 1980b; McDonald et al. 1978; Perry 1978; Tao 1978c, 1980a, 1980b). Generally, results of these referees have shown that seed-testing laboratories can reproduce their own results on the same sample of seed within acceptable confidence limits. However, when the same tests are conducted by *different* laboratories, the amount of variability is often unacceptable.

Many possibilities exist to explain the lack of standardization among laboratories (AOSA 1983). For example, most vigor tests used to date require a degree of subjectivity. The seedling vigor classification and tetrazolium vigor tests possibly involve the most subjective interpretations because seeds and seedlings or both are separated into categories based on characteristics that are difficult to describe precisely. Even the cold test and accelerated aging test require the classification of seedlings as normal or abnormal.

Variations in temperature, moisture, and other environmental conditions are more critical for tests in which rate of growth or rate of a biochemical process is measured than for those such as the standard germination test which measures the completion of a process. For example, a 1°C difference in temperature during the course of the standard laboratory germination test probably has little effect on the final percent germination, but a 1°C temperature difference may have a considerable effect on results in the seedling growth rate test or seed deterioration in the accelerated aging test, as would minor variations in the moisture content of germination substrata or the relative humidity of the air. Consequently, conditions and equipment that are suitable for the standard laboratory germination test may not be suitable for vigor tests.

The cold test would appear to be difficult to standardize because it exposes seeds to soil microorganisms to measure their ability to resist attack under cold, wet conditions. Thus, the standardization of the cold test procedure in its present state still requires standardization of the substrate (soil) microflora, a task that would be difficult, if not impossible. The use of sterile media inoculated with specific microorganisms has been suggested, an approach that is probably too simplistic, since it is difficult to culture microorganisms and maintain a constant level of pathogenicity. Also, pathogens behave differently in soil, where there is a complex population of microorganisms, from how they behave in the absence of such an interacting population.

A recent study by Byrum and Copeland (1995) indicated that the cold test for corn in the U.S. corn belt is as repeatable as the standard germination test. This study was conducted on four corn seed lots with varying levels of quality and each sample was tested by 10 different laboratories throughout the Midwest and Midsouth representing official, commercial, and crop improvement associations. This study illustrates the present state of the art of cold testing in the United States and demonstrates the potential for the eventual standardization of all vigor tests.

It may be important to determine whether such factors as dormancy affect vigor test results. Many species (e.g., legumes) have seeds that are impermeable or only slowly permeable to water. This can affect the percentage of seeds that germinate as well as the length of seedling growth in stress tests and can cause a biased index of vigor in a seedling growth rate test. It can also affect the leaching of electrolytes from seeds in a conductivity test.

Dormancy can also confound the interpretation of vigor test results. For example, dormancy can cause lettuce and cereal seeds to germinate poorly in tests conducted at high temperatures. This difficulty is often overcome in the brick grit test by exposing cereal seeds to low temperatures in the moist brick grit prior to the germination period. If dormancy is suspected of affecting vigor results, the seed should be checked by a standard warm germination test or a tetrazolium test.

Clearly, vigor testing for all species and regions has not yet achieved the same level of standardization possessed by the standard germination test. However, both seed-testing organizations and the seed trade are focusing on improving the reproducibility of vigor tests. Considering the important role that vigor tests can have in plant breeding, seed production, quality control, and marketing programs, standardization is badly needed. Although future research and testing are required before vigor testing becomes a routine phase of seed testing, the promise of vigor testing in the future is bright.

Use of Seed Vigor Tests

The AOSA Seed Vigor Testing Handbook (1983) has included a consideration of the many possible uses of seed vigor tests. For example, many farmers know from experience that seed lots with equal germination levels may emerge from the soil quite differently, resulting in erratic field stands, replanting (on some occasions), or both. These same farmers recognize that this problem is usually more severe when adverse environmental conditions occur at or immediately after planting. Thus, if farmers know that they will be planting seed in adverse soil conditions or can predict an adverse field environment following planting, high-vigor seeds should provide higher field emergence than low-vigor seeds. If this prevents replanting, and the associated delayed maturity or yield reductions due to poor stands, additional money spent on high-vigor seeds may be worthwhile. It cannot be assumed, however, that high-vigor seeds will produce excellent emergence and stands in *any* soil environment. It should improve the chances for satisfactory emergence, *but it will not guarantee it*. Conditions prevailing after planting may be so severe that even the most vigorous seeds cannot perform satisfactorily.

Many seed companies in the central United States conduct a series of vigor tests on each soybean seed lot and combine this information into a seed vigor index. They establish an acceptable vigor level for a specific production season and seed lots are evaluated (in-house) prior to conditioning, treatment, and marketing. In a few cases, those seed lots that exceed a specified vigor level after conditioning are advertised and promoted in general terms as vigor-proven, vigor-rated, or high-vigor seed. If the seed is sold with strong emphasis on vigor, a buyer should ask the dealer for specific information on how many vigor tests were conducted and the criteria used to assess vigor potential. *One should not assume that the statement "vigor-tested" automatically means a sound vigor testing program or implies higher vigor than other seed which is not promoted as vigor-tested.*

Does high-vigor seed mean higher yields? This question is asked by many farmers following implications concerning yield by some seed producers. Unfortunately, there have been fewer comparisons of seed vigor to final yield than to field emergence. As with other investigations relating to seed vigor, the yield comparisons are variable depending on the crop planted, the stand achieved, and the original vigor level of the seed. Many factors that may not relate to the vigor of the seed can influence yield during the growing season. Also, certain crops and genotypes have the ability to compensate for minor stand differences following emergence, and this may result in little difference in final yield. As would be expected, if adequate field stands resulted from high-vigor seeds, higher yields should occur. However, less evidence supports improved yields using high-vigor seeds than supports equal stands achieved from planting medium- or low-vigor seeds. To date, insufficient evidence is available to show a strong relationship between yield and vigor in the absence of stand differences. Future research on plant growth and development or the physiological or biochemical basis for seed vigor may improve our understanding of this relationship.

Questions

1. What are some evidences of genetic influence on seedling vigor?
2. How much loss of vigor do you feel results from harvesting before complete seed maturity? Which crops are more likely to be affected?
3. Which preharvest environmental factors are most influential in affecting seed vigor?
4. The seed corn industry traditionally has not emphasized the different performance of the different seed grades. Is this consistent with the documented effect of seed size on seedling vigor?
5. Mechanical damage to seed is more serious in certain crops. Name several species in which damage is most likely.
6. Should all carryover seed by tested for vigor?

7. List some advantages which might accrue if it were agronomically feasible to presoak seed prior to planting to avoid chilling injury.

8. Name several ways of seed stimulation. Which are agronomically feasible?

9. Name several seed vigor tests. Which are most valuable for measuring seed quality?

10. Do you think cold tests would be valuable for measuring seed vigor in crops other than corn and soybeans?

11. Do physical growth rate tests or physiological measurements most truthfully measure seed quality?

General References

Abdul-Baki, A. A. 1969. Relationship of glucose metabolism to germinability and vigor in barley and wheat seeds. *Crop Science* 9:732–737.

Aguiar, P. A. A. 1974. Some relationships between seed diameter and quality in soybeans. Master's Thesis, Mississippi State University.

Anderson, J. D. 1970. Physiological and biochemical difference in deteriorating barley seed. *Crop Science* 10:36–39.

Association of Official Seed Analysts. 1983. *Seed Vigor Testing Handbook*. No. 32.

Association of Official Seed Analysts. 1991. Rules for testing seeds. *Journal of Seed Technology*. 12:1–109.

Atkin, J. D. 1958. Relative susceptibility of snap bean varieties to mechanical injury of seed. *Proceedings of the American Society of Horticultural Science*. 72:370–373.

Austin, R. B. 1972. Effects of environment before harvesting. In: *Viability of Seeds,* ed. E. H. Roberts, pp. 104–109. Syracuse, N.Y.: Syracuse University Press.

Barriga, C. 1961. Effects of mechanical abuse of Navy bean seed at various moisture levels. *Agronomy Journal* 53:250–251.

Baskin, C. C. 1977. Vigor test methods—Accelerated aging. *Association of Official Seed Analysts Newsletter* 51:42–52.

Beighley, D. H., and N. W. Hopper. 1981. The relationship of chemical composition and electrical conductivity of cowpea seed to field performance. *Agronomy Abstracts* 1981:117.

Belcher, E. W., and L. Miller. 1974. Influence of substrate moisture level on the germination of sweet gum and sand pine seed. *Proceedings of the Association of Official Seed Analysts* 65:88–89.

Bonner, F. T. 1974. Tests for vigor in cherrybark oat acorns. *Proceedings of the Association of Official Seed Analysts* 64:109–114.

Burris, J. S. 1975. Seedling vigor and its effect on field production of corn. *Proceedings of the 30th Annual Corn and Sorghum Research Conference*. 185–193.

———. 1976. Seed/seedling vigor and field performance. *Journal of Seed Technology* 1:58–74.

Burris, J. S., O. T. Edje, and A. H. Wahab. 1969. Evaluation of various indices of seed and seedling vigor in soybean, *Glycine max* (L.) Merr. *Proceeding of the Association of Official Seed Analysts* 59:73–81.

———. 1973. Effect of seed size on seedling performance in soybeans. II. Seedling growth and photosynthesis, and field performance. *Crop Science* 13:207–210.

Burris, J. S., A. H. Wahab, and O. T. Edje. 1971. Effects of seed size on seedling performance in soybeans. I. Seedling growth and respiration in the dark. *Crop Science* 11:429–436.

Byrd, H. W., and J. C. Delouche. 1971. Deterioration of soybean seeds in storage. *Proceedings of the Association of Official Seed Analysts* 61:42–57.

Byrum, J. R., and L. O. Copeland. 1995. Variability in vigor testing of maize (*Zea mays* L.) seed. *Seed Science and Technology* (in press).

Caldwell, W. P. 1972. Relationship of preharvest environmental factors to seed deterioration in cotton. Ph.D. Dissertation, Mississippi State University.

Cantliffe, D. J. 1981. Vigor in vegetable seeds. *Acta Horticulturae* 11:219–226.

Ching, T. M. 1973. Adenosine triphosphate content and seed vigor. *Plant Physiology* 51:400–402.

Cox, F. R., and P. H. Reid. 1964. Calcium-boron nutrition as related to concealed damage in peanuts. *Agronomy Journal* 56:173–176.

Czabator, F. J. 1962. Germination value: An index combining speed and completeness of pine seed germination. *Forest Science* 8:386–396.

Delouche, J. C. 1974. Maintaining soybean seed quality. In: *Soybean: Production, Marketing, and Use*. Muscle Shoals, Ala.: NFDC, TVA, Bull. Y-69:46–62.

———. 1980. Environmental effects on seed development and seed quality. *Horticultural Science* 15:775–780.

Delouche, J. C., and C. C. Baskin. 1973. Accelerated aging techniques for predicting the relative storability of seed lots. *Seed Science & Technology* 1:427–452.

Delouche, J. C., and W. P. Caldwell. 1960. Seed vigor and vigor tests. *Proceedings of the Association of Official Seed Analysts* 50:124–129.

Djavanshir, K., and H. Pourbeik. 1976. Germination value—A new formula. *Silvae Genetica* 25 (2):79–83.

Egli, D. B., and D. M. TeKrony. 1979. Relationship between soybean seed vigor and yield. *Agronomy Journal* 71:755–759.

El-Sharkawi, H. M., and I. Springuel. 1977. Germination of some crop plant seeds under reduced water potential. *Seed Science and Technology* 5:677–688.

Fernandez, R. G., and R. J. Laird. 1959. Yield and protein content of wheat in central Mexico as affected by available soil moisture and nitrogen fertilization. *Agronomy Journal* 51:33–36.

Fontes, J. A. N., and A. J. Ohlrogge. 1972. Influence of seed size and population on yield and other characteristics of soybeans. (*Glycine max* (L.) Merr.). *Agronomy Journal* 64:833–836.

Fox, R. L., and W. A. Albrecht. 1957. Soil fertility and the quality of seeds. *Missouri Agricultural Experiment Station Research Bulletin* 619.

Franck, W. J. 1950. Introductory remarks concerning a modified working of the international rules for seed testing on the basis of experience gained after the World War. *Proceedings of the International Seed Testing Association* 16:405–430.

Fritz, T. 1965. Germination and vigor tests of cereal seed. *Proceedings of the International Seed Testing Association* 39:923–927.

Gloyer, W. C. 1932. Percentage of hard shell in pea and bean varieties. *New York Agricultural Experiment Station Bulletin* 195.

Green, D. E., L. E. Cavanah, and E. L. Pinnell. 1966. Effect of seed moisture content, field weathering, and combine cylinder speed on soybean seed quality. *Crop Science* 6:7–10.

Green, D. E., E. L. Pinnell. 1968. Inheritance of soybean seed quality. II. Heritability of visual ratings of soybean seed quality. *Crop Science* 8:11–15.

Green, D. E., E. L. Pinnell, L. E. Cavanah, and L. F. Williams. 1965. Effect of planting date and maturity date on soybean seed quality. *Agronomy Journal* 57:165–168.

Hadas, A. 1977. A suggested method for testing seed vigor under water stress in simulated arid conditions. *Seed Science and Technology* 5:519–525.

Halmer, P., and J. D. Bewley. 1984. A physiological perspective on seed vigor testing. *Seed Science and Technology* 12:561–575.

Harman, G. E., and L. R. Mattick, 1976. Association of lipid oxidation with seed aging and death. *Nature* 260:323–324.

Harris, H. B., M. B. Parker, and B. J. Johnson. 1965. Influence of molybdenum content of soybean seed and other factors associated with seed source on progeny response to applied molybdenum. *Agronomy Journal* 57:397–399.

Hawkins, R. C. and P. J. M. Cooper. 1979. Effects of seed size on growth and yield of maize in Kenya highlands. *Experimental Agriculture* 15:73–79.

Heydecker, W. 1972. Vigor. *Viability of Seeds*, ed. E. H. Roberts, pp. 209–252. Syracuse, N.Y.: Syracuse University Press.

Heydecker, W. 1977. Stress and seed germination: An agronomic view. In: *The Physiology and Biochemistry of Seed Dormancy and Germination,* ed. A. A. Khan, pp. 237–276. Amsterdam: Elsevier Press.

Hicks, D. R., R. H. Peterson, W. E. Lueschen, and J. H. Ford. 1976. Seed grade effect on corn performance *Agronomy Journal* 68:819–820.

Hiltner, L., and G. Ihssen. 1911. Uber das schlechte Auflaufen und die Auswinterung des Getreides infolge Befalls des Saatgutes durch Fusarium. Landw. *Jahrbuch fur Bayern* 1:20–60, 315–362.

Hopper, N. W., and H. R. Hinton. 1980. The use of electrical conductivity as a measure of cottonseed quality. *Agronomy Abstracts* 1980:109.

Hunter, R. B., and L. W. Kannenberg. 1972. Effects of seed size on emergence, grain yield, and plant height in corn. *Canadian Journal of Plant Science* 52:252–256.

Isely, D. 1957. Vigor tests. *Proceedings of the Association of Official Seed Analysts* 47:176–182.

Johnson, D. R., and V. D. Leudders. 1974. Effects of planted seed size on emergence and yield in soybeans (*Glycine max* (L.) Merr.). *Agronomy Journal* 66:117–118.

Johnson, R. R., and L. M. Wax. 1978. Relationship of soybean germination and vigor tests to field performance. *Agronomy Journal* 70:273–278.

Joo, P. K., B. A. Orman, A. M. Moustafa, and M. Hafdahl. 1980. Can leachate electro-conductivity be a useful tool for corn seed emergence potential evaluation? *Agronomy Abstracts* 1980:109.

Kannenberg, L. W., and R. W. Allard. 1964. An association between pigment and lignin formation in the seed coat of lima bean. *Crop Science* 4:621–622.

Koostra, P. T., and J. F. Harrington. 1969. Biochemical effects of age on membranal lipids of *Cucumis sativus* L. seed. *Proceedings of the International Seed Testing Association* 34:329–340.

Kyle, J. H., and T. F. Randall. 1963. A new concept of the hard seed character in *Phaseolus vulgaris* L. and its use in breeding and inheritance studies. *Proceedings of the American Society of Horticultural Science* 83:461–475.

Lebedeff, G. A. 1947. Studies of the inheritance of hard seed in beans. *Journal of Agricultural Research* 74:205–215.

Leggatt, C. W. 1948. Germination of boron-deficient peas. *Scientific Agriculture* 28:131–139.

Lowe, L. B., G. S. Ayers, and S. K. Ries. 1972. The relationship of seed protein and amino acid composition to seedling vigor and yield of wheat. *Agronomy Journal* 64:608–611.

Lowe, L. B., and S. K. Ries. 1973. Endosperm protein of wheat seed as a determinant of seedling growth. *Plant Physiology* 51:57–60.

Luedders, V. D., and J. S. Burris. 1979. Effect of broken seed coats on field emergence of soybeans. *Agronomy Journal* 71:877–879.

Maguire, J. D. 1962. Speed of germination-aid in selection and evaluation for seedling emergence and vigor. *Crop Science* 2:176–177.

Manohar, M. S. 1966. Effect of osmotic systems on germination of peas (*Pisum sativum* L.). *Planta* 71:81–88.

McDaniel, R. G. 1969. Relationship of seed weight, seedling vigor, and mitochondrial metabolism in barley. *Crop Science* 9:823–827.

McDaniel, R. G., and I. V. Sarkissian. 1968. Mitochondrial heterosis in maize. *Genetics* 59:465–475.

McDonald, M. B., Jr. 1975. A review and evaluation of seed vigor tests. *Proceedings of the Association of Official Seed Analysts* 65:109–139.

———. 1977a. AOSA vigor subcommittee report: 1977 vigor test "referee" program. *Association of Official Seed Analysts Newsletter* 51(5):14–41.

———. 1977b. The influence of seed moisture on the accelerated aging vigor test. *Journal of Seed Technology* 2(1):19–28.

———. 1980a. Assessment of seed quality. *Horticultural Science* 15:784–788.

———. 1980b. Vigor test subcommittee report. *Association of Official Seed Analysts Newsletter* 54(1):37–40.

McDonald, M. B. 1994. The history of seed vigor testing. *Journal of Seed Technology*.

McDonald, M. B., Jr., and B. R. Phaneendranath. 1978. A modified accelerated aging seed vigor test for soybean. *Journal of Seed Technology* 3(1):27–37.

McDonald, M. B., Jr., and D. O. Wilson. 1979. An assessment of the standardization and ability of the ASA-610 to rapidly predict soybean germination. *Journal of Seed Technology* 4(2):1–12.

———. 1980. ASA-610 ability to detect changes in soybean seed quality. *Journal of Seed Technology* 5(1):56–66.

McDonald, M. B., Jr., K. L. Tao, C. C. Baskin, D. F. Grabe, and J. F. Harrington. 1978. AOSA vigor seed committee report. 1978 vigor test "referee" program. *Association of Official Seed Analysts Newsletter* 52(4):31–42.

McNeal, F. H., M. A. Berg, P. L. Brauig, and C. F. McGuire. 1971. Productivity and quality response of five spring wheat genotypes (*T. aestivum*) to nitrogen fertilizer. *Agronomy Journal* 63:908–910.

McWilliams, J. R., and P. J. Phillips. 1970. Effect of osmotic and matric potentials on the availability of water for seed germination. *Australian Journal of Biological Science* 24:423–431.

Michel, B. E., and M. R. Kaufmann. 1973. The osmotic potential of polyethylene glycol 6000. *Plant Physiology* 51:914–916.

Miles, D. F., Jr., and L. O. Copeland. 1980. The relationship of vigor tests and field performance in soybeans (*Glycine* max (L.) Merr.), *Agronomy Abstracts* 1980:111.

Muchena, S. S., and C. O. Grogan. 1977. Effects of seed size on germination of corn (*Zea mays*) under simulated water stress conditions. *Canadian Journal of Plant Science* 57:921–923.

Nass, H. G., and P. L. Crane. 1970. Effect of endosperm mutants on germination and early seedling growth rate in maize (*Zea mays* L.). *Crop Science* 10:139–140.

Nichols, M. A., and W. Heydecker. 1968. Two approaches to the study of germination data. *Proceedings of the International Seed Testing Association* 33:531–540.

Nicholson, J. F., O. D. Dhingra, and J. B. Sinclair. 1972. Internal seed-borne nature of *Sclerotinia sclerotiorum* and *Phomopsis* spp. and their effects on soybean seed quality. *Phytopathology* 61:1261–1263.

Nobbe, F. 1876. *Handbuch der samenkunde*. Berlin: Wiegandt-Hempel-Parey.

Parmar, M. T., and R. P. Moore. 1968. Carbowax 6000, mannitol, and sodium chloride for simulating drought conditions in germination studies of corn (*Zea mays* L.) of strong and weak vigor. *Agronomy Journal* 60:192–195.

Perry, D. A. 1973. Seed vigour and stand establishment. *Horticultural Abstracts* 42:334–342.

Perry, D. A. 1977. A vigor test for seeds of barley (*Hordeum vulgare*) based on measurement of plumule growth. *Seed Science and Technology* 5:709–719.

————. 1978. Report of the vigor test committee 1974–1977. *Seed Science and Technology* 6:159–181.

Perry, D. A., ed. 1981. *Handbook of Vigor Test Methods.* Zurich: International Seed Testing Association.

Pinthus, M. J., and U. Kimel. 1979. Speed of germination as a criterion of seed vigor in soybeans. *Crop Science* 19:291–292.

Pollock, B. M., and E. E. Roos. 1972. Seed and seedling vigor. In: *Seed Biology,* Vol. 1, ed. T. T. Kozlowski, pp. 314–318. New York: Academic Press.

Potts, H. C., J. D. Duangpatra, W. G. Hairston, and J. C. Delouche. 1978. Some influences of hard-seededness on soybean seed quality. *Crop Science* 18:221–224.

Priestley, D. A., and A. C. Leopold. 1979. Absence of lipid oxidation during accelerated aging of soybean seeds. *Plant Physiology* 63:726–729.

Ries, S. K., O. Moreno, W. F. Meggitt, C. J. Schweizer, and S. A. Ashkar. 1970. Wheat seed protein-chemical influence on and relationship to subsequent growth and yield in Michigan and Mexico. *Agronomy Journal* 62:746–748.

Roncadori, R. W., O. L. Brooks, and C. E. Perry, 1972. Effect of field exposure on fungal invasion and deterioration of cotton seed. *Phytopathology* 62:1137–1139.

Schoorel, A. F. 1960. Report on the activities of the vigor test committee. *Proceedings of the International Seed Testing Association* 25:519–525.

Sharma, M. L. 1973. Simulation of drought and its effect on germination of five pasture species. *Agronomy Journal* 65:982–987.

Sherf, A. F. 1953. Correlation of germination data of corn and soybean seed lots under laboratory, greenhouse, and field conditions. *Proceedings of the Association of Official Seed Analysts* 43:127–130.

Shieh, W. J., and M. B. McDonald. 1982. The influence of seed size, shape, and treatment on inbred seed corn quality. *Seed Science and Technology* 10:307–313.

Simon, E. W. 1974. Phospholipids and plant membrane permeability. *New Phytology* 73:377–420.

Simon, E. W., and H. H. Wiebe. 1975. Leakage during imbibition, resistance to damage at low temperature, and the water content of peas. *New Phytology* 74:407–411.

Simpson, D. M., and B. M. Stone. 1935. Viability of cottonseed as affected by field conditions. *Journal of Agricultural Research* 50:435–447.

Singh, J. N., S. K. Tripathi, and P. S. Negi. 1972. Note on the effect of seed size on germination, growth, and yield of soybeans. *Indian Journal of Agricultural Science* 42:83–86.

Spencer, G. F., F. R. Earle, I. A. Wolff, and W. H. Tallent. 1973. Oxygenation of unsaturated fatty acids in seeds during storage. *Chemistry and Physiology of Lipids* 10:191–202.

Stewart, R. R. C., and J. D. Bewley. 1980. Lipid peroxidation associated with accelerated aging of soybean axes. *Plant Physiology* 65:245–249.

Streeter, J. G. 1965. Possible mechanisms in the loss of seed viability. *Association of Official Seed Analysts Newsletter* 39(3):27–35.

Sullivan, G. A. 1973. Effects of dolomitic limestone, gypsum, and potassium on planting seed quality, yield, and grade of peanuts. *Arachis hypogea* L. Ph.D. Dissertation, North Carolina State University, Raleigh.

Suryatmana, G., L. O. Copeland, and D. F. Miles. 1980. Comparison of laboratory indices of seed vigor with field performance of Navy bean (*Phaseolus vulgaris* L.). *Agronomy Abstracts* 1980:113.

Tao, K. J. 1978a. Effect of soil water holding capacity on the cold test for soybeans. *Crop Science* 18:979–982.

————. 1978c. The 1978 "referee" test for soybean and corn. *Association of Official Seed Analysts Newsletter* 52(4):43–66.

————. 1978b. Factors causing variations in the conductivity test for soybean seeds. *Journal of Seed Technology* 3(1):10–18.

————. 1980a. The 1979 Vigor "referee" test for soybean and corn. *Association of Official Seed Analysts Newsletter* 54(1):40–58.

————. 1980b. The 1980 vigor "referee" test for soybean and corn seed. *Association of Official Seed Analysts Newsletter* 54(3):53–68.

TeKrony, D. M., and D. B. Egli. 1977. Relationship between laboratory indices of soybean seed vigor and field emergence. *Crop Science* 17:573–577.

TeKrony, D. M., and D. B. Egli, 1991. Relationship of seed vigor to crop yield: A review. *Crop Science* 31:816–822.

TeKrony, D. M., D. B. Egli, and J. Balles. 1980a. The effect of the field production environment on soybean seed quality. In: *Seed Production,* ed. P. D. Hebblethwaite. London: Butterworth and Co., Ltd.

TeKrony, D. M., A. D. Phillips, and D. B. Egli. 1980b. The effect of field weathering on soybean seed viability and vigor. *Crop Science* 72:749–753.

Tempe, de J. 1963. The use of correlation coefficients in comparing methods for seed vigor testing. *Proceedings of the International Seed Testing Association* 28:167–172.

Timson, J. 1965. New method of recording germination data. *Nature* 207:216–217.

Tucker, H., and L. N. Wright. 1965. Estimating rapidity of germination. *Crop Science* 5:398–399.

Ullrich, S. E., and R. F. Eslick. 1978. Lysine and protein characterization of spontaneous shrunken endosperm mutant of barley. *Crop Science* 18:809–812.

Van Onckelen, H. A., R. Verbeek, M. A. Stone, and A. A. Khan. 1973. The metabolism of barley embryo under accelerated aging conditions and its relation to alpha amylase synthesis and RNA metabolism of aleurone. *Plant Physiology* 51:S–39.

Villiers, T. A. 1973. Aging and the longevity of seeds in field conditions. In: *Seed Ecology,* ed. W. H. Heydecker, pp. 265–288. University Park, PA: The Pennsylvania University Press.

Walters, H. J. and C. E. Caviness. 1973. Breeding for improved soybean seed quality. *Arkansas Farm Research* 22 (5):5.

Wang, B. S. P. 1973. Laboratory germination criteria for red pine (*Pinus resinosa* Ait.) seed. *Proceedings of the Association of Official Seed Analysts* 63:94–101.

Wester, R. E. 1970. Nonbleached, a genetic marker for quality seed in lima beans. *Seed World* 160(7):24.

Whaley, W. G. 1950. The growth of inbred and hybrid maize. *Growth* 14:123–154.

Woodruff, J. M., F. S. McCain, and C. S. Hoveland. 1967. Effect of relative humidity, temperature, and light intensity during boll opening on cotton seed quality. *Agronomy Journal* 59:441–444.

Woodstock, L. W. 1965. Seed vigor. *Seed World* 97(5):6.

Woodstock, L. W. 1968. Relationship between respiration during imbibition and subsequent growth rates in germinating seeds. In: *3rd International Symposium on Quantitative Biology of Metabolism,* ed. A. Locker, pp. 136–146.

————. 1969. Biochemical tests for seed vigor. *Proceedings of the International Seed Testing Association* 34:253–263.

————. 1973. Physiological and biochemical tests for seed vigor. *Seed Science and Technology* 1:127–157.

Woodstock, L. W., and M. F. Combs. 1965. Effects of gamma-irradiation of corn seed on the respiration and growth of the seedling. *American Journal of Botany* 52:563–569.

Woodstock, L. W., and D. F. Grabe. 1967. Relationship between seed respiration during imbibition and subsequent seedlings growth in *Zea mays* L. *Plant Physiology* 42:1071–1076.

Woodstock, L. W., and B. M. Pollock. 1965. Physiological predetermination: Imbibition, respiration, and growth of lima bean seed. *Science* 150:1031–1032.

Woodstock, L. W., B. Reiss, and M. F. Combs. 1967. Inhibition of respiration and seedling growth by chilling treatments in *Cacao theobroma*. *Plant and Cell Physiology* 8:339–342.

Yaklich, R. W., and M. M. Kulik, 1979. Evaluation of vigor tests in soybean seeds. Relationship of the standard germination test, seedling vigor classification, seedling length and tetrazolium staining to field performance. *Crop Science* 19:247–252.

8

Seed Longevity and Deterioration

Seeds are uniquely equipped to survive as viable regenerative organisms until the time and place are right for the beginning of a new generation; however, like any other form of life, they cannot retain their viability indefinitely and eventually deteriorate and die. Fortunately, neither nature nor agricultural practice ordinarily requires seeds to survive longer than the next growing season, though seeds of most species are able to survive much longer under the proper conditions.

THE LIFE SPAN OF SEEDS

Long-Lived Seeds

Museum botanists in Canada reported the germination of lupin seeds that had been buried deep in a Canadian peat bog for an estimated 10,000 years (Porsild and Harrington 1967). This is the longest known record for safe seed storage even with a literature that abounds with evidences of long seed storage (Justice and Bann 1978). Germinating Indian lotus seeds from a Manchurian lake bed were first estimated to be 120 to 400 years old (Ohga 1926), but were later found, using radiocarbon dating, to be over 1000 years old (Libby 1951). At the National Museum of Paris, viable seeds of several species that had been collected 100 to 160 years previously were found (Becquerel 1934). Frequently, one hears of reports of germination of the so-called "mummy" seed from Egyptian tombs; however, these reports have not been substantiated.

Many long-term storage studies are in progress in the United States. These have already provided considerable data on the seed longevity of many species. The two oldest and best-known studies are those founded by Beal in 1879 at Michigan State University and Duvel of the United States Department of Agriculture in 1902. In both studies, seeds buried in soil continue to germinate demonstrating the remarkable life span of many seeds (Figure 8.1).

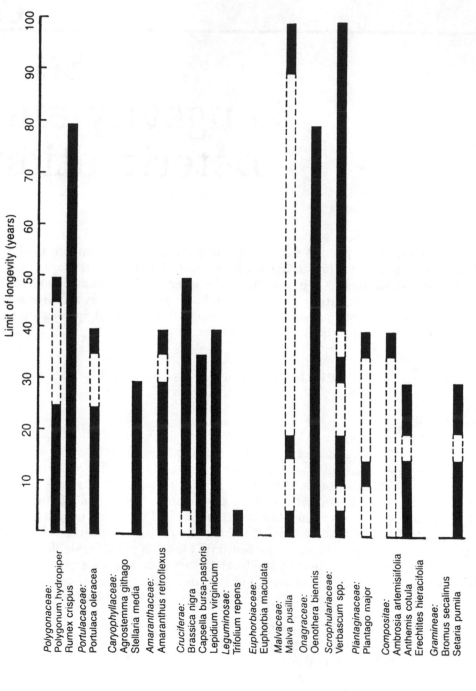

Figure 8.1. *Limits of longevity for seeds in Beal's burial experiment. Germination tests were performed every five years for the first 40 years and every 10 years thereafter. Dashed lines indicate test periods when no positive germinations were noted. "Verbascum spp." indicates* V. thapsus *and/or* V. blattaria. *(From Priestley, 1986).*

Short-Lived Seeds

In contrast to evidence on long-lived seeds, some species have seeds that are remarkably short-lived, particularly when stored in open air. To distinguish between these types of long- and short-lived seeds, Roberts (1973) proposed the terms *orthodox* and *recalcitrant* seeds. Orthodox seeds are long-lived seeds. They can be successfully dried to moisture contents as low as 5% without injury and are able to tolerate freezing temperatures. Recalcitrant seeds are those seeds which cannot be dried to moisture contents below 30% without injury and are unable to tolerate freezing. As a result, recalcitrant seeds live for only short durations and are difficult to successfully store because their high moisture content encourages microbial contamination and results in more rapid seed deterioration. Furthermore, storage of recalcitrant seeds at subzero temperatures causes the formation of ice crystals which disrupts cell membranes and causes freezing injury.

Recalcitrant and orthodox seeds differ greatly in their ecology and seed morphology (Chin et al. 1989). Recalcitrant seeds are primarily from perennial trees in the moist tropics such as coconut, coffee, and cacao. In some cases, they also come from temperate trees such as citrus. Other species with remarkably short life spans that lose viability after a few months in the open air are willow, poplar, cottonwood, and elm. Seeds of river maple (*Acer saccharinum*) reportedly last only a few days under natural conditions (Jones 1920). Survival in nature apparently depends on the ability of the seeds to germinate and become established quickly after they have fallen from the tree. Seeds of wild rice (*Zizania aquatica*) also lose viability rapidly in open-air storage and must be kept in water at low temperatures for longer survival (Duvel 1905). Most orthodox seeds come from annual temperate species adapted to open fields.

Recalcitrant seeds also differ from orthodox seeds in their seed morphology. Most recalcitrant seeds mature and exist in their fruits and are covered with fleshy or juicy arriloid layers and impermeable testa (Chin et al. 1989). At physiological maturity, they are generally much higher (50–70%) in moisture content than orthodox seeds (30–50%). At the same time, they are larger than most orthodox seeds, even though their embryos are only about 15% the size of an orthodox seed embryo (Table 8.1). These differences have made successful storage of recalcitrant seeds difficult and an active area of research study. It is generally believed that recalcitrant seeds never go into dormancy but, instead, continue their development and progress toward germination (Berjak et al. 1990). Most attempts at storing these seeds have focused on using endogenous seed inhibitors such as abscisic acid (Goldbach 1979) or replacing the high water content with other substances such as sugars or ethylene glycol to permit successful storage even under low temperatures without inducing ice-crystal formation and subsequent seed damage (Crowe and Crowe 1986). A more detailed consideration of recalcitrant seeds is provided by Chin and Roberts (1980). It should be noted that some seeds now appear to fit neither an orthodox nor recalcitrant category and, for these seeds, an intermediate category has been suggested (Ellis 1991). Citrus and coffee seeds may fit this intermediate classification.

Table 8.1. Seed Size, 1000 Seed wt., and Moisture Content of Typical Recalcitrant and Orthodox Seeds

Crop species	Seed size, length × width	1000 Seed wt.	Moisture content
	mm	g	$gH_2O\ g^{-1}fw$
Recalcitrant			
Nephelium lappaceum L.	28 × 16	3,555	0.49
Artocarpus heterophyllus Lam.	35 × 24	8,520	0.52
Artocarpus champeden (Lour.) Spreng.	30 × 20	5,814	0.71
Lansium domesticum Corr.	17 × 13	2,335	0.52
Bouea ganadaria	22 × 15	3,530	0.46
Durio zibethinus Murr.	42 × 25	14,783	0.50
Theobroma cacao L.	25 × 25	1,995	0.36
Orthodox			
Hibiscus esculentus L.	6 × 4	146	0.18
Vigna sesquipedalis (L.) Fruw.	12 × 5	192	0.16

From Chin et al. (1989).

CONCEPTS OF SEED DETERIORATION

Seed deterioration is a major problem in agricultural production. In the United States alone, seed sales amounted to over $2 billion in 1984. It has been estimated that about 25% of that value, or approximately $500 million in revenue, was lost due to poor seed quality (McDonald and Nelson 1986). Worldwide, these losses are even greater, particularly in lesser-developed countries and in geographic regions where high temperatures and high relative humidities prevail during seed maturation and storage. Although the significance of these losses is readily apparent, the overall importance of seed deterioration becomes even more manifest when it is realized that over $800 million is lost annually due to deterioration and as a consequence of breakage and microorganism spoilage during production, storage, and shipping of feed grains (Salunkhe et al. 1985).

Seed deterioration can be characterized by the following three general concepts (Delouche 1973):

1. **Seed deterioration is an inexorable process.** All living things must eventually deteriorate and die. Although death still remains an inevitable result of life, it is possible to retard the rate of deterioration through optimum storage practices.
2. **Seed deterioration is an irreversible process.** Once seed deterioration has occurred, this anabolic process cannot be reversed. Simply stated, low-quality seeds cannot be made into high-quality seeds. Some mechanisms for preconditioning or treating seeds with fungicides improve field emergence. However, these treatments only allow the optimum expression of seed potential; they do not alter the basic physiological quality of the seed.
3. **Seed deterioration varies among seed populations.** It is now well established that certain varieties exhibit less deterioration than others. Even within a variety, the storage potential of individual lots varies, and even within a seed lot, individual seeds have differing storage potential.

FACTORS INFLUENCING THE LIFE SPAN OF SEEDS

Internal Factors

The physical condition and physiological state of seeds greatly influence their life span. Seeds that have been broken, cracked, or even bruised deteriorate more rapidly than undamaged seeds (McDonald 1985; Priestley 1986). Even in the absence of physical symptoms, seeds may be physiologically impaired and become susceptible to rapid deterioration. Several kinds of environmental stresses during seed development, and prior to physiological maturity, can reduce the longevity of seeds—for example, deficiency of minerals (N, K, Ca) (Harrington 1960b), water (Haferkamp et al. 1953), and temperature extremes (Justice and Bass 1978). Immature and small seeds within a seed lot do not store as well as mature and large seeds within a seed lot (Wien and Kueneman 1981; Minor and Paschal 1982). Hard-seededness also extends seed longevity (Flood 1978; Patil and Andrews 1985).

Relative Humidity and Temperature

The two most important factors that influence the life span of seeds are relative humidity and temperature. The effects of relative humidity (and its subsequent effect on seed moisture) and temperature of the storage environment are highly interdependent. Most crop seeds lose their viability at relative humidities approaching 80% and temperatures of 25 to 30°C but can be kept 10 years or longer at relative humidities of 50% or less and a temperature of 5°C or lower (Toole 1950). According to Harrington (1973), because of this interdependency, the sum of the percentage of relative humidity plus the temperature in degrees Farenheit should not exceed 100 for safe storage. Another report indicates that for safe storage from one to three years, this combined total may be as high as 120 as long as the temperature contributes no more than half the total (Bass 1967). It has been suggested that the relative humidity should be no higher than 60% for seeds at 21°C and no higher than 70% for seeds 4 to 10°C; however, at 5°C and 45 to 50% relative humidity, many crop seeds can be safely stored for 10 years or longer (Toole 1957).

These data reflect the intimate association among seed moisture, storage temperature, and seed longevity. This interaction was recognized by Harrington (1972) when he suggested the following two "rules of thumb" regarding optimum seed storage: (1) each 1% reduction in seed moisture doubles the life of the seed, and (2) each 5°C reduction in seed temperature doubles the life of the seed. Let's examine each of these important variables critically.

Seed Moisture

Harrington (1972) recognized that there must be some qualifications to the above rules of thumb. First, the rule regarding seed moisture does not apply above 14 or below 5% seed moisture. Seeds stored at moisture contents above 14% begin to exhibit increased respiration, heating, and fungal invasion that destroy seed viability more rapidly than that indicated by the first rule of thumb. Below 5% seed moisture, a breakdown of membrane structure hastens seed deterioration. (This is probably a consequence of reorientation of hydrophyllic cell membranes due to the loss of the water molecules necessary to retain their configuration.) Thus, storage of most seeds between 5 and 6% seed moisture appears to be ideal for maximum longevity.

Water in seeds is a complex system and is only partially understood (Leopold 1986; Stanwood and McDonald 1989). The probability of various structural features existing have been derived through statistical thermodynamical approaches (Nementhy and Sheraga 1962; Rahman and Stillinger 1971). Empirical physical properties of water such as viscosity, density, and thermodynamical properties were compiled by the United States Bureau of Statistics in the 1920s and are valid for bulk water but not that associated with plant substances. The way in which water interacts with plant substances is more complex (Kavanau 1964). In much of the literature, we find references to bound water, adsorbed water, and free water which have been characterized into three phases or zones (Vertucci and Leopold 1986) depending on the way it is held by the plant substances. Bound water is tightly held to ionic groups such as amino or carboxyl groups and exists as a monolayer around macromolecules of the seed. Adsorbed water is considered to exist in multilayers, loosely held by bonding to hydroxyl and amide groups above the monolayer of bound water. Free water is considered as capillary or solution water held only by capillary forces to the seed tissues. However, these concepts about seed moisture may be oversimplifications.

Even though much is not known about water-plant substance relationships, it is known that water is associated with the seed system in several patterns. In some cases, it is actually part of the chemical structure of other molecules of the seed tissue, held by hydrogen bonding (vectorized polar bonds), and does not exist as discrete water molecules. In other cases, water is held as discrete molecules in bonding interactions with seed tissue molecules, though the arrangement and stability of this type of water is highly variable. These interactions may extend into the surrounding liquid, forming gradient patterns of structure in a dynamic state of turnover. There may be water, but always in association with other systems, and at which point water is bound and free is difficult to ascertain.

Moisture Equilibrium.   The hygroscopic nature of seeds allows them to maintain equilibrium moisture content with any given relative humidity (see Table 8.2). Equilibrium is attained when the seed has no further tendency to absorb or lose moisture. Hygroscopic equilibrium curves, also called absorption isotherms, are graphic expressions of the relationship between the moisture content of seeds and their ambient relative humidity at constant temperatures. They are established by measuring the absorption or desorption at successive relative humidities and can be used to predict seed moisture content at any given relative humidity.

The hygroscopic moisture equilibrium curve in Figure 8.2 is a sigmoidlike curve, with three rather distinct phases representing different stages of water absorption or desorption. Phase one represents very tightly held water that may actually be a part of the chemical structure of the seed. This kind of water cannot be removed without destruction of the seed tissue. This phase may also include some water held as discrete molecules in bonding interactions with the seed tissue molecules.

Phase two of the moisture equilibrium curve represents water that is more loosely

Table 8.2. Approximate Moisture Content of Vegetable and Field Crop Seeds in Equilibrium with Air at Different Relative Humidities at Room Temperature (Approximately 77°F) (Moisture Content Wet Basis, in Percentages)

Vegetable seeds

Species	% Relative Humidity					
	10	20	30	45	60	75
Bean, Broad	4.2	5.8	7.2	9.3	11.1	14.5
Bean, Lima	4.6	6.6	7.7	9.2	11.0	13.8
Bean, Snap	3.0	4.8	6.8	9.4	12.0	15.0
Beet, Garden	2.1	4.0	5.8	7.6	9.4	11.2
Cabbage	3.2	4.6	5.4	6.4	7.6	9.6
Cabbage, Chinese	2.4	3.4	4.6	6.3	7.8	9.4
Carrot	4.5	5.9	6.8	7.9	9.2	11.6
Celery	5.8	7.0	7.8	9.0	10.4	12.4
Corn, Sweet	3.8	5.8	7.0	9.0	10.6	12.8
Cucumber	2.6	4.3	5.6	7.1	8.4	10.1
Eggplant	3.1	4.9	6.3	8.0	9.8	11.9
Lettuce	2.8	4.2	5.1	5.9	7.1	9.6
Mustard, Leaf	1.8	3.2	4.6	6.3	7.8	9.4
Okra	3.8	7.2	8.3	10.0	11.2	13.1
Onion	4.6	6.8	8.0	9.5	11.2	13.4
Onion Welsh	3.4	5.1	6.9	9.4	11.8	14.0
Parsnip	5.0	6.1	7.0	8.2	9.5	11.2
Pea	5.4	7.3	8.6	10.1	11.9	15.0
Pepper	2.8	4.5	6.0	7.8	9.2	11.0
Radish	2.6	3.8	5.1	6.8	8.3	10.2
Spinach	4.6	6.5	7.8	9.5	11.1	13.2
Squash, Winter	3.0	4.3	5.6	7.4	9.0	10.8
Tomato	3.2	5.0	6.3	7.8	9.2	11.1
Turnip	2.6	4.0	5.1	6.3	7.4	9.0
Watermelon	3.0	4.8	6.1	7.6	8.8	10.4

Field crop seeds

Species	% Relative Humidity						
	15	30	45	60	75	90	100
Barley	6.0	8.4	10.0	12.1	14.4	19.5	26.8
Buckwheat	6.7	9.1	10.8	12.7	15.0	19.1	24.5
Shelled Corn, yd	6.4	8.4	10.5	12.9	14.8	19.1	23.8
Shelled Corn, wd	6.6	8.4	10.4	12.9	14.7	18.9	24.6
Shelled Corn, pop	6.8	8.5	9.8	12.2	13.6	18.3	23.0
Flaxseed	4.4	5.6	6.3	7.9	10.0	15.2	21.4
Oats	5.7	8.0	9.6	11.8	13.8	18.5	24.1
Peanut	2.6	4.2	5.6	7.2	9.8	13.0	
Rice, Milled	6.8	9.0	10.7	12.6	14.4	18.1	23.6
Rye	7.0	8.7	10.5	12.2	14.8	20.6	26.7
Sorghum	6.4	8.6	10.5	12.0	15.2	18.8	21.9
Soybeans	4.3	6.5	7.4	9.3	13.1	18.8	
Wheat, White	6.7	8.6	9.9	11.8	15.0	19.7	26.3
Wheat, Durum	6.6	8.5	10.0	11.5	14.1	19.3	26.6
Wheat, Soft Red Winter	6.3	8.6	10.6	11.9	14.6	19.7	25.6
Wheat, Hard Red Winter	6.4	8.5	10.5	12.5	14.6	19.7	25.9
Wheat, Hard Red Spring	6.8	8.5	10.1	11.8	14.8	19.7	25.0

Data from Harrington (1960a).

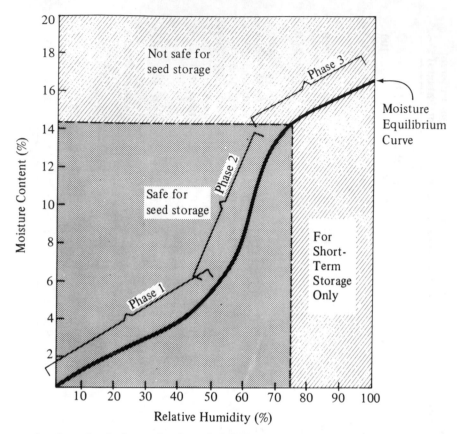

Figure 8.2. *An absorption isotherm, showing relationship of seed moisture content with relative humidity of air, at a given temperature.*

held than that of phase one. For most seeds, this portion of the moisture equilibrium is represented by a straight-line relationship between relative humidity and moisture content. Water represented by the upper portion of phase two is easily removed by drying; however, the lower portion representing strong bonding is difficult to remove. Water in the upper portion of phase two contributes significantly to seed deterioration during storage.

Water in phase three represents water loosely held by very weak bonding and free water in the intercellular and intertissue spaces. It is easily eliminated during drying; but if not eliminated, it contributes to rapid seed deterioration.

Effect of Temperature. The moisture equilibrium curves are only slightly influenced by temperature; an increase in temperature causes a slight reduction in moisture content at a fixed relative humidity (Haynes 1961). Most equilibrium moisture curves are established at 77°F; however, a formula for predicting the effect of temperature at any relative humidity has been devised (Hibbard and Miller 1928).

The Hysteresis Phenomenon. The desorption equilibrium curve will usually be slightly higher than the adsorption curve. The difference in the equilibrium moisture content during adsorption and desorption is known as *hysteresis*. The higher desorption isotherm has been attributed to the appearance of additional points of attachment (polar sites) for bound water as a result of tissue swelling (Abdul-Baki and Anderson 1970). On desorption, it was suggested that the disappearance of these polar sites was delayed by their tendency to hold and keep the bound water that intervenes to block the collapse of the seed tissue. The study concluded that hysteresis could only occur in absorptive substances with a high degree of structural rigidity.

Temperature

At temperatures of 0°C, formation of intracellular ice crystals can disrupt membrane integrity and contribute to seed deterioration. Seeds with moisture levels below 14% do not form ice crystals. It should be noted, however, that at 14% initial moisture, seeds stored in cold rooms below 0°C will likely gain moisture. Most cold rooms have a high relative humidity, and seeds achieve equilibrium with that relative humidity after a brief period of storage. Thus, seeds stored at low temperatures must be in conditions in which the relative humidity is controlled or placed in moisture-proof containers to avoid increases in moisture content and increased deterioration.

Association Between Seed Moisture Content and Temperature. The axiom that the sum of the percentage of relative humidity and temperature in degrees Farenheit should not exceed 100 for safe storage implies an equivalence of these effects of temperature and humidity on seed longevity. We now know that this is not true. Both parameters influence seed metabolism. High relative humidities increase seed moisture content, which results in biochemical events such as increased hydrolytic enzyme activity, enhanced respiration, and increases in free fatty acids. High temperatures serve to enhance the rate at which many enzymatic and metabolic reactions occur, causing a more rapid rate of deterioration. However, seed moisture content is considered to be the most critical factor in maintaining seed longevity. Although seed moisture content and high temperatures are interrelated, high temperatures hasten the deterioration of high-moisture seeds by increasing the metabolic activity of hydrolyzed substrates and enzymes. High temperatures exert only a minimal deteriorative effect on low-moisture seeds. It has been shown that low-moisture seeds store well at temperatures up to 25°C but high-moisture seeds will store well only if the temperature is reduced to 10°C or less. Thus, although temperature and relative humidity interact to determine seed longevity, the control of relative humidity and its subsequent effect on seed moisture content is more critical than storage temperature in achieving optimum storage conditions.

Genetic Factors

Seeds of some species are genetically and chemically equipped for longer storability than others under similar conditions. Most long-lived seeds belong to species possessing hard, impermeable seed coats. Seeds of *Canna* (Sivori et al. 1968), *Lotus* (Wester 1973), and *Lupinus* (Porsild and Harrington 1967) have been reported to be viable even after 500 years. Other hard-seeded genera reported by Harrington (1972) to be germinable after 100 years include *Albizia*, *Cassia*, *Goodia*, and *Trifolium*.

Seeds of other species are characteristically short-lived. These include vegetables such as lettuce, onion, and parsnip and agronomic crops such as rye (Table 8.3). Wheat exhibits intermediate storability compared to alsike clover which has excellent storability (Table 8.3). Generally, seed species possessing high oil content do not store as well as those with low oil content. However, such total compositional analyses may be misleading. What is more important is the quantity of oil present in the portion of the seed responsible for germination. For example, whole wheat seeds contain only about 3% oil, but their embryo portion has about 27% oil. Seeds of different species may also be chemically similar but have greatly different storability due to differences in genetic potential. For example, Chewings fescue and annual ryegrass seeds are similar in appearance and chemical composition; however, ryegrass seeds have much better storability under comparable conditions. These genetic factors affect seed stora-

Table 8.3. Relative Storability Index (1 = 50% of the Seeds are Expected to Germinate After 1 to 2 Years Storage, 2 = 50% of the Seeds are Expected to Germinate After 3 to 5 Years Storage, 3 = 50% of the Seeds are Expected to Germinate After 5 or More Years) of Important Crop Seeds (from Justice and Bass (1978)).

Crop	Relative Storability Index	Crop	Relative Storability Index
Alfalfa	3	Bahiagrass	2
Barley	2	Field bean	2
Sugarbeet	3	Creeping bentgrass	3
Velvet bentgrass	3	Bermuda grass	1
Kentucky bluegrass	2	Big bluestem	2
Mountain brome	2	Smooth brome	2
Buckwheat	2	Buffalograss	3
Reed canarygrass	1	Carpetgrass	3
Alsike clover	3	Berseem clover	2
Crimson clover	2	Subterraneum clover	2
Field corn	1	Cotton	1
Cowpea	1	Dallisgrass	1
Chewings fescue	2	Tall fescue	2
Flax	2	Blue grama	2
Hardinggrass	1	Korean lespedeza	1
Yellow lupine	1	Foxtail millet	1
Pearl millet	1	Oats	2
Orchardgrass	1	Peanut	1
Rape	2	Redtop	1
Rice	2	Rye	1
Perennial ryegrass	2	Sorghum	1
Soybean	1	Sunflower	1
White sweetclover	3	Timothy	2
Birdsfoot trefoil	2	Hairy vetch	3
Wheat	2	Broccoli	2
Cabbage	2	Carrot	2
Sweet corn	2	Cucumber	2
Lettuce	1	Onion	1
Parsley	1	Parsnip	1
Pumpkin	2	New Zealand spinach	2
Tomato	3	Turnip	2

bility and have led to the development of tables where species differences are divided into three broad categories according to their relative storability (Table 8.3).

Genetic differences in storage potential are not limited to seeds of different species. Differences in seed storability may also occur among cultivars. For example, the bean cultivar Black Valentine stores better than Brittle Wax (Toole and Toole 1953). Similarly, some inbred lines of corn have been shown to germinate 90% after 12 years of storage while others were completely dead following the same storage period (Lindstrom 1942). Kueneman (1983) demonstrated by means of reciprocal soybean crosses that the maternal plant was responsible for increased storability of F1 seeds perhaps by influencing the permeability of the seed coat. Thus, inheritance clearly exerts a marked effect on seed longevity and can be a focus of breeding programs. It should not be forgotten, however, that the environment strongly alters the genetic potential for seed longevity.

Presence of Microflora

Two types of fungi invade seeds: field fungi and storage fungi. Field fungi infect seeds that are developing on the mother plant and typically require high relative humidity (90 to 95%) or high seed moisture content (30 to 35%). Since these conditions occur only during seed maturation or imbibition, field fungi seldom contribute to seed deterioration during storage.

In contrast, storage fungi have the capacity to grow without free water. In general, they grow at seed moisture contents in equilibrium with relative humidities from 65 to 90%. The optimum temperature for growth of storage fungi is about 30 to 33°C, with a maximum at 55°C and a minimum of 0°C. Most storage fungi belong to one of two principal genera, *Aspergillus* and *Penicillium*. Species from these genera are saprophytes and survive on dead tissue. Most storage fungi ultimately invade the embryo of the seed. They cause seed deterioration not only through invasion, but also by producing toxic metabolites that destroy cells that provide the dead tissue on which they subsist. An excellent treatment of the importance of microflora has been prepared by Christensen and Meromick (1986) and been the subject of symposia (West 1986; APS 1983). Interestingly, bacteria do not substantially reduce the germination of stored seeds since they require free water for growth.

Mechanical Damage

Seed production practices such as harvesting, cleaning, and handling inevitably lead to mechanical damage. Although the immediate effect of such damage on seed quality is generally not serious, the delayed effects of mechanical damage on seed longevity are much more troublesome and of much greater economic significance (McDonald 1985). Seed deterioration is inexorable and progressive; thus, small mechanically damaged areas that initially have little impact on seed performance may later increase in size and cause deterioration of vital embryonic tissues, resulting in poor seed quality (Moore 1972). Direct injuries to embryonic tissues are much more detrimental to seed longevity than are large injuries to nonembryonic tissues. Mechanical damage also promotes invasion by storage fungi, which can enter the seed through cracks in the seed coat (Beattie and Boswell 1939; Mamiepic and Caldwell 1963).

Another important production process that permits the germination of many hard seeds is scarification. However, this mechanical or chemical treatment invariably causes some mechanical damage. Studies have also shown that scarified seeds do not store as well as nonscarified seeds (Battle 1948; Brett 1952). Delinting of cotton seeds also reduces storability (Simpson 1946) as does hulling of cool-season grasses (Canode 1972).

Seed Maturity

Factors such as temperature, moisture, variety, and nutrient status influence seed maturity, which, in turn, influence seed storability. The greatest storage potential is attained at the time of physiological maturity, or maximum dry weight of the seed. Although this appears to be a simple principle for individual seeds, many plants (e.g., carrot, grasses) have an indeterminate flowering pattern; that is, the most mature flowers grow at the base of the inflorescence with more immature flowers formed on the newer branches. Thus, varying stages of seed maturity, from recently fertilized ovules to mature seeds, occur on the same plant. Harvested seeds from such plants show varying degrees of maturity and different storage potential.

Although immature (or small) seeds have been shown in a number of studies to be inferior to mature seeds in viability and vigor (Ries 1971; Lopez and Grabe 1973; Clark and Peck 1968; Austin and Longden 1967; Fontes and Ohlrogge 1972), other factors such as nutrient status can also influence seed longevity. Harrington (1960b) demonstrated that carrot and pepper plants grown in nutrient solutions deficient in nitrogen, potassium, or calcium produced seeds that did not store well over an eight-year period. Seeds deficient in phosphorus, however, did not exhibit decreased storage potential. A similar nutrient deficiency symptom that affects seed quality is "marsh spot" of pea. These seeds have a magnesium deficiency that expresses itself as a brown necrotic area on the adaxial surface of the cotyledons and results in low-quality pea seeds.

PREDICTING SEED DETERIORATION

The ability to predict or forecast seed deterioration would be extremely valuable to seed companies and germplasm repositories since the loss of seed quality could be anticipated and seed stocks replenished in orderly fashion. Since we know that the type of seed, its initial germinability and moisture content, and the temperature and relative humidity conditions of the storage environment greatly affect the degree of seed deterioration, it has been suggested that seed deterioration can be modeled by mathematical equations (Roberts 1973, 1986). The general principle is that low-moisture, high-quality seeds stored under cool, dry conditions maintain seed quality better than high-moisture, low-quality seeds under hot, humid conditions. To utilize this principle, Roberts (1986) developed the following equation:

$$K_E - C_w \log m - C_H t - C_Q t^2 \qquad (1)$$
$$v = K_i - p/10$$

where v equals the probit of percent germinability after a storage period of p days; K_i is the probit of initial germinability of the seed lot; K_E, C_W, C_H, and C_Q are species constants; m is seed moisture on a fresh weight basis; and t is the storage temperature in degrees celsius. This assumes that loss in seed germination is normally distributed in time under a constant environment. This is useful for two reasons. First, the sigmoidal curve typical of the decline in germination during storage can be made a straight line using the probit transformation (Figure 8.3). Thus, the curve is completely described

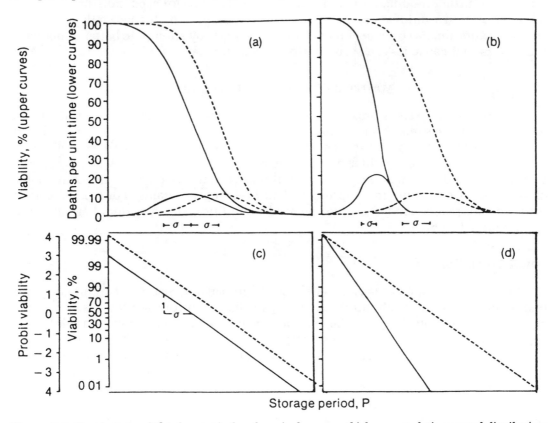

Figure 8.3. *Graphs (a) and (b) show typical seed survival curves, which are cumulative normal distributions of negative slope, under conditions when the temperature and moisture content remain constant during storage. The frequency distribution of lifespans which give rise to these survival curves are also shown. The standard deviation, (σ, is indicated below each curve. Graphs (c) and (d) show the same survival curves as (a) and (b), respectively, when percentage viability is transformed to probit values; these curves are described by Eq. (1) and the slope, 1/σ, is indicated on one of the curves in (c). Graphs (a) and (c) show two survival curves representing two different seed lots of the same species stored under identical conditions which, therefore, have identical slopes but where the seed lot constant, K_i, of one seed lot (---) has a greater value than the other seed lot (—). Graphs (b) and (d) show two survival curves of the same seed lot (therefore the same K_i-value, stored in two different environments so that the slope, 1/σ, in the less deleterious environment (---) has half the value of the other environment). After Roberts and Ellis (1982).*

by two parameters, slope and intercept. Second, germination following long-term storage may be anticipated from short-term results by extrapolation of the line. This is the method by which the four species constants are determined.

Considerable research is needed to validate this mathematical test of seed deterioration rate. Its limitations are associated with precisely characterizing the quality of the seed being stored and the storage conditions. Bewley and Black (1982) identified these factors as (1) cultivar and harvest variability, (2) pre- and postharvest conditions, (3) oxygen pressure effects during storage, and (4) fluctuating environmental conditions. Wilson and McDonald (1989) tested the predictions using *Phaseolus vulgaris* seeds and found that further modification of the equation to account for seed moisture influence on deterioration would be useful. Even so, this attempt at mathematically describing seed deterioration has been useful to physiologists as they attempt to better understand the factors that cause loss in seed quality during storage.

MAINTAINING SEEDS IN STORAGE

The purpose of seed storage is to preserve planting stocks from one season to the next. In some cases (e.g., seed companies), the objective of seed storage is to maintain seed quality for the longest duration possible. This approach creates a greater diversity in seed inventory and provides a guarantee of seed supply in years when acceptable seed quality and production is low. In addition, seed storage enables the maintenance of germplasm over time for improved plant breeding programs. Four principal approaches are recommended for storing seeds: conditioned, cryogenic, hermetic, and containerized storage.

Conditioned Storage

Seeds of most species may be safely stored for several years by careful control of temperature and relative humidity. Although such conditions are too costly for most agricultural seed lots, they may be extremely valuable for preserving germ plasm and certain high-value seed stocks. In some parts of the world, especially in the tropics, conditioned storage is necessary in order to maintain high viability of some seeds from harvest to planting (Harrington 1973). At least four factors should be considered when evaluating the economics of seed storage (Welch and Delouche 1974).

Type of Seed To Be Stored. Seed storage requires a large investment in construction, maintenance, and power. However, the relatively high cost of seed, particularly that which must be carried over, may justify such an investment. Factors that should be considered in making conditioned storage decisions include: (a) the differential between seed and grain prices, (b) the prevailing price of seed to be stored and its storage potential, (c) the weight per volume of seed (costs of construction, conditioning, and storage capacity will reflect storage volume rather than seed weight), (d) the extent of loss resulting from disposing the seed in alternative markets (e.g., selling as grain, or dumping in the case of treated seed), and (e) the projected seed inventory (volume). Although all seeds benefit from conditioned storage, it becomes practical only when income derived from seed storage exceeds the costs involved.

Length of Storage. Costs of installation and maintenance of a conditioned store-room can be substantial over time and will vary with geographical location. The costs of operating a conditioned storeroom will be less in the Great Lakes area where the prevailing environment is generally cool and dry compared to hot and humid areas of the Gulf of Mexico. In most instances, it is only essential to maintain conditioned storage for seeds from one growing season through the following summer to the next spring planting, a period of 18 to 20 months.

Quality of Seed Stored. Many people believe that conditioned storage is more beneficial for low-quality than high-quality seeds. Figure 8.4 demonstrates that low-quality garden bean and sorghum seeds store poorly compared to high-quality seeds even though all lots have the same initial germination. Seed quality cannot be improved by seed storage. Even the best storage conditions can only maintain the quality of the seed placed in storage, and low-quality seeds deteriorate more rapidly under conditioned storage than high-quality seeds.

Loss of Seed Weight During Conditioned Storage. High-moisture seeds that are placed in a conditioned storeroom at 40% relative humidity will gradually decrease to 10% moisture content as moisture equilibrium is attained. Thus, the seed will weigh less following conditioned storage as a result of water loss. Consequently, conditioned storage will eventually result in higher-density seed lots with more seed per unit volume and may become an important economic consideration for seed producers.

Requirements of Conditioned Storage. Conditioned storage involves placing seeds in a dry and cool environment for extended periods. But, how dry and how cool? Seeds of most grain crops (e.g., corn, wheat, barley, sorghum, and oats) will maintain satisfactory germination and vigor for about one year at moisture contents of 12 to 13% under normal warehouse temperatures. When longer storage is needed, seed moisture content should be less than 11% and the temperature should not exceed 20°C. Similarly, seeds of most temperate zone grass and legume crops can be stored safely for one year at moisture contents of 10 to 11% under normal warehouse temperatures. When longer storage is needed, 10% seed moisture content and temperatures of 20°C or less are recommended.

Seed of other species (e.g., soybean and peanut) may lose viability even over one year if kept at 11 to 12% moisture content at temperatures of 20°C or less. Although carryover of soybean seed lots is generally not recommended, they may be kept safely for 18 to 20 months in conditioned storage facilities of 20°C and 50% relative humidity. While most orthodox seeds can be successfully stored under conditioned storage facilities, operation of the facilities is costly and mechanical breakdowns of the cooling units are common. As a result, more reliable approaches to long-term preservation of seeds have been sought.

To achieve these storage conditions, conditioned seed storage rooms typically contain refrigeration and dehumidification equipment. The refrigeration is necessary to lower the temperature of the seed storage room and dehumidification is required because lower storage air temperatures hold less water causing the relative humidity of the air to increase. In general, the temperature and relative humidity conditions of the storage environment are determined to meet satisfactory storage conditions identified by Harrington (1972) while satisfying the objectives of the seed storage manager.

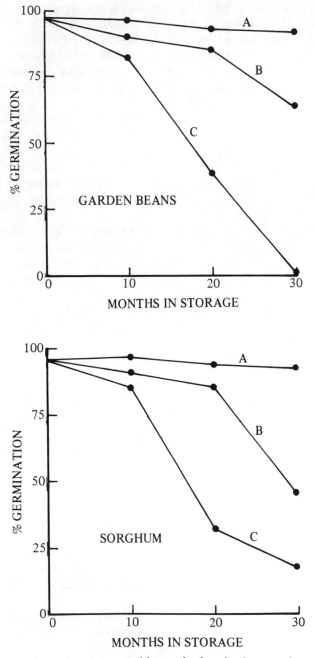

Figure 8.4. *Differences in longevity of three seed lots each of garden beans and sorghum under open storage conditions. (From Delouche (1973).)*

Storage of seed for maintaining germplasm requires conditions that are not feasible for seed industry quality control programs. In this instance, seeds are often stored at temperatures ranging from $-15°$ to $-20°C$.

Cryogenic Storage

While most orthodox seeds can be successfully stored in conditioned storage facilities, routine operation of such facilities is costly and mechanical breakdowns are common. Another approach to minimize these difficulties is cryogenic storage. This storage method places seeds into liquid nitrogen at a temperature of $-196°C$ (seeds are actually placed into the gaseous phase of the liquid nitrogen, at approximately $-150°C$, to facilitate handling and safety. The differential in seed temperature is not considered significant in influencing seed longevity.).

The advantage of this approach is that seeds are placed at a temperature where little detrimental physiological activity occurs, thereby prolonging storage life. From a practical perspective, the cost of the liquid nitrogen is minimal compared to maintaining conditioned storerooms. In addition, no working parts are necessary so repair of equipment is not required. Liquid nitrogen is also an inert gas that volatilizes easily. Therefore, it remains relatively safe, although air circulation in the storage room is necessary to guard against the room being filled with nitrogen gas and to prevent asphyxiation. Finally, this seed storage method is limited in capacity to the amount of storage space available in the cryogenic tanks. It is not practical for most commercial seed, although it may be useful for maintaining valuable seed germplasm over prolonged periods.

Many studies are now under way on the ability to safely store a variety of seeds using cryogenic storage (Stanwood 1985). The general conclusion is that most agronomic crops can be successfully stored in liquid nitrogen. It seems clear that under such circumstances, there is little question of metabolism or even conventional chemical activity ensuring successful seed storage for prolonged periods. Yet, not all seeds are able to tolerate liquid nitrogen storage. The reasons for this are still unknown although seed moisture content is certainly involved. In seeds, there is a distinct moisture content below which freezable water which kills cells does not occur. Sesame seeds, for example, are able to tolerate liquid nitrogen freezing up to 12% moisture, at which point germination declines rapidly. A central theme of cryogenic storage has been to replace freezable water in seeds with protective cryogenic compounds such as sugars and glycols. The objective is to replace the water in the seeds by infiltrating the protective compounds into the seeds without inducing subsequent seed injury.

Hermetic Storage

In recent years, packaging seeds in moisture-resistant or *hermetically sealed* containers for storage and marketing has been explored. The purpose of such containers is to maintain seeds at safe storage moisture levels. Figures 8.5 and 8.6 show the effectiveness of several types of packaging materials in preventing moisture uptake and maintaining seed viability. Ordinary paper and cloth containers were least effective, while various laminate and polyethylene materials were moderately effective. Metal cans were com-

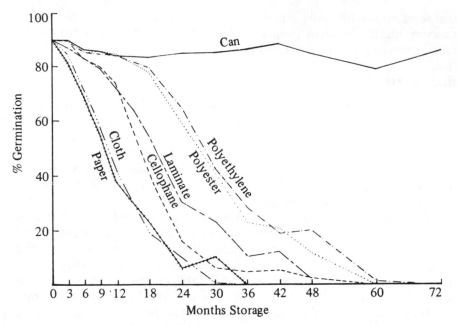

Figure 8.5. *The effect of different packaging materials on the germination of creeping red fescue seed. (From Grabe and Isely (1969).)*

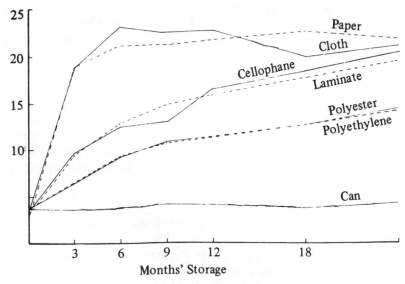

Figure 8.6. *The effect of different packaging materials on the moisture content of creeping red fescue seed. (From Grabe and Isely (1969).)*

pletely effective in maintaining seed moisture at the initial 5% level. Such completely moisture-proof containers hermetically seal the seed and are effective for long-term seed storage up to 10 years or more. The effectiveness of other materials was directly associated with their ability to resist moisture.

The moisture content of seeds placed in hemetic storage must be lower (2 to 3%) than that at which seeds are normally packaged in nonsealed storage. Most observations indicate that starchy seeds above 12% and oily seeds above 9% deteriorate faster in sealed storage than in nonsealed storage (Harrington 1973). It is believed that the atmosphere in a moisture-proof container holding seed at 13% moisture will equilibrate at a relative humidity of about 65%. Some seed storage fungi that are detrimental to seed quality multiply under such conditions. In contrast, the relative humidity surrounding corn seed packaged in porous containers such as paper bags at 13% moisture will rise to nearly 100% and decrease below 65% at other times. As a result, seeds placed in open containers will generally decrease in moisture during the winter when relative humidities are the lowest and thus inhibit the growth of storage fungi.

One of the central benefits of hermetic seed storage is the ability to remove ambient air from the seeds and replace it with specific gases known to prolong seed storage life. This is necessary because the composition of the gaseous environment in hermetic storage changes with time and the effect is more pronounced with high seed moisture content. For example, Roberts and Abdalla (1968) demonstrated that pea seeds stored at 18.4% moisture content at 25°C in ambient air had a decrease in oxygen from 21 to 1.4% and an increase in carbon dioxide from 0.03 to 12% after only 11 weeks. This decrease in oxygen concentration and/or increase in carbon dioxide level may differentially affect the storage capability of the seeds.

The most extreme case of hermetic storage is to store seeds *in vacuo* at low moisture contents. While some reports have indicated that this approach extends seed storage life (Bockholt et al. 1969; Lougheed et al. 1976), the majority of reports suggest no benefit from vacuum storage (Justice and Bass 1978; Bass and Stanwood 1978). Studies have also been conducted which replace ambient air with a pure gas in sealed containers. Some advantage has been reported for pure carbon dioxide (Bennici et al. 1984; Harrison 1966) and the inert gases nitrogen, argon, and helium (Harrison 1966; Quaglia et al. 1980), but most reports show little or no advantage to this approach (Justice and Bass 1978). Interestingly, a consensus is emerging that seeds stored in oxygen at low moisture contents deteriorate more rapidly (Harrison 1966; Roberts and Abdalla 1968; Ohlrogge and Kernan 1982). This effect may be associated with the physical/physiological mechanisms occurring during lipid peroxidation that will be described later.

Containerized Seed Storage

Absolute humidity control in seed storage areas requires considerable investment in specially constructed rooms and dehumidifying equipment. Most seed stock organizations and commercial seed companies have special facilities for storing germ plasm and high-value seed stocks where relative humidity and temperature are closely regulated. The National Seed Storage Laboratory operated by the United States Department of Agriculture at Fort Collins, Colorado, as a permanent repository for valuable germ plasm of all kinds of plants, provides temperature and relative humidity conditions satisfactory for long-term storage of all kinds of seeds.

Where elaborate facilities are not available, humidity can be closely regulated in a closed container by use of chemical desiccants that have known moisture equilibrium values. Either saturated salt solutions or acid solutions may be used. Sulfuric acid (H_2SO_4) is a common acid desiccant and is diluted with water to provide increasing relative humidity levels (Table 8.4). Since H_2SO_4 is very caustic and corrosive, it should be handled with extreme caution to prevent injury to personnel as well as to the seeds. Saturated salt solutions are less hazardous to both seed and personnel; however, different salts must be used to attain various humidity levels (Table 8.5). Complete saturation is attained when a given quantity of water dissolves its full capacity of salt crystals. To maintain complete saturation, some additional salt should be added so that the relative humidity is held constant. Care must be used to keep the seed separate from the liquid desiccant. This can be accomplished by using metal platforms, cheesecloth, or wire containers for holding the seed. Depending on the size of the container, small fans may be used to keep the relative humidity distributed uniformly throughout. Exposure of seed to uniform relative humidity can also be aided by use of small seed containers or appropriate use of wire screens throughout the seed mass.

Table 8.4. Relative Humidity in Equilibrium with Different Concentrations of Aqueous Acid Solutions at Various Temperatures

Acid	Temperature °F	Acid by weight, percent			
		20	40	60	80
H_2SO_4 (Sulfuric)	0	87.3	55.7	15.0	3.14
	50	87.4	56.6	15.8	3.88
	68	87.7	56.7	16.3	4.76
	86	87.5	56.6	17.0	5.75
	104	87.6	57.5	17.8	6.88
	112	88.8	58.2	18.8	8.2
		Acid by weight, percent			
		20	30	40	50
HNO_3 (Nitric)	0	89.2	78.4	65.3	45.7
	50	86.7	77.0	63.0	45.6
	68	86.6	75.2	61.5	
	86	86.6	74.9	61.3	
	104	85.9	74.1	60.5	
	112	86.5	74.6		
	140	86.9	75.6		
		Acid by weight, percent			
		10	20	30	40
HCl (Hydrochloric)	0	83.5	56.0	27.4	8.9
	50	83.5			
	68	83.2			
	86	84.2			

Calculated by $\dfrac{p}{p_0}$ from basic data.

From Hall (1957).

Table 8.5. Relative Humidity in Equilibrium with Saturated Salt Solutions at Different Temperatures

Salt	Temp. (°F)	RH (Percent)	Salt	Temp. (°F)	RH (Percent)
$BaCl_2 \cdot 2H_2O$	77	91.2	$LiCl \cdot H_2O$	50	13.3
(Barium chloride)	86	90.8	(Lithium chloride)	68	12.4
	95	90.2		86	11.8
	104	89.7		104	11.6
$CaCl_2$	20	44	$MgCl_2$	73	32.9
(Calcium chloride)	32	41	(Magnesium chloride)	86	32.4
	50	40		100	31.9
	70	35			
$Ca(NO_3)_2$	73	51.8	$Mg(NO_3)_2$	73	53.5
(Calcium nitrate)	86	46.6	(Magnesium nitrate)	86	51.4
	100	38.9		100	49.0
$CuCl_2 \cdot 2H_2O$	68	68.7	$Na_2Cr_2O_7 \cdot 2H_2O$	50	57.9
(Cupric chloride)	77	68.7	(Sodium dichromate)	68	55.2
	86	68.3		86	52.5
	104	67.4		104	49.8
KCl	68	89.2	NH_4Cl	20	82
(Potassium chloride)	77	87.2	(Ammonium chloride)	32	83
	86	85.3		50	81
	95	83.8		70	75
KNO_2	68	49.0	$NaC_2H_3O_2 \cdot 3H_2O$	68	76.0
(Potassium nitrite)	77	48.2	(Sodium acetate)	77	73.7
	86	47.2		86	71.3
	100	45.9		95	68.8
KNO_3	32	97.6	$NaNO_2$	68	65.3
(Potassium nitrate)	50	95.5	(Sodium nitrate)	77	64.3
	68	93.2		86	63.3
	86	90.7		100	61.8
	104	87.9			
K_2CrO_4	68	86.6	$NaCl$	32	74.9
(Potassium chromate)	77	86.5	(Sodium chloride)	50	75.2
	86	86.3		68	75.5
	100	85.6		86	75.6
				104	75.4
				122	74.7

From Hall (1957).

A common seed desiccant is silica gel treated with cobalt chloride which serves as an indicator dye and turns from blue to pink when the relative humidity exceeds 45%. The silica gel-cobalt chloride granules are placed in a metal box with the seeds in a proportion of 1 kg desiccant per 10 kg of seeds. Seeds can thus be stored safely for several years because the relative humidity within the box is kept below 45%. The advantages of this desiccant storage system are: (1) construction of conditioned storage facilities is not necessary, (2) maintenance and operation costs are minimal, (3) the metal storage boxes are insect-, rodent-, and moisture-proof, and (4) the seeds are not damaged by storage fungi since they are maintained at 45% relative humidity. The only care required for desiccant seed storage is frequent inspection of the silica gel to insure that the indicator silica gel remains blue.

SYMPTOMS OF SEED DETERIORATION

The process of seed deterioration is complex; therefore, the conclusions from seed deterioration studies are often difficult to critically evaluate. There are at least two reasons for this. First, most seed deterioration studies have focused on whole seed germination response or enzyme analyses without considering the fact that seed deterioration probably does not occur uniformly throughout a seed. For example, young vigorous embryos that are excised and transplanted onto deteriorated endosperms do not grow as well as those planted onto nondeteriorated endosperms. This suggests that endosperm deterioration markedly alters the performance of a vigorous embryo (Floris 1970; Mandal and Basu 1981). Similarly, Bulat (1963) showed that unique seed deterioration patterns occurred in 41 species. In many cases, the necrosis was first detected in the embryo and usually around the radicle tip. Thus, it should be recognized that seed parts vary in their chemistry and susceptibility to deterioration. Studies examining this differential susceptibility of seed parts which collectively influence the whole seed response should be useful in understanding the mechanisms of seed deterioration.

Second, all seed lots are composed of individual seeds, each possessing its own unique capability to perform in the field. A hypothetical distribution of loss in seed quality over time, therefore, can be represented by Figure 8.7. Two observations can

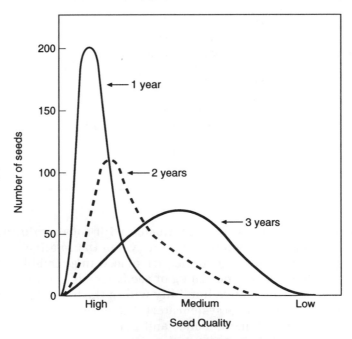

Figure 8.7. *Hypothetical distribution of loss in soybean seed quality over a three-year period. (From McDonald and Wilson (1980).)*

be made regarding this deterioration (McDonald and Wilson 1980). First, the proportion of high-quality seeds within a population decreases with increasing storage time. Second, the curves shift from high to low seed quality and the range becomes wider with increased storage time. As seeds age, overall quality level of the seed lot declines. Since some seeds are inherently stronger than others, a remnant of high-quality seeds remains even after a three-year storage period. The point of this example is that seeds in storage do not die simultaneously but age at differing rates. Yet, most studies of seed deterioration are expressed as an average of some subsample greater than one seed, e.g., conductivity of 20 seeds, or as some average value of a subsample that represents the entire population of seeds, e.g., percentage germination. Thus, total population studies of seed deterioration do not represent what is occurring at the individual seed level.

Despite these difficulties, the most visible symptoms of seed deterioration are observed first at the whole seed morphological level and then during germination and seedling growth. However, these are preceded by numerous ultrastructural and physiological changes whose symptoms are not as readily apparent but can be detected by sophisticated monitoring techniques that attempt to identify changes in the deteriorating seed at the physiological level.

Seed Symptoms

Morphological Changes. Seed coat color often provides an indication of seed deterioration, particularly for legumes. Darkening of the seed coat in deteriorating clover (Vaughan and Delouche 1968), peanut (Marzke et al. 1976), and soybean (Saio et al. 1980) seeds have been reported. Such color changes are presumably due to oxidative reactions in the seed coat which are accelerated under high temperature and relative humidity conditions (Hughes and Sandsted 1975). Beyond these seed coat effects, other morphological changes in deteriorating seeds have been reported. Lettuce seeds develop red necrotic lesions in the cotyledons known as cotyledonary necrosis (Bass 1970) and lentil seeds characteristically turn yellow after prolonged storage (Nozzolillo and DeBezada 1984).

Ultrastructural Changes. Deteriorated dry seeds have been examined for ultrastructural changes using electron microscopy and two general patterns of coalescence of lipid bodies and plasmalemma withdrawal associated with deterioration have been observed. Coalescence of lipid bodies in the embryo has been found in a broad group of species, including wheat (Anderson et al. 1970), peas (Harman and Granett 1972), and pine (Fernandez Gracia de Castro and Martinez-Honduvilla 1984). In lettuce, this coalescence of lipid bodies has been detected only in the embryonic axis but not the cotyledons (Smith 1983). Withdrawal of the plasmalemma has also been detected in these species (above) as well as in rye (Hallam et al. 1973). It is significant that both of these events influence cell membrane integrity.

Cell Membranes. One common facet of deteriorating seeds is their inability to retain cellular constituents which leak out during imbibition. This has three important seed quality implications: (1) many of these cellular constituents are essential for normal, vigorous germination, (2) some of the exuded compounds are necessary for maintenance of internal osmotic potential which is responsible for normal water uptake and provides the turgor pressure required for radicle protrusion, and (3) the external leakage of these substances encourages the growth of pathogenic microflora.

The reasons for this increased leakage have been attributed to cell membrane disruptions associated with the loss of membrane phospholipids. Phospholipid decrease has been reported in deteriorating cucumber (Koostra and Harrington 1969), peanut (Pearce and Abdel-Samad 1980), pea (Powell and Matthews 1981), soybean (Priestley and Leopold 1983), tomato (Francis and Coolbear 1984), and sunflower (Halder et al. 1983) seeds. The loss of phospholipids in deteriorating seeds is generally considered to be due to either phospholipase enzyme activity or lipid peroxidation. Interestingly, some (Koostra and Harrington 1969; Petruzzelli and Taranto 1984) have suggested that the decline in phospholipids occurs only under conditions of high relative humidity (as encountered in an accelerated aging test) but not under long-term dry storage. Since phospholipases are hydrolytic enzymes, they function most actively under conditions of high seed moisture content. Thus, the decline in phospholipids reported in the literature under accelerated aging conditions may be attributed to these enzymes and could be one reason that the rate of seed deterioration is accelerated compared to that in dry seed storage. In the latter case, lipid peroxidation may be more responsible for the observed declines in membrane integrity.

Loss of Enzyme Activity. The most sensitive tests for measuring incipient seed deterioration are those that measure activity of certain enzymes associated with breakdown of food reserves or biosynthesis of new tissue during germination. Examples include amylases (Saxena and Maheshwari 1980), proteinases (Nowak and Mierzwinski 1978), cytochrome c oxidase (Ching 1972), and glyceraldehyde phosphate dehydrogenase (Harman et al. 1976).

Among the biochemical tests that have been used to measure loss of enzyme activity are the tetrazolium test (dehydrogenase) (see Chapter 5) and the glutamic acid decarboxylase (GADA) activity test. Other enzymes that have been correlated with seed deterioration are the oxidases such as catalase (Crocker and Harrington 1918), peroxidase (McHargue 1920), amylase (Anderson 1970), and cytochrome oxidase (Throneberry and Smith 1954). Although these positive associations have been reported, other studies have shown that high levels of peroxidase activity existed in old wheat seeds (Brocq-Rosseu and Gain 1908) and high levels of dehydrogenases have been found in heat-damaged barley seeds (MacLeod 1952). Changes in the levels of specific enzymes may not always provide an accurate indication of seed deterioration. Many of these studies were conducted under accelerated aging conditions which increase seed moisture content and contamination by microflora—both conditions which are seldom encountered in long-term dry seed storage.

Reduced Respiration. Respiration is a composite expression of activity of a large group of enzymes that react together in breaking down food reserves. As seeds deteriorate, respiration becomes progressively weaker, and ultimately leads to loss of germination. However, prior to loss of germinability, the respiration level during the early stages of germination has been correlated with subsequent seedling vigor (Woodstock and Feeley 1965).

Changes in respiration have been studied under natural aging (Woodstock and

Grabe 1967; Kittock and Law 1968; Abdul-Baki 1969; Anderson 1970), accelerated aging (Woodstock and Feeley 1965; Abdul-Baki 1969), chilling injury (Throneberry and Smith 1954; Woodstock and Pollock 1965, Woodstock and Feeley 1965), and irradiation (Woodstock and Combs 1965; Woodstock and Justice 1967) conditions. Although the association between oxygen utilization and seed deterioration has been demonstrated, a more sensitive index of deterioration may be the respiratory quotient (RQ), which represents the CO_2 evolved divided by the oxygen utilized. High RQ values (1.5 or higher) have been reported in deteriorated seeds (Woodstock and Grabe 1967; Anderson 1970; Woodstock et al. 1984) and may be due to increases in CO_2 evolution, reductions in oxygen uptake, or both. Regardless of the cause, it appears that reductions in the rate of respiration are closely associated with seed deterioration. This may be causally related to a breakdown in membrane structure, particularly in the mitochondrial cristae, which would reduce total respiration.

Increases in Seed Leachates. A frequently observed symptom of deteriorated seeds is their increased leachate content when soaked in water. The degree of deterioration is associated with the concentration of seed exudates that may be found in the steep solution. These exudates are a reflection of the amount of membrane degradation that has occurred. The leachate concentration has been measured by electrical conductance methods (Hibbard and Miller 1928), and also by determining the soluble sugar content of the leachate (Abdul-Baki and Anderson 1970).

Increase in Free Fatty Acid Content. The hydrolysis of phospholipids leads to the release of glycerol and fatty acids, and this reaction accelerates with increasing seed moisture content (Harrington 1973). The continual accumulation of free fatty acids culminates in a reduction in cellular pH and is detrimental to normal cellular metabolism (Earnshaw et al. 1970). Furthermore, it denatures enzymes, resulting in a loss of activity (Tortora et al. 1978). Individual cotton seeds containing 1% or more of free fatty acids usually will not germinate (Hoffpauir et al. 1947).

There is debate as to whether the increase in free fatty acids is due to production of lipases by microflora or by the seed itself. Hummel et al. (1954) concluded that the increase in free fatty acids in wheat was entirely attributable to mold. Fungal invasion is also thought to be a major cause of the breakdown of lipids to fatty acids (Christensen et al. 1949). Others (Priestley 1986) have indicated that lipase activity can be detected in seeds which are at moisture contents too low for fungal growth.

Performance Symptoms

Eventually the deterioration of seeds is observable in their lowered performance during germination. Delayed seedling emergence is among the first noticeable symptoms, followed by a slower rate of seedling growth and development and decreased germination. As seeds deteriorate, the environmental conditions under which they will germinate become narrower (Heydecker 1969); this symptom also occurs early in seed deterioration. Loss of field emergence potential is another frequently observed symptom of deterioration (Grabe 1965).

Another symptom of deteriorated seeds is decreased resistance to environmental stresses during germination and early seedling growth (Isely 1957; Woodstock and

Pollock 1965). Still another is reduced yield potential. This may occur even in the absence of the more obvious symptoms accompanying germination and seedling establishment.

As some areas of the seed lose their viability, the seed may still be able to produce a seedling, although it may be morphologically abnormal due to malfunction of the deteriorated areas. Such a condition in lettuce has been called red cotyledon malfunction, causing lack of hypocotyl elongation and stunting of the radicle, resulting in a stunted seedling incapable of survival (Dempsey and Harrington 1951). Aged onion seeds do not develop the cotyledonary knee necessary for emergence through the soil, and aged seeds of cucurbits do not develop the hypocotyl peg that normally cracks open the seed coat, allowing the cotyledons to emerge (Harrington 1973).

The ultimate performance symptom of seed deterioration is the complete loss of germinability and death of the seed.

POSSIBLE CAUSES OF SEED DETERIORATION

An understanding of the fundamental factors that induce aging is essential in a study of seed deterioration. The following discussion cites several theories that have been suggested as basic causes of deterioration. Some of the theories included are entirely speculative. It is fairly certain that seed deterioration occurs from a combination of several causes.

Lipid Peroxidation

Of all the models presented to explain seed deterioration, the lipid peroxidation model has stimulated the greatest interest (Wilson and McDonald 1986b; Bewley 1986). A free radical is an atom or a group of atoms with an unpaired electron. They can be produced either through autoxidation or enzymatically by lipoxygenase which is present in many seeds. The autoxidation mechanism is often initiated by oxygen around unsaturated or polyunsaturated fatty acids such as oleic and linoleic acids which are most common in seed membranes (Figure 8.8). The result is the release of a free radical, often hydrogen (H•) from a methylene group of the fatty acid that is adjacent to a double bond. In other cases, the free radical hydrogen may combine with other free radicals from carboxyl groups (ROOH) leaving a peroxyfree radical (ROO•). Once these free radicals are initiated, they create profound damage to membranes, particularly those where electron transport is most frequent, and continue to propagate other free radicals until they combine with free radicals which terminates the reaction. The result is the loss of membrane integrity in the case of phospholipids.

It has been noted that lipid autoxidation occurs in all cells; but in fully imbibed cells, water acts as a buffer between the reactive compounds and the macromolecules, thus preventing enzyme inactivation (Harrington 1973). Lipid autoxidation is accelerated at high temperature and increased oxygen concentration. Harrington (1973) considered this to be a cause of seed deterioration only at moisture contents below 6%, since moisture contents from 6% to 12% maintain seed viability, and above 12% other factors are responsible for deterioration.

Lipoxygenase enzymes also generate free radicals. However, their activity is great-

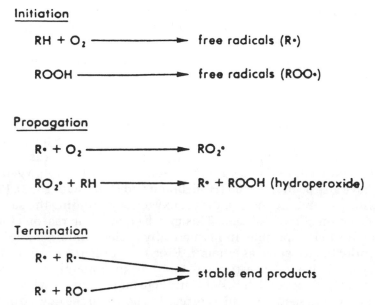

Initiation

$$RH + O_2 \longrightarrow \text{free radicals } (R\cdot)$$

$$ROOH \longrightarrow \text{free radicals } (ROO\cdot)$$

Propagation

$$R\cdot + O_2 \longrightarrow RO_2\cdot$$

$$RO_2\cdot + RH \longrightarrow R\cdot + ROOH \text{ (hydroperoxide)}$$

Termination

$$R\cdot + R\cdot$$
$$R\cdot + RO\cdot \longrightarrow \text{stable end products}$$

Figure 8.8. *Free radical chain reaction resulting in autoxidation. (From Wilson and McDonald (1986).)*

est when the seed moisture content exceeds 14%, while autoxidation is believed to occur primarily at lower seed moisture contents. Thus, the mechanism of lipid peroxidation may be different under accelerated aging (lipoxygenase) compared to long-term aging (autoxidation) conditions. It should also be noted that oxygen is deleterious to seed storage based on this proposal, which is consistent with the success of hermetic seed storage and that lipid peroxidation causes loss of membrane integrity.

Free radicals also attack compounds other than fatty acids. Changes in protein structure of seeds have also been observed and attributed to free radicals. Free radicals of lipid peroxides damage cytochrome c by changing its physical and catalytic properties, suggesting that free radicals attach themselves to the protein by covalent linkages (Tappel 1962). Sulfhydryl levels decrease in wheat flour with increasing oxygen content (Yoneyama et al. 1970). The "sick wheat syndrome" which results in a discoloration of the embryo with increasing storage time has been attributed to a condensation reaction between lysine or methionine protein residues and reducing sugars (Feeney and Whitaker 1982). Even more importantly, free radicals are also suspected of assault on chromosomal DNA. As seeds age, the propensity for genetic mutations increases. Many of these mutations can be detected as chromosomal aberrations (Ghosal and Mondal 1978; Murata et al. 1981) which delay the onset of mitosis necessary for germination (Murata et al. 1980). While these chromosomal aberrations delay seedling growth, continued development of the seedling results in fewer cells with these chromosomal irregularities, presumably because abnormal cells are not able to compete with normal ones (Murata et al. 1984). As a result, Roos (1982) has argued that mitotic lesions are unlikely to affect the genetic integrity of stored germ plasm.

Noteworthy products of lipid peroxidation include the volatile aldehydes. Harman et al. (1978) reported that deteriorated seeds produced 20 times more volatile aldehydes during imbibition than did fresh seeds. They further suggested that these compounds served to stimulate fungal spore germination and provided directive signals for growth (Harman et al. 1982). Wilson and McDonald (1986a) simplified the analysis of volatile aldehydes from imbibing seeds and proposed that their levels could be used as an index of seed vigor.

Four approaches have been described to minimize lipid peroxidation in seeds (Wilson and McDonald 1986b): (1) Lipid modification—breeders have focussed on changing the proportion of saturated/unsaturated fatty acids to favor an increase in saturated fatty acids which are less prone to lipid peroxidation. This has been accomplished in certain oil seeds (Knowles 1969; Wilson et al. 1981). Lipoxygenase-deficient soybean lines have also been reported (Hildebrand and Hymowitz 1982). (2) Regulation of oxygen pressure—reducing the quantity of oxygen surrounding the seeds might also decrease the initiation of free radicals. This may be one of the reasons for the success of storing seeds for longer periods in hermetically sealed containers. (3) Antioxidant treatments—Antioxidants such as vitamin E or α-tocopherol are known to be free radical scavengers. It has been estimated that one tocopherol molecule can afford antioxidant protection to several thousand fatty acid molecules (Bewley 1986). Tocopherol is known to occur in seed oils and a correlation between the degree of unsaturation in the oil and the tocopherol content has been reported (Hove and Harris 1951). To date, however, attempts to eliminate lipid peroxidation using antioxidants have not been successful (Yang and Yu 1982; Parrish and Bahler 1983). Part of the difficulty may be the infusion of these compounds into the seed since they are water insoluble. Woodstock et al. (1983) showed that freeze-drying of tocopherol impregnated seeds prolonged the storage life of parsley and onion seeds. (4) Hydration/dehydration treatments—hydration/dehydration treatments, also known as priming (Chapter 11), do not appear to extend seed storage life but do lead to improved seedling performance during germination. One possible explanation of this observation has been that repair of free radical damage to membranes and other compounds occurs during the hydration phase (Ward and Powell 1983). Studies supporting this notion have shown that deteriorated seeds perform better following imbibition (Goldsworthy et al. 1982), placing them in high humidity (Sanchez and de Miguel 1983), or placing seeds in an osmoticum followed by a drying treatment (Brocklehurst and Dearman 1983; Burgass and Powell 1984). This practical approach to repair of lipid peroxidation-induced membrane damage has been implemented in many commercial seed quality control programs.

Degradation of Functional Structures

As seed deterioration progresses, cellular membranes lose their selective permeability, permitting the cytoplasmic metabolites to leach into the intercellular spaces. Membrane degradation occurs from both hydrolysis of phospholipids by phospholipase and phospholipid autoxidation; therefore, in the strictest sense, it might be considered a result of aging rather than a cause.

Mitochondrial degradation and functional changes appear to play a major role in seed deterioration. A comprehensive review of the literature on aging in both plants and animals shows that many mitochondrial changes undoubtedly play a role in seed deterioration (Abu-Shakra 1963; Priestley 1986). Such changes decrease in number as deterioration proceeds; mitochondria become permanently swollen and lose their natural swelling-contracting ability (Cowdry 1956). Later they become pigmented and fragmented (Payne 1946; Weiss and Lansing 1953; Andrews 1956). Degradation of the mitochondrial membranes also occurs, leading to loss of function and eventual fragmentation (Abu-Shakra 1963). Two important aspects of mitochondrial deterioration are an increase in ATPase (Kielley and Kielley 1951; Sacktor 1953) and a decline in oxidative phosphorylation ability necessary to complete its respiratory function (Lehninger and Remmert 1959; Wojtczak and Wojtczak 1960; Weinbach et al. 1963). Since ATPase catalyzes the breakdown of ATP and ADP, it depletes energy available in the mitochondria. Levels of ATP are lower in aged seeds of a number of crops (Lunn and Madsen 1981; Banerjee et al. 1981). ATP also helps restore contraction ability to aged, swollen mitochondria; thus, the loss of ATP accelerates mitochondrial disintegration (Weinbach et al. 1963).

Functional changes in the mitochondria may be repaired partly or in whole by the addition of unsaturated fatty acids, phospholipids, chelating agents, and fatty acid binding compounds (albumin) (Chefurka 1963; Wojtczak and Wojtczak 1960; Rossi et al. 1964). It is believed that certain growth regulators contribute to protection of mitochondria and other cytoplasmic membranes and thus help maintain their integrity and selective permeability.

Inability of Ribosomes to Dissociate

Associated with the degradation of function in deteriorating seeds is the dissociation of ribosomes. Recent evidence indicates that the dissociation of polyribosomes must occur before attachment of preformed mRNA occurs, leading to protein synthesis in germinating seedlings (App et al. 1971). In nonviable seeds, the ribosomes fail to dissociate (Bray and Chow 1976) and protein synthesis is retarded. Such declines in protein synthesis are a measurable symptom of aging. There is also evidence that long-lived mRNA is lost during extended seed storage (Osborne 1983; Ghosh and Choudhuri 1984). It has also been suggested that aging depresses the synthesis of newly formed mRNA (Osborne et al. 1977; Weidner and Zalewski 1982).

Enzyme Degradation and Inactivation

Decreased activity of enzymes such as catalase, dehydrogenase, and glutamic acid decarboxylase in deteriorating seeds is well documented. The general decrease in enzyme activity in the seed lowers its respiratory potential, which in turn lowers both the energy (ATP) and food supply to the germinating seed. Several changes in the enzyme macromolecular structure may contribute to their lowered effectiveness (Walter 1963). They may undergo compositional changes by losing or gaining certain functional groups, by oxidation of sulfhydryl groups, or by conversion of amino acids within the protein structure. The enzymes may also undergo configurational changes such as: (1) partial folding or unfolding of ultrastructure, (2) condensation to form polymers, or (3) degradation to subunits.

Formation and Activation of Hydrolytic Enzymes

As seed moisture content approaches levels necessary for germination, hydrolytic enzymes are activated. If the seed moisture content remains high or reaches higher levels, normal germination may occur; however, if moisture levels for germination are not attained, the seed deteriorates because of energy expenditure or accumulation of breakdown products. The increase in free fatty acids, a symptom of deterioration caused by activation of lipase enzymes in oil-containing seeds, has already been discussed. A related group of enzymes, the phospholipases, hydrolyzes the phospholipids and thus destroys the membrane structure of the seed. Activation of phosphatase enzymes converts ATP to ADP, resulting in an energy loss accompanied by increased phosphate acidity. Other hydrolytic enzymes activated by high moisture levels are amylases and proteolases. This kind of deterioration is rapid and is important only over short periods at moisture levels around 20% and above. Below 20% moisture, other kinds of deterioration predominate.

Breakdown in Mechanisms for Triggering Germination

Harrington (1973) has made a strong case for the idea that the breakdown of various triggering mechanisms also causes seed deterioration. Evidence has been cited on the role of gibberellins (Paleg 1960; Penner and Ashton 1965) and cytokinins in triggering enzyme activity leading to germination. Further evidence for this theory is the improved germination in aging seeds after exposure to growth hormones (Harrington 1973). For example, the exposure of partially aged rape seed to ethylene gas enabled them to produce normal seedlings (Takayanagi and Harrington 1971). It has been noted that gibberellic acid improved the germination and vigor of partially aged celery seeds (Harrington 1973) and can protect seeds from age-induced damage (Petruzzelli and Taranto, 1985) although other studies on aged barley seeds have shown no effect by gibberellic acid (Huber and McDonald 1982).

Genetic Degradation

A great deal of indirect evidence supports the view that seed deterioration is associated with random somatic mutations that impair the cellular functions of vital seed tissues. The increase in chromosomal aberrations in deteriorating seeds as a result of somatic mutation has been observed in many species. Fusion and fragmentation aberrations that were observed in both x-rayed and aged barley seeds occurred at higher rates than in fresh seeds (Gustafsson 1937). Two general observations have been made on work with barley seeds (James 1960): (1) the properties of chromosomes change gradually with age, and (2) mutations in seeds are highly correlated with seed age and decline in viability in aging seeds. The area of greatest sensitivity to x-ray damage and mortality is in the central root nuclei. The following observations have been made in support of the genetic theory: (1) extracts from aged seeds retard the germination of fresh seeds, (2) mutations in seeds are highly correlated with seed age and decline in viability, (3) spontaneous mutations arise in aging seeds and become evident by chromosome breakage in the presplit phase, (4) differences in the shoot and root tips in aged seed closely parallel those in irradiated seeds (D'Amato and Hoffman-Ostenhof 1956). Earlier, D'Amato (1952) suggested that mutations in aged seeds were induced by mutagens formed during the aging process and probably were decomposition products resulting from seed metabolism.

One observer (Harrington 1973) discounts genetic degradation as a primary cause of aging. Harrington cites evidence that although more chromosomal aberrations occur with increasing age, animal tissues rid themselves of the altered cells (Kohn 1963). This notion has been supported by Roos (1982). Further, radiation-induced injury in bacterial cells is reduced by quick repair of damaged DNA (Witkin 1966). If enough cells in the germinating seedlings are damaged, slower growth and seedling abnormality can occur; however, ordinarily radiation-killed cells are quickly crushed or replaced by normal tissue (Harrington 1973).

Depletion of Food Reserves

Depletion of food reserves is one of the oldest theories on deterioration; however, it has not survived critical scrutiny. In fact, most seeds contain enough food materials to last thousands of years (James 1960). Even 1000- to 2000-year-old wheat seeds that have been found in ancient tombs still retain most of their food reserves; thus, even nonviable seeds contain enough food reserves for seedling growth and development (Barton 1961). We know that the biochemical degradation processes in dry seeds are almost imperceptibly small and could not account for depleting the food reserves within the life span of most seeds.

Starvation of Meristematic Cells

The theory of starvation of meristematic cells was introduced at the USDA-ARS Seed Quality Research Symposium in 1971 (Harrington 1973); however, in principle, it had been implied earlier in an evaluation of the reserve food depletion theory (Harrington 1973). It was noted that respiration may deplete the tissues involved in the transfer of nutrition from reserve storage areas and thus prevent them from reaching the embryo. Another study elaborated on this, noting that meristematic cells, even though only a few cells away from abundant reserves of energy, may die from lack of food or from injury (Harrington 1973). It was speculated that perhaps the meristematic cells exhausted their energy supply, with no way to convert ADP to ATP.

Accumulation of Toxic Compounds

Under low-moisture storage, the reduced respiration and enzyme activity may be responsible for accumulation of toxic substances that reduce seed viability (Harrington 1973). When aged wheat embryos were transplanted onto young endosperms and young embryos onto aged endosperms, a progressive decline in germination and vigor of both transplants has been observed (Floris 1970), strongly indicating a gradual accumulation of toxic metabolites. It has been suggested that the presence of abscisic acid, a germination inhibitor, in several seeds supports this theory as a probable cause of aging (Harrington 1973; Ryugo 1969; Martin et al. 1969) as well as phenolic compounds (Sreeramulu 1983) and polyamines (Mukhopodhyay et al. 1983).

Questions

1. Name some crop and weed species noted for their short-lived seed.
2. What internal and external factors influence seed longevity?
3. Why do some seeds have higher moisture equilibrium curves (isotherms) than others?
4. How do the relative amounts of fats, carbohydrates, and proteins affect seed storability?
5. If it were possible to regulate temperature and relative humidity independently, which control would be most important for preserving seed viability in storage? Why?
6. Name several symptoms of seed deterioration. List them in order of their relative appearance in decline of germination in seed storage.
7. What are some possible causes of seed deterioration?
8. How would you realistically provide for adequate seed storage in the midwestern United States? Alaska? the Andes mountains in South America? Indonesia?
9. Describe Harrington's "rules of thumb," emphasizing their relationship to seed moisture and temperature during storage.
10. Explain why seeds placed in sealed storage at 13% moisture will deteriorate more rapidly than seeds in open storage.

General References

Abdul-Baki, A. A. 1969. Metabolism of barley seed during early hours of germination. *Plant Physiology* 44:733–738.

Abdul-Baki, A. A., and J. D. Anderson. 1970. Viability and leaching of sugars for germinating barley. *Crop Science* 10:31–34.

Abu-Shakra, S. 1963. Biochemical study of aging in seeds. Ph.D. Dissertation. Oregon State University, Corvallis.

American Phytopathological Society. 1983. Deterioration mechanisms in seeds. *Phytopathology* 73:313–339.

Anderson, J. D. 1970. Physiological and biochemical differences in deteriorating barley seed. *Crop Science* 10:36–39.

Anderson, J. D., J. E. Baker, and E. K. Worthington. 1970. Ultrastructural changes of embryos in wheat infected with storage fungi. *Plant Physiology* 46:857–859.

Andrews, W. 1956. The mitochondria of neuron. *International Review of Cytology* 5:147–170.

App, A. A., M. G. Bulis, and W. J. McCarthy. 1971. Dissociation of ribosomes and seed germination. *Plant Physiology* 47:81–86.

Austin, R. B., and P. C. Longden. 1967. Some effects of seed size and maturity on the yield of carrot crops. *Journal of Horticultural Science* 42:339–353.

Banerjee, A., M. M. Choudhuri, and B. Ghosh. 1981. Changes in nucleotide content and histone phosphorylation of ageing rice seeds. *Zeitschrift fur Pflanzenphysiologie* 102:33–36.

Barton, L. V. 1961. *Seed Preservation and Longevity*. London: Leonard Hill.

Bass, L. N. 1967. Controlled atmosphere and seed storage. *Seed Science and Technology* 1:463–492.

Bass, L. N. 1970. Prevention of physiological necrosis (red cotyledons) in lettuce seeds (*Lactuca sativa* L.). *Journal of American Society of Horticulture Science* 95:550–553.

Bass, L. N., and P. C. Stanwood. 1978. Long-term preservation of sorghum seed as affected by seed moisture, temperature, and atmospheric environment. *Crop Science* 18:575–577.

Battle, W. R. 1948. Effect of scarification on longevity of alfalfa seed. *Journal of the American Society of Agronomy* 40:758–759.

Beattie, J. H., and V. R. Boswell. 1939. Longevity of onion seed in relation to storage conditions. *U.S. Department of Agriculture Circular No. 512.*

Becquerel, P. 1934. La longevite des graines macrobiotiques. *Comptes Rendus Academie des Sciences* (Paris) 199:1662–1664.

Bennici, A., M. B. Bitonti, C. Floris, D. Gennai, and A. M. Innocenti. 1984. Ageing in *Triticum durum* wheat seeds: Early storage in carbon dioxide prolongs longevity. *Environmental and Experimental Botany* 24:159–165.

Berjak, P., J. M. Farrant, D. J. Mycock, and N. W. Pammenter. 1990. Recalcitrant (homoiohydrous) seeds: The enigma of their desiccation sensitivity. *Seed Science and Technology* 18:297–310.

Bewley, J. D. 1986. Membrane changes in seeds as related to germination and the perturbations resulting from deterioration in storage. In: *Physiology of Seed Deterioration*, eds. M. B. McDonald and C. J. Nelson, Madison, WI, pp. 27–47. Crop Science Society of America.

Bewley, J. D., and M. Black. 1982. *Physiology and Biochemistry of Seeds in Relation to Germination*, Vol. II. New York: Springer-Verlag.

Bockholt, A. J., J. S. Rogers, and T. R. Richmond. 1969. Effects of various storage conditions on longevity of cotton, corn, and sorghum seeds. *Crop Science* 9:151–153.

Bray, C. M., and T. Y. Chow. 1976. Lesions in ribosomes of nonviable pea (*Pisum arvense*) embryonic axis tissue. *Biochim Biophys Acta* 442:14–23.

Brett, C. C. 1952. Factors affecting the viability of grass and legume seed in and during shipment. *Proceedings of the International Grassland Congress* 6:878–884.

Brocklehurst, P. A., and J. Dearman. 1983. Interactions between seed priming treatments and nine seed lots of carrot, celery, and onion. I. Laboratory germination. *Annals Applied Biology* 102:577–584.

Brocq-Rosseu, D., and E. Gain. 1908. Sur la duree desperoxy distases des graines. *Comptes Rendus de L'Academie des Sciences* 146:545–548.

Bulat, H. 1963. Das allmahliche, durch ungunstige Largerungsbedingungen beschleunigte Absterben der Samen bzw. Ruckgang der Keimfahigkeit im Bilde des topografischen Tetrazoliumverfahrens. *Proceedings of International Seed Testing Association* 28:713–751.

Burgass, R. W., and A. A. Powell. 1984. Evidence for repair processes in the invigoration of seeds by hydration. *Annals of Botany* 53:753–757.

Canode, C. L. 1972. Germination of grass seed as influenced by storage conditions. *Crop Science* 12:79–80.

Chefurka, W. 1963. Comparative study of the dinitrophenol-induced ATPase activity in relation to mitochondrial aging. *Life Sciences* 6:399–406.

Chin, H. F., B. Krishnapillay, and P. C. Stanwood. 1989. Seed moisture: Recalcitrant vs. orthodox seeds. In: *Physiology of Seed Deterioration*, eds. M. B. McDonald and C. J. Nelson, pp. 15–22. Crop Science Society of America, Madison, WI.

Chin, H. F., and E. H. Roberts. 1980. *Recalcitrant Crop Seeds*. Kuala Lumpur, Malaysia: Tropical Press.

Ching, T. M. 1972. Aging stresses on physiological and biochemical activities of crimson clover (*Trifolium incarnatum* L. var. Dixie) seeds. *Crop Science* 12:415–418.

Christensen, C. M., and R. A. Meronuck. 1986. *Quality Maintenance in Stored Grains and Seeds*. Minneapolis, Minn.: University of Minnesota Press.

Christensen, C. M., J. H. Olafson, and W. F. Geddes. 1949. Grain storage studies. III. Relation of molds in moist stored cottonseed to increased production of carbon dioxide, fatty acids, and heat. *Cereal Chemistry* 26:109–128.

Clark, B. E., and N. H. Peck. 1968. Relationship between size and performance of snap bean seeds. *New York State Agricultural Experiment Station Bulletin 819*, pp. 4–30.

Cowdry, E. V. 1956. E. W. Dempsey's variations in the structure of mitochondria. *Journal of Biophysical and Biochemical Cytology* 2(4-Suppl.):305–310.

Crocker, W., and G. T. Harrington. 1918. Catalase and oxidase content of seeds in relation to their dormancy, age, vitality, and respiration. *Journal of Agricultural Research* 15:137–174.

Crowe, J. H., and L. M. Crowe. 1986. Stabilization of membranes in anhydrobiotic organisms. In: *Membranes, Metabolism, and Dry Organisms,* ed. A. C. Leopold, pp. 188–209. Ithaca, New York: Cornell University Press.

D'Amato, F. 1952. The problem of the origin of spontaneous mutations. *Caryologia* 5:1–13.

D'Amato, F., and O. Hoffmann-Ostenhof. 1956. Metabolism and spontaneous mutations in plants. *Advances in Genetics* 8:1–28.

Delouche, J. C. 1973. Precepts of seed storage. *Proceedings of the Mississippi State Seed Processors Shortcourse,* 1973:93–122.

Dempsey, W. H., and J. F. Harrington. 1951. Red cotyledon of lettuce. *California Agriculture* 5(7):4.

Duvel, J. W. T. 1905. The storage and germination of wild rice seed. *U.S. Department of Agriculture Plant Industry Bulletin 90, Part I.*

Earnshaw, M. J., B. Truelove, and R. D. Butler. 1970. Swelling of *Phaseolus vulgaris* mitochondria in relation to free fatty acid levels. *Plant Physiology* 45:318–321.

Ellis, R. H. 1991. The longevity of seeds. *Horticultural Science* 26:1119–1125.

Feeney, R. E., and J. R. Whitaker. 1982. The Maillard reaction and its prevention. In: *Food Protein Deterioration,* ed. J. P. Cherry, pp. 201–229. Washington, D.C.: American Chemical Society.

Fernandez Garcia de Castro, M., and C. J. Martinez-Honduvilla. 1984. Ultrastructural changes in naturally aged *Pinus pinea* seeds. *Physiologia Plantarum* 62:581–588.

Flood, R. G. 1978. Contribution of impermeable seed to longevity in *Trifolium subterraneum* (subterranean clover). *Seed Science and Technology* 6:647–654.

Floris, C. 1970. Aging *Triticum durum* seeds: Behavior of embryo and endosperms from aged seeds as revealed by the embryo-transplantation technique. *Journal of Experimental Botany* 21:462–468.

Fontes, L. A. N., and A. J. Ohlrogge. 1972. Influence of seed size and population on yield and other characteristics of soybeans [*Glycine max* (L) Merr.]. *Agronomy Journal* 64:833–836.

Francis, A., and P. Coolbear. 1984. Changes in the membrane phospholipid composition of tomato seeds accompanying loss of germination capacity caused by controlled deterioration. *Journal of Experimental Botany* 35:1764–1770.

Ghosal, K. K., and J. L. Mondal. 1978. Nature and consequence of chromosomal damage in aged seeds of tetraploid and hexaploid wheat. *Seed Research* 6:129–134.

Ghosh, B., and M. M. Chaudhuri. 1984. Ribonucleic acid breakdown and loss of protein synthetic capacity with loss of viability of rice embryos (*Oryza sativa*). *Seed Science and Technology* 12:669–677.

Goldbach, H. 1979. Imbibed storage of *Melicoccus bijugatus* and *Eugenia brasiliensis* (*E. dombeyi*) using abscisic acid as a germination inhibitor. *Seed Science and Technology* 7:403–406.

Goldsworthy, A., J. L. Fielding, and M. B. J. Dover. 1982. "Flash imbibition:" A method for the reinvigoration of aged wheat seed. *Seed Science and Technology* 10:55–65.

Grabe, D. F. 1965. Prediction of relative storability of corn seed lots. *Proceedings of the Association of Official Seed Analysts* 55:92–96.

Grabe, D. F., and D. Isely. 1969. Seed storage in moisture-resistant packages. *Seed World* 104(2):4.

Gustafsson, A. 1937. Der Tod Als Ein Nuklearer Prozess. *Hereditas* 23:1–37.

Haferkamp, M. E., L. Smith, and R. A. Nilan. 1953. Studies of aged seeds. I. Relation of age of seed to germination and longevity. *Agronomy Journal* 45:434–437.

Halder, S., S. Kole, and K. Gupta. 1983. On the mechanism of sunflower seed deterioration under two different types of accelerated aging. *Seed Science and Technology* 11:331–339.

Hall, C. W. 1957. *Drying Farm Crops*. Reynoldsburg, Ohio: Agricultural Consulting Associates, Inc.

Hallam, N. D., B. E. Roberts, and D. J. Osborne. 1973. Embryogenesis and germination in rye (*Secale cereale* L.). III. Fine structure and biochemistry of the nonviable embryo. *Planta* 110:279–290.

Harman, G. E., and A. L. Granett. 1972. Deterioration of stored pea seed: Changes in germination, membrane permeability, and ultrastructure resulting from infection by *Aspergillus ruber* and from aging. *Physiological Plant Pathology* 2:271–278.

Harman, G. E., A. A. Khan, and K. L. Tao. 1976. Physiological changes in the early stages of germination of pea seeds induced by aging and by infection by a storage fungus, *Aspergillus ruber*. *Canadian Journal of Botany* 54:39–44.

Harman, G. E., B. Nedrow, and G. Nash. 1978. Stimulation of fungal spore germination by volatiles from aged seeds. *Canadian Journal of Botany* 56:2124–2127.

Harman, G. E., B. L. Nedrow, B. E. Clark, and L. R. Mattick. 1982. Association of volatile aldehyde production during germination with poor soybean and pea seed quality. *Crop Science* 22:712–716.

Harmon, D. 1956. Aging: A theory based on free radical and radiation chemistry. *Journal of Gerontology* 11:298–300.

Harrington, J. F. 1960a. Drying, storing, and packaging seed to maintain germination and vigor. *Seedsmen's Digest* 11(1):16.

———. 1960b. Germination of seeds from carrot, lettuce, and pepper plants grown under severe nutrient deficiencies. *Hilgardia* 30:219–235.

———. 1972. Seed storage and longevity. In: *Seed Biology, Vol. 3*, ed. T. T. Kozlowski, pp. 145–240. New York: Academic Press.

———. 1973. Biochemical basis of seed longevity. *Seed Science and Technology* 1:453–461.

Harrison, B. J. 1966. Seed deterioration in relation to storage conditions and its influence upon germination, chromosomal damage, and plant performance. *Journal of National Institute of Agricultural Botany* 10:644–663.

Haynes, B. C., Jr. 1961. Vapor pressure determination of seed hygroscopicity. *U.S. Department of Agriculture Technical Bulletin 1229*.

Heydecker, W. 1969. The vigor of seeds—a review. *Proceedings of the International Seed Testing Association* 4(2):201–219.

Hibbard, R. P., and E. V. Miller. 1928. Biochemical studies on seed viability. I. Measurements of conductance and reduction. *Plant Physiology* 3:335–352.

Hildebrand, D. F., and T. Hymowitz. 1982. Inheritance of lipoxygenase-1 activity in soybean seeds. *Crop Science* 22:851–855.

Hoffpauir, C. I., D. J. Petty, and J. D. Guthrie. 1947. Germination and free fatty acid in individual cotton seeds. *Science* 106:344–345.

Hove, E. L., and P. L. Harris. 1951. Note on the linoleic acid-tocopherol relationship in fats and oils. *Journal of American Oil Chemical Society* 28:405.

Huber, T. A., and M. B. McDonald. 1982. Gibberellic acid influence on aged and unaged barley seed germination and vigor. *Agronomy Journal* 74:386–389.

Hughes, P. A., and R. F. Sandsted. 1975. Effect of temperatures, relative humidity, and light on the color of 'California Light Red Kidney' bean seed during storage. *Horticultural Science* 10:421–423.

Hummel, B. C. W., L. S. Cuendet, C. M. Christensen, and W. F. Geddes. 1954. Grain storage studies. XIII. Comparative changes in respiration, viability, and chemical composition of mold-free and mold-contaminated wheat upon storage. *Cereal Chemistry* 31:143–150.

Isely, D. 1957. Vigor tests. *Proceedings of the Association of Official Seed Analysts.* 47:176–182.

James, E. 1960. Seed deterioration. *5th Farm Seed Research Conference Proceedings,* pp. 31–39. Published by the American Seed Trade Association, Washington, D.C.

Jones, H. A. 1920. Physiological study of maple seeds. *Botanical Gazette* 69:127–152.

Justice, O. L., and L. N. Bass. 1978. Principles and Practices of Seed Storage. *USDA Agricultural Handbook 506.*

Kavanau, J. L. 1964. *Water and Solute-Water Interactions.* San Francisco: Holden-Day.

Kielley, W., and R. Kielley, 1951. Myokinase and adenosinetriophosphatase in oxidative phosphorylation. *Journal of Biological Chemistry* 191:485–500.

Kittock, D. L., and A. G. Law. 1968. Relationship of tetrazolium chloride reduction by germinating wheat seeds. *Agronomy Journal* 60:286–288.

Knowles, P. F. 1969. Modification of quantity and quality of safflower oil through plant breeding. *Journal of American Oil Chemical Society* 46:130–133.

Kohn, R. R. 1963. Mutations and aging. *Science* 142–540.

Koostra, P. T., and J. F. Harrington. 1969. Biochemical effects of age on membranal lipids of *Cucumis sativus* L. seed. *Proceedings of International Seed Testing Association* 34:329–340.

Kueneman, E. A. 1983. Genetic control of seed longevity in soybeans. *Crop Science* 23:5–8.

Lehninger, A. L., and L. F. Remmert. 1959. An endogenous uncoupling and swelling agent in liver mitochondria and its enzymatic formation. *Journal of Biological Chemistry* 234:2458–2464.

Leopold, A. C., ed. 1986. *Membranes, Metabolism, and Dry Organisms.* Ithaca, New York: Cornell University Press.

Libby, W. F. 1951. Radiocarbon dates II. *Science* 114:291–296.

Lindstrom, E. W. 1942. Inheritance of seed longevity in maize inbreds and hybrids. *Genetics* 26:154.

Lopez, A., and D. F. Grabe. 1973. Effect of protein content on seed performance in wheat (*Triticum aestivum* L.). *Proceedings of the Association of Official Seed Analysts* 63:106–116.

Lougheed, E. C., D. P. Murr, P. M. Harney, and J. T. Sykes. 1976. Low-pressure storage of seeds. *Experientia* 32:1159–1161.

Lunn, G., and E. Madsen. 1981. ATP-levels of germinating seeds in relation to vigor. *Physiologia Plantarum* 53:164–169.

MacLeod, A. M. 1952. Enzyme activity in relation to barley viability. *Transactions of the Botanical Society of Edinburgh* 36:18–33.

Mamiepic, N. G., and W. P. Caldwell. 1963. Effects of mechanical damage and moisture content upon viability of soybeans in sealed storage. *Proceedings of the Association of Official Seed Analysts* 53:215–220.

Mandal, A. K., and R. N. Basu. 1981. Role of embryo and endosperm in rice seed deterioration. *Proceedings of Indian National Science Academy, Part B, Biological Science* 47:109–114.

Mandal, A. K., and R. N. Basu. 1983. Maintenance of vigour, viability, and yield potential of stored wheat seed. *Indian Journal of Agricultural Science* 53:905–912.

Martin, G. C., M. I. R. Mason, and H. I. Forde. 1969. Changes in endogenous growth substances in the embryos of *Fuglans regia* during stratification. *Journal of the American Society for Horticultural Science* 94:13–17.

Marzke, F. O., S. R. Cecil, A. F. Press, and P. K. Harein. 1976. Effects of controlled storage atmospheres on the quality, processing, and germination of peanuts. *U.S. Department of Agriculture, Agricultural Research Service* 114:1–12.

McDonald, M. B. 1985. Physical seed quality of soybean. *Seed Science and Technology* 13:601–628.

McDonald, M. B., and C. J. Nelson, eds. 1986. *Physiology of Seed Deterioration.* Crop Science Society of America, Madison, WI.

McDonald, M. B., and D. O. Wilson. 1980. ASA-610 ability to detect changes in soybean seed quality. *Journal of Seed Technology* 5:56–66.

McHargue, J. S. 1920. The significance of the peroxidase reaction with reference to the viability of seeds. *Journal of the American Chemical Society* 42:612–615.

Minor, H. C., and E. H. Paschal. 1982. Variation in storability of soybeans under simulated tropical conditions. *Seed Science and Technology* 10:131–139.

Moore, R. P. 1972. Effects of mechanical injuries on viability. In: *Viability of Seeds*, ed. E. H. Roberts, pp. 94–114. Syracuse, N.Y.; Syracuse University Press.

Mukhopadhyay, A., M. M. Choudhuri, K. Sen, and B. Ghosh. 1983. Changes in polyamines and related enzymes with loss of viability in rice seeds. *Phytochemistry* 22:1547–1551.

Murata, M., E. E. Roos, and T. Tsuchiya. 1980. Mitotic delay in root tips of peas induced by artificial seed aging. *Botanical Gazette* 141:19–23.

Murata, M., E. E. Roos, and T. Tsuchiya. 1981. Chromosome damage induced by artificial seed aging in barley. I. Germinability and frequency of aberrant anaphases at first mitosis. *Canadian Journal of Genetic Cytology* 23:267–280.

Murata, M., E. E. Roos, and T. Tsuchiya. 1984. Chromosome damage induced by artificial seed aging in barley. III. Behavior of chromosomal aberrations during plant growth. *Theoretical and Applied Genetics* 67:161–170.

Nementhy, G., and H. A. Scheraga. 1962. Structure of water and hydrophobic bonding in proteins. I. A model for the thermodynamic properties of liquid water. *Journal of Chemical Physics* 63:3382–3400.

Nowak, J., and T. Mierzwinski. 1978. Activity of proteolytic enzymes in rye seeds of different ages. *Zeitschrift fur Pflanzenphysiologie* 86:15–22.

Nozzolillo, C., and M. DeBezada. 1984. Browning of lentil seeds, concomitant loss of viability, and the possible role of soluble tannins in both phenomena. *Canadian Journal of Plant Science* 64:815–824.

Ohga, I. 1926. The germination of century-old and recently harvested Indian *Lotus* fruits, with special reference to the effect of oxygen supply. *American Journal of Botany* 13:754–759.

Ohlrogge, J. B., and T. P. Kernan. 1982. Oxygen-dependent aging of seeds. *Plant Physiology* 70:791–794.

Osborne, D. J. 1983. Biochemical control systems operating in the early hours of germination. *Canadian Journal of Botany* 61:3568–3577.

Osborne, D. J., M. Dobrzanska, and S. Sen. 1977. Factors determining nucleic acid and protein synthesis in the early hours of germination. *Symp. Soc. Exp. Biol.* 31:177–194.

Paleg, L. G. 1960. Physiological effects of gibberellic acid: I. On carbohydrate metabolism and amylase activity of barley endosperm. *Plant Physiology* 35:293–299.

Parrish, D. J., and C. C. Bahler. 1983. Maintaining vigor of soybean seeds with lipid anitoxidants. *Proceedings of the Plant Growth Regulator Society of America* 10:165–170.

Patil, V. N., and C. H. Andrews. 1985. Cotton seeds resistant to water absorption and seed deterioration. *Seed Science and Technology* 13:193–199.

Payne, F. 1946. The cellular picture of the anterior pituitary of normal fowls from embryo to old age. *Anatomical Record* 96:77–91.

Penner, D., and F. M. Ashton. 1965. Effect of benzyladenine on the proteolytic activity of germinating squash seeds. *Plant Physiology* 40(suppl.):lxxix.

Petruzzelli, L., and G. Taranto. 1984. Phospholipid changes in wheat embryos aged under different storage conditions. *Journal of Experimental Botany* 35:517–520.

Petruzzelli, L., and G. Taranto. 1985. Effects of permeations with plant growth regulators via acetone on seed viability during accelerated aging. *Seed Science and Technology* 13:183–191.

Porsild, A. E., and C. R. Harrington. 1967. *Lupinus articus* Wats. grown from seeds of the Pleistocene Age. *Science* 158:113–114.

Powell, A. A., and S. Matthews. 1981. Association of phospholipid changes with early stages of seed aging. *Annals of Botany* 47:709–712.

Priestley, D. A. 1986. *Seed Aging.* Ithaca, New York: Cornell University Press.

Priestley, D. A., and A. C. Leopold. 1983. Lipid changes during natural aging of soybean seeds. *Plant Physiology* 63:726–729.

Quaglia, G., R. Cavaioli, P. Catani, J. Shejbal, and M. Lombardi. 1980. Preservation of chemical parameters in cereal grains stored in nitrogen. In: *Controlled Atmosphere Storage of Grains,* ed. J. Shejbal, pp. 319–333. New York: Elsevier.

Rahman, A., and F. H. Stillinger. 1971. Molecular dynamics study of liquid water. *Journal of Chemical Physics* 55:3331–3359.

Ries, S. K. 1971. The relationship of size and protein content of bean seed with growth and yield. *Proceedings of the American Society for Horticultural Science* 96:557–560.

Roberts, E. H. 1973. Predicting the storage life of seeds. *Seed Science and Technology* 1:499–514.

Roberts, E. H. 1986. Quantifying seed deterioration. In: *Physiology of Seed Deterioration,* eds. M. B. McDonald and C. J. Nelson, pp. 101–123. Crop Sci. Soc. Amer. Spec. Publ. 11. Madison, WI.

Roberts, E. H., and F. H. Abdalla. 1968. The influence of temperature, moisture, and oxygen on period of seed viability in barley, broad beans, and peas. *Annals of Botany* 32:97–117.

Roberts, E. H., and R. H. Ellis. 1982. Physiological, ultrastructural and metabolic aspects of seed viability. In: *The Physiology and Biochemistry of Seed Development, Dormancy and Germination,* ed. A. A. Khan, pp. 465–485. Amsterdam: Elsevier Biomedical Press.

Roos, E. E. 1982. Induced genetic changes in seed germplasm during storage. In: *The Physiology and Biochemistry of Seed Development, Dormancy, and Germination,* ed. A. A. Khan, pp. 409–434. New York: Elsevier.

Rossi, C. R., L. Sartorelli, L. Tato, and N. Siliprandi. 1964. Relationship between oxidative phosphorylation efficiency and phospholipid content in rat liver mitochondria. *Archives of Biochemistry and Biophysics* 107:170–175.

Ryugo, K. 1969. Abscisic acid, a component of the beta-inhibitor complex in the *Prunus* endocarp. *Journal of the American Society for Horticultural Science* 94(1):5–8.

Sacktor, B. 1953. Investigations on the mitochondria of the house fly, *Musca domestica* L. *Journal of General Physiology* 36:371–387.

Sacktor, B., J. J. O'Neill, and D. G. Cochran. 1958. The requirement for serum albumin in oxidative phosphorylation of flight muscle mitochondria. *Journal of Biological Chemistry* 233:1233–1235.

Saio, K., I. Nikkuni, Y. Ando, M. Otsuru, Y. Terauchi, and M. Kito. 1980. Soybean quality changes during model storage studies. *Cereal Chemistry* 57:77–82.

Salunkhe, D. K., J. K. Chavan, and S. S. Kadam. 1985. *Postharvest Biotechnology of Cereals.* Boca Raton, Fla.: CRC Press.

Sanchez, R. A., and L. C. de Miguel. 1983. Aging of *Datura ferox* seed embryos during dry storage and its reversal during imbibition. *Zeitschrift fur Pflanzenphysiologie* 110:319–329.

Saxens, O. P., and D. C. Maheshwari. 1980. Biochemical aspects of viability in soybean. *Acta Botanica Indica* 8:229–234.

Simpson, D. M. 1946. The longevity of cottonseed as affected by climate and seed treatments. *Agronomy Journal* 38:32–45.

Sinex, F. M. 1960. Aging and the lability of irreplaceable molecules—II. The amide groups of collagen. *Journal of Gerontology* 15:15–18.

Sivori, E., F. Nakayama, and E. Cigliano. 1968. Germination of Achirs seed (*Canna* sp.) approximately 550 years old. *Nature* 219:1269–1270.

Smith, M. T. 1983. Cotyledonary necrosis in aged lettuce seeds. *Proceedings of the Electron Microscopy Society of South Africa* 13:129–130.

Sreeramulu, N. 1983. Germination and food reserves in bambarra groundnut seeds (*Voandzeia subterranea* Thouars) after different periods of storage. *Annals of Botany* 51:209–216.

Stanwood, P. C. 1985. Cryopreservation of seed germplasm for genetic conservation. In: *Plant Cryopreservation*, ed. K. Kartha, pp. 199–225. Boca Raton, Fla.: CRC Press.

Stanwood, P. C. 1986. Dehydration problems associated with the preservation of seed and plant germplasm. In: *Membranes, Metabolism, and Dry Organisms*, ed. A. C. Leopold, pp. 327–310. Ithaca, N.Y.: Cornell University Press.

Stanwood, P. C., and M. B. McDonald, eds. 1989. *Seed Moisture*. Madison, Wisc.; Crop Science Society of America.

Takayanagi, K., and J. F. Harrington. 1971. Enhancement of germination rate of aged seeds by ethylene. *Plant Physiology* 47:521–524.

Tappel, A. L. 1962. Hematin compounds and lipoxidase as biocatalysts. In: *Symposium on Foods: Lipids and Their Oxidation*, pp. 122–138. Corvallis, Oregon: Oregon State University.

Throneberry, G. O., and F. G. Smith. 1954. Seed viability in relation to respiration and enzymatic activity. *Proceedings of the Association of Official Seed Analysts* 44:91–95.

Toole, E. H. 1950. Relation of seed processing and of conditions during storage on seed germination. *Proceedings of the International Seed Testing Association* 16:214–227.

———. 1957. *Storage of vegetable seeds*. U.S. Department of Agriculture Leaflet 220.

Toole, E. H., and E. Brown. 1946. Final results of the Duvel buried seed experiment. *Journal of Agricultural Research* 72:201–210.

Toole, E. H., and V. K. Toole. 1953. Relation of storage conditions to germination and to abnormal seedlings of bean. *Proceedings of the International Seed Testing Association* 18:123–129.

Tortora, P., G. M. Hanozet, A. Guerritoe, M. T. Vincenzini, and P. Vanni. 1978. Selective denaturation of several yeast enzymes by free fatty acids. *Biochim Biophys Acta* 522:297–306.

Vaughan, C. E., and J. C. Delouche. 1968. Physical properties of seeds associated with viability in small-seeded legumes. *Proceedings of the Association of Seed Analysts* 58:128–141.

Vertucci, C. W., and A. C. Leopold. 1986. Physiological activities associated with hydration level in seeds. In: *Membranes, Metabolism, and Dry Organisms*, ed. A. C. Leopold, pp. 35–50. Ithaca, N.Y.; Cornell University Press.

Walter, H. 1963. Chemical reactivity of a macromolecule as a function of its age. *Biochemica et Biophysica Acta* 69:410–411.

Ward, F. A., and A. A. Powell. 1983. Evidence for repair processes in onion seeds during storage at high seed moisture contents. *Journal of Experimental Botany* 34:277–282.

Weidner, S., and K. Zalewski. 1982. Ribonucleic acids and ribosomal protein synthesis during germination of unripe and aged wheat caryopses. *Acta Society Botanica Poland* 51:291–300.

Weinbach, E. C., H. Sheffield, and J. Garbus. 1963. Restoration of oxidative phosphorylation and morphological integrity of swollen, uncoupled rat liver mitochondria. *Proceedings of the National Academy of Science* 50:561–568.

Weiss, J., and A. I. Lansing. 1953. Age changes in the time structure of the anterior pituitary of the mouse. *Proceedings of the Society of Experimental Biology and Medicine Journal* 82:460–466.

Welch, G. B., and J. C. Delouche. 1974. Conditioned storage of seed. *Proceedings and Reports of the Southern Seedsmen's Association* 1974:30–40.

West, S. H., ed. 1986. *Physiological-Pathological Interactions Affecting Seed Deterioration.* Madison, Wisc.; Crop Science Society of America.

Wester, H. V. 1973. Further evidence of age of ancient viable Lotus seeds from Pulantien Deposit. Manchuria. *Horticultural Science* 5:371–377.

Wien, H. C., and E. A. Kueneman. 1981. Soybean seed deterioration in the tropics. II. Varietal differences and techniques for screening. *Field Crops Research* 4:123–132.

Wilson, D. O., Jr., and M. B. McDonald. 1986a. A convenient volatile aldehyde assay for measuring seed vigour. *Seed Science and Technology* 14:259–268.

Wilson, D. O., Jr., and M. B. McDonald. 1986b. The lipid peroxidation model of seed aging. *Seed Science and Technology* 14:269–300.

Wilson, D. O., Jr., and M. B. McDonald. 1989. A probit planes method for analyzing seed deterioration data. *Crop Science* 29:471–476.

Wilson, R. F., J. W. Burton, and C. A. Brim. 1981. Progress in selection for altered fatty acid composition in soybeans. *Crop Science* 21:788–791.

Witkin, E. M. 1966. Radiation-induced mutations and their repair. *Science* 152:1345–1353.

Wojtczak, L., and A. B. Wojtczak. 1960. Uncoupling of oxidative phosphorylation and inhibition of ATP-Pi exchange by a substance from insect mitochondria. *Biochemica and Biophysica Acta* 39:277–289.

Woodstock, L. W., and M. F. Combs. 1965. Effects of gamma-irradiation of corn seed on the respiration and growth of the seedlings. *American Journal of Botany* 52:563–569.

Woodstock, L. W. and J. Feeley. 1965. Early seedling growth and initial respiration rates as potential indicators of seed vigor in corn. *Proceedings of the Association of Official Seed Analysts.* 55:131–139.

Woodstock, L. W., K. Furman, and T. Solomos. 1984. Changes in respiratory metabolism during aging in seeds and isolated axes of soybean. *Plant Cell Physiology* 25:15–26.

Woodstock, L. W., and D. F. Grabe. 1967. Relationship between seed respiration during imbibition and subsequent seedling growth in *Zea mays* L. *Plant Physiology* 42:1071–1076.

Woodstock, L. W., and O. L. Justice. 1967. Radiation-induced changes in respiration of corn, wheat, sorghum, and radish seeds during initial stages of germination in relation to subsequent seedling growth. *Radiation Botany* 7:129–136.

Woodstock, L. W., S. Maxon, K. Faul, and L. Bass. 1983. Use of freeze-drying and acetone impregnations with natural and synthetic anti-oxidants to improve storability of onion, pepper, and parsley seeds. *Journal of American Society of Horticultural Science* 108:692–696.

Woodstock, L. W., and B. M. Pollock. 1965. Physiological predetermination: imbibition, respiration, and growth of lima bean seed. *Science* 150:1031–1032.

Yang, S. F., and Y. B. Yu. 1982. Lipid peroxidation in relation to aging and loss of seed viability. *Search* 16(1):2–7.

Yoneyama, T., I. Suzuki, and M. Murohashi. 1970. Natural maturing of wheat flour. I. Changes in some chemical components and in farinograph and extensigraph properties. *Cereal Chemistry* 47:19–26.

9
Seed Production

During the last 50 years, the availability of high-quality seed of improved crop varieties, along with modern power equipment, improved fertilizers, and better methods of weed and insect control, have revolutionized farming. The seed industry has played a vital role in this modern revolution, with its expanding production capability, efficiency in rapid seed increase of new varieties, and effective maintenance of genetic purity. The quantity of seed needed by farmers each year is enormous. It is estimated that North American farmers alone use over 12 billion pounds annually of field crop, vegetable, flower, and tree seeds.

Prior to the era of modern agriculture, seeds were usually a by-product of grain or hay production. Frequently, the poorest part of the crop was reserved as seed, which was often gathered from haylofts, along with chaff, weed seeds, and other incidental material. Occasionally, farmers would have enough excess seed to share with a neighbor, and a farm-to-farm exchange resulted. Although seed was sometimes available on the commercial market, the supply was sporadic, and often of questionable quality. Unscrupulous marketing gimmicks were sometimes used to take advantage of the customers' inability to identify seeds and their quality. This led to skepticism about the quality of seeds bought off the farm and created a climate of distrust that acted against the development of a legitimate seed industry.

DEVELOPMENT OF THE SEED INDUSTRY

Several factors are responsible for the development of the seed industry in North America: (1) an increased number of new, available varieties, (2) the development of seed certification and seed law enforcement programs, (3) the development of a cleaning and conditioning technology, (4) a better knowledge of seed quality, and (5) the emergence of the seed grower as a specialist.

The Hatch Act of 1875 accelerated the development of the U.S. seed industry. This Act established a network of experiment stations throughout the country and stimulated the development of improved crop varieties. Seed certification was a direct outgrowth of the experiment stations of the land grant colleges. Certification programs were established in most states between 1915 and 1930, and most of the new "college-bred" field crop varieties first became available as certified seed. During this period, many farmers became accustomed to purchasing seed from recognized seed producers rather than planting their own "bin-run" seed.

With the availability of new varieties, a strong seed industry was necessary with the ability for rapid seed increase and distribution with adequate safeguards for varietal purity. Otherwise, the impact of new improved varieties would have been lost. The National Foundation Seed Project, another outgrowth of the experiment station network, assured the success of many new varieties by providing for rapid increase of basic seed stocks; however, it is no longer an active program.

Since the early 1900s, the technology of seed cleaning and conditioning has made tremendous strides. Machinery has been designed that can condition the seed more efficiently and improve the quality over "bin-run" seed by removing contaminating weed seeds, other crop seeds, and inert material. Machines have also been developed to improve germination by eliminating poor-quality crop seeds. Facilities have become available to chemically treat and inoculate the seed and package it in attractive containers for merchandising appeal. Seed quality testing has been developed as a service to both seed producers and farmers. Such testing has provided new insights about seed quality and has done much to increase quality consciousness among farmers and seed producers.

The development of the seed producer as a specialist and the growth of the seed industry have been closely related. Since seed production is no longer a by-product of grain or hay production, a large seed industry has developed in areas widely removed from where the seed is used and a seed increase and merchandising chain has developed to market seed over wide geographical areas. Seed stocks of a new variety are sent to a seed production area for increase. Quantities of commercial seed are returned to the area of adaptation for forage or turf production. Thus, seed production outside the area of use has contributed in a major way to the development of the seed producer as a specialist and has helped the growth of the seed industry. Seed production is a complex business involving many different and integrated operations. Several reviews of these operations have been captured by Hebblethwaite (1980), George (1985), Kelly (1988), and McDonald and Copeland (1995).

SEED PRODUCTION IN AREA OF USE

The seeds of many field crops are produced in the same area where they are planted for commercial production. As a general rule, when adequate quantities of high-quality seed can be produced in the area of use, it is best to do so. Crop varieties usually produce higher seed yields in the area where they are adapted. Also, transportation and marketing costs from producers to consumers are minimized. This is particularly important for seed of cereals, soybean, and certain other crops that involve larger seed volumes and greater planting rates than small-seeded grasses and legumes. Other than small-seeded grasses and legumes and certain specialty crops, seed of most field crops is produced in the area of principal adaptation and use.

SEED PRODUCTION OUTSIDE THE AREA OF USE

A large part of the seed planted in the United States and Canada is produced outside its major area of use, including almost all seeds of certain grass and legume crops, and many vegetable and ornamental crops. Most of this specialized seed production is in the western regions of North America, where a unique combination of climate and cultural factors has contributed to the specialized seed production capability. A closer look at this western seed industry indicates why it has developed and flourished far removed from areas where most of the seed is used.

Advantages—Climatic Factors

The probability of dry harvesting weather is the greatest single reason for the success of the western seed production region. Weather records for these areas show that the probability of rain during the harvest season is minimal. In the western valleys of the Pacific Northwest, the relative assurance of favorable harvesting weather, combined with plentiful natural precipitation during the spring and early summer, provides a unique competitive advantage in turf and forage seed production. In the drier western areas, water is provided during the growing season from vast irrigation systems.

Nonirrigated Production. Although the inland valleys of the Pacific Northwest are particularly well suited for nonirrigated grass seed production, seeds of forage legumes, sugar beet, and garden vegetables are also produced in this area. Few geographical regions are more climatically suited for grass seed production. Rainfall is plentiful and almost continuous throughout the mild winters and spring, providing ideal conditions for plant development and seed head production. In early summer the rainy season ends, and the weather becomes dry with a low probability of rain from mid-June until September. Weather records show many summers without precipitation for 30–60 days. Such conditions favor maturation of seed heads and threshing at harvest, without the continuous threat of rainy weather that often plagues seed harvest in other areas.

Irrigated Production. Irrigated seed production in the western United States is principally for both large- and small-seeded legumes, although there are areas of highly successful grass seed production also. Although pockets of irrigated legume seed production are found throughout the West, major production is centered in California, eastern Oregon, and Washington eastward to Idaho and Utah. These areas have a low probability of rainfall through the harvesting season and have adequate water for irrigation. Another major advantage in legume seed production is the relative abundance of natural insect pollinators. Great success has been achieved in building up native bee pollinators and introducing various types of wild bees from other areas. Dry weather during the flowering period and an abundance of insect pollinators usually assures good pollination and seed set.

The warm, dry summer climate of the western United States is also ideally suited for production of disease-free seed of vegetable and field crops that are highly susceptible to seedborne diseases. Practically all of the seed of garden pea and snap bean and much of the dry edible bean types are produced in Idaho, Washington, and California. The warm, dry summer weather combined with stringent phytosanitary seed production regulations prevent buildup of many of the bacterial, fungal, and viral diseases that are highly destructive in the warm, humid summers of the central and eastern regions where much of the seed is planted.

Disadvantages

Transportation and marketing costs are major disadvantages of producing seeds outside their area of use. However, this cost is minimal compared to the advantage provided.

Perhaps the greatest potential danger of producing seed outside the major area of crop adaptation is the likelihood of a *genetic drift* in the varieties caused by the different environmental stresses (such as day length, temperature, soil type) to which the plants are exposed during seed production. The potential for a drift in germ plasm composition is especially critical in cross-pollinated forage species; environmental stresses in the seed production area may cause the failure of the individual plants in the varietal population to survive or to produce their proportionate share of seed. As a result, the germ plasm represented by their segment of the population decreases, while other segments of the population increase proportionately. Thus, a gradual drift in the genetic balance of germ plasm occurs, which may change the varietal characters. Although this drift may be gradual and hardly noticeable in the region of seed production, after a few years of production it may seriously affect the varietal performance in its area of adaptation.

Safeguards against a serious genetic drift are provided through a limited-germination certification system, in which the number of generations of seed increase outside its area of adaptation is carefully controlled. The certification system provides these limitations and minimizes the likelihood of genetic drift.

GRASS SEED PRODUCTION

Seed production of grasses can be appropriately discussed in terms of cool season grasses, Great Plains grasses, and southern grasses (see Table 9.1).

Cool Season Grasses

Cool season grasses are those adapted to the northern regions of North America. They include most of the lawn and turfgrass species in the United States except for those of the most southerly regions. These grasses grow actively at cool temperatures during the spring and fall. Depending on the severity of the temperatures, their winter response varies. During the mild winters of the Pacific Northwest, their growth is minimal but they maintain a green color. Under more severe winter conditions, cool season grasses go dormant and lose their green color. Dormancy is followed by very rapid spring growth until flowering in early June. Seed is mature about the first of July.

Most seed production of the cool season grasses is located in nonirrigated areas of the Pacific Northwest; however, considerable amounts of seed are also produced under irrigation in the Pacific intermountain region. Relatively small but important

Table 9.1. Classification of Grasses of the United States

Cool Season Grasses	Warm Season Grasses	
	Great Plains Grasses	**Southern Grasses**
tall fescue	gramagrass	Bermudagrass
red fescue	bluestems	Dallisgrass
ryegrass	buffalograss	rescuegrass
tall oatgrass	Indian ricegrass	Bahiagrass
Kentucky bluegrass	lovegrass	Zoysia grass
smooth brome grass	blue panicum	napier grass
orchard grass	switchgrass	pangola grass
Reed canarygrass	dropseeds	centipedegrass
timothy	buffelgrass	Rhodegrass
bentgrass	Texas wintergrass	carpetgrass
Russian wild rye	Indiangrass	Vaseygrass
wheatgrasses	vine mesquite	St. Augustine grass
	curly mesquite grass	
	thatchgrass	

pockets of production exist in midwestern and central states, particularly in Minnesota, South Dakota, Kentucky, and Missouri.

Seed fields may be established in rows or in solid plantings, by using a grain drill. In recent years, row spacings of 24 to 48 in. have become more popular, although most seed production is still from solid seedlings. In the Great Plains, Russian wild rye has been planted in 84-in. rows. In addition to providing assurance of cleaner, weedfree fields, row seedlings offer several other advantages. Row plantings generally produce higher yields than solid seedings and require lower seeding rates. It is thought that row plantings offer more economical use of fertilizer and prolong stand productivity. Solid stands, especially in the sod-forming grasses, tend to become *sod bound* after a few years and their productivity declines. This condition seems to be a result of overpopulation and excessive competition for available nutrients, as well as the physical crowding caused by overpopulation. The adverse effects of sod binding can be minimized by increased nitrogen applications and by postharvest field burning.

Seedling rates lower than those used for forage or turf production have been found best for grass seed production. Small-seeded grasses, such as bentgrass or bluegrass, are often seeded at ¼–1 lb/ac, while larger-seeded grasses, such as tall fescue and Russian wild rye, may be seeded at 2–4 lb/ac.

The ability of grass seed producers in the Pacific Northwest to increase small quantities of seeds of new forage and turf varieties for widespread availability in the United States and in the world market has gained them international recognition. These seed producers use techniques of stand establishment, weed control, culture, harvesting, and conditioning that are geared for high varietal purity, freedom from weed and other crop seed and higher germination standards. Since much of their seed is used for establishment of quality lawns and high-grade pastures, their attention to seed quality is important.

One method of stand establishment uses a special drill that places a charcoal barrier directly over the seed immediately after planting (Figure 9.1). A nonselective

A

B

Figure 9.1. *Establishing a grass seed field by use of a protective charcoal barrier: (A) laying down the barrier and planting the seed, (B) results following seedling emergence. (Courtesy of Richard Bailey.)*

herbicide (e.g., Karmex), is then applied to the entire field. The herbicide directly above the drilled seed is absorbed by the charcoal layer, allowing seed germination without injury while all seedlings between the bands are eliminated. Once the crop is established, the use of selective herbicides and hand roguing keeps weed contamination to a minimum.

Seed is harvested (Figure 9.2) by swathing the crop into windrows and allowing it to dry a few days before threshing. Much of the seed is taken directly to farm cleaners where it may be cleaned completely or given a rough scalping to remove most of the foreign material, such as straw, stones, and other large and easily separated contaminants. If cleaned completely, it is sold to a wholesaler on a pure seed basis. Alternatively, it may be sold "in the rough" to a wholesaler who finishes the cleaning procedure. In the latter case, the grower is paid on a pure seed basis. Much of the grass seed is grown on a contract basis, so commitments are made in advance concerning quality of seed to be delivered, the degree of cleaning, and the price to be paid.

Field Burning—A Dilemma. Grass seed fields in the Pacific Northwest traditionally have been burned after the seed is removed. This practice has been widely recognized as a valuable cultural practice to the grass seed grower, since burning significantly increases yields, and controls destructive insects and diseases that overwinter in unburned plant residues. Burning also aids in weed control, delays the development of "sod binding," and returns potash and other mineral nutrients to the soil.

Although the benefits are significant and conspicuous, so is the air pollution from the burning fields. In recent years, ecologists and the nonagricultural population have become increasingly critical of this practice and have influenced state legislatures to ban open-field burning. Considerable research is under way to find alternatives to open-field burning, but none has yet proved satisfactory, including the use of portable burners using straw or propane as fuel. Future possibilities include: (1) removal of all stubble and chaff from the field by mechanical methods, and (2) chemical control of diseases and insects. If satisfactory alternatives are not employed, growers may have to accept lower yields, which could increase the cost of forage and turf grass seed throughout the world.

Warm-Season Grasses

Warm-season grasses make their maximum growth during the summer. This growth is aided by the adequate summer rainfall in their area of adaptation. These grasses start their spring growth about three weeks later than cool-season grasses and cease growing with the first hard frost in the fall. In the winter they are completely dormant. Warm-season grasses predominate in the central and southern Great Plains area and throughout the South.

Great Plains Grasses. These are mostly native grasses of the high plains of western and southern Texas and western Oklahoma, Kansas, Nebraska, and the surrounding areas. Seeds of these grasses were originally harvested from native pastures for use in converting croplands to grass and for improving rangelands. Even today much of the seed of these grasses is harvested from native stands. However, a specialized seed production industry is developing due to at least two reasons: (1) seed production

A

B

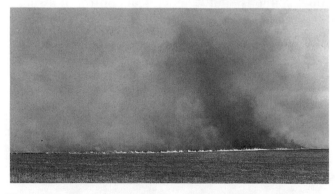

C

Figure 9.2. *Grass seed production in Oregon: (A) general view of a fescue seed field, (B) swathing (windrowing) a ryegrass seed field, and (C) postharvest field burning. (B and C, Courtesy of Harold Youngberg.)*

from native stands is sporadic and undependable, occurring only when seasonable moisture conditions are unusually favorable, and (2) as improved varieties of native grasses become available, the need for a program for maintaining seed availability intensifies.

Most seed fields of the Great Plains grasses are established in rows, fertilized as needed for optimum seed yields, and perhaps irrigated in periods of drought. Harvesting is usually performed by direct combining, although some of the grasses are harvested by a stripper of the type previously used for harvesting native bluegrass stands. The harvested seeds of many of the Great Plains grasses are extremely chaffy and are not free flowing. This creates extreme difficulty in cleaning and handling the harvested seed and has even discouraged production of certain species. However, handling techniques are becoming available that will remove much of the chaff and increase the ease of conditioning. Thus, a seed industry for these species is emerging.

Southern Grasses. The adaptation of a plant to its environment is reflected in its development and seed yield. Consequently, seed of the warm-season southern grasses (for example, Bermuda grass) must be produced in southern regions of the United States rather than in the Pacific Northwest. Almost all Bermuda grass seed production is limited to Arizona and southern California. Bermuda grass seed fields may be established either vegetatively by "spriggings" (planting stolons or rhizomes), or by seeding (rows or broadcast). Regardless of the method used, the plants soon spread into solid stands. Zoysia grass is another southern species in which seed fields are established by vegetative plantings. Both Bermuda grass and Zoysia grass are harvested for turf purposes in vegetative plugs or sprigs, from which lawns are established (though Bermuda grass seed may also be harvested).

Much of the southern forage grass seed production is harvested from pastures that have been grazed during the summer, fall, and early spring. This dual use and early removal of grass material appears to stimulate seed production relative to ungrazed stands. Recent availability of improved varieties has stimulated more seed production in rows rather than as a byproduct of pasture production. Harvesting may be done by combining directly from a windrow, or occasionally by shocking and threshing.

LEGUME SEED PRODUCTION

Seed of the forage legumes is produced under two distinct types of management systems and climatic regions. One kind is the specialized seed production industry that has developed in the western United States and Canada. The second, often a byproduct of forage production, is not centered in any geographic area, but is scattered throughout North America.

Western Legume Seed Production

The greatest proportion of high-quality legume seed is produced in the western region of North America, particularly in western Canada, California, Oregon, Washington, and eastward to Colorado and Oklahoma. This area has great natural advantages for legume seed production, especially alfalfa seed. The dry climate, adapted soils, abundance of water for irrigation, and availability of insect pollinators have raised seed production of alfalfa and other legumes to levels unattainable in other areas of North America. Occasionally in this region, alfalfa yields 2000 lb of seed per acre.

Cultural Practices. Alfalfa seed fields are usually established in 21- to 41-in. rows, although some solid stands are used. Seeding rates vary from ⅔–4 lb/ac for rows to as much as 15 lb/ac for solid stands. Research has shown that within certain limits, lower seeding rates tend to give the highest seed yields. Alfalfa stands respond to lower plant populations by producing more seed per plant, thus increasing total production. Spacing the plants individually or in hills within the row often increases seed yields, although it may also increase weed problems. In the San Joaquin Valley of California, highest yields are obtained in 40-in. rows planted at only 1 lb/ac. Row planting also offers advantages over solid stands through more efficient use of seed and fertilizer and in facilitating weed control.

The most serious weed in alfalfa seed production is field dodder (*Cuscuta spp.*). Dodder (see Figure 9.3) is a parasitic weed which grows particularly well on alfalfa, though several other crops and weeds may also serve as hosts. It is particularly troublesome for alfalfa seed production because its similar seed size makes it difficult to separate from alfalfa seed lots. Although the seed is similar in size, its rough seed coat texture allows it to be separated by means of a velvet roll machine (see Chapter 10). Dodder is controlled in alfalfa seed fields by a combination of chemical and manual methods. Most states have added dodder to their noxious weed lists to prevent incoming alfalfa and clover seed lots from becoming sources of dodder contamination.

Insect Pollinators. The availability of insect pollinators gives the western seed production area a great advantage for alfalfa and other insect-pollinated legumes. Several important insect pollinators used are: (1) honeybees, (2) leaf-cutting bees, (3) alkali bees, and (4) bumblebees. Both wild and domesticated honeybees are used, although neither are efficient in "tripping" (pollinating) the alfalfa flowers (see Figure 9.4). When the honeybees visit the flowers for either nectar or pollen, they "trip" the alfalfa flower by dislodging the sexual column, which is concealed by the keel petals. The tripping exposes the stigma to the pollen carried by the bee from other flowers. When the sexual column is released it snaps upward with considerable force, striking the bee on the underside. Honeybees soon learn how to visit flowers without tripping the flower and thus become less effective pollinators. Some honeybees are primarily nectar gatherers, while others gather pollen and are more effective as pollinators. In spite of their rather low effectiveness, the large numbers of honeybees and the availability of domesticated colonies make them the most important insect pollinator for alfalfa.

Leaf-cutting bees (*Megachile spp*) are a wild species that are more effective pollinators than honeybees, although they rarely occur naturally in sufficient numbers to pollinate commercial seed fields alone. They are small, dark-colored bees (see Figure 9.5) that build their nests in colonies in small holes in wood. They may bore the holes themselves or use existing holes made by other insects. Oblong leaf cuttings are used to make a cell into which the eggs are laid. Each cell is provided with pollen moistened with nectar. After the eggs are laid, the cell is sealed with circular leaf cuttings. If the hole is long enough, several cells may be placed in tandem within the same hole.

In recent years, alfalfa seed growers have increased the number of leaf-cutting bees by providing artificial egg-laying sites where colonies can be established. These colonies

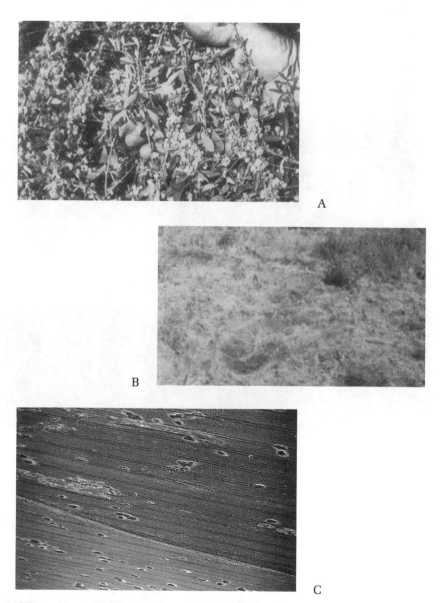

Figure 9.3. *Three views of dodder and its control in alfalfa seed production: (A) closeup of dodder in the flowering stage growing on alfalfa, (B) a severe dodder infestation of alfalfa, and (C) an aerial view of an alfalfa seed field in which dodder has been sprayed with a contact herbicide, allowed to dry, and subsequently burned (this is a common control method). (Courtesy of Howard Roylance).*

standard
petal

wing
petal

keel
petal

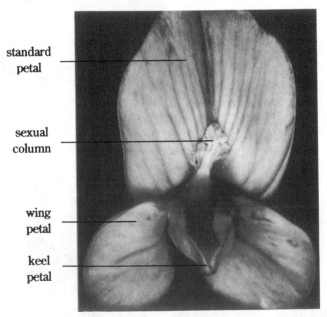

standard
petal

sexual
column

wing
petal

keel
petal

Figure 9.4. *The parts of an alfalfa flower important in insect pollination: above, untripped; below, tripped. (Courtesy of William P. Nye, Logan, Utah.)*

A

B

C

Figure 9.5. *Leaf-cutting bees, their habitat and use in alfalfa seed production: (A) leaf-cutting bee about to enter nest cell, (B) artificial leaf-cutting bee cells in which the eggs are laid, (C) portable leaf-cutting bee colonies. (A, Courtesy of William P. Nye; B and C, Courtesy of Howard Roylance.)*

are often portable and can be moved to different parts of the seed field as needed. One can observe the effectiveness of leaf-cutting bees in the conspicuous circles of well-pollinated plants (with wilted flowers) around their colonies. Leaf-cutting bees have two disadvantages that lower their value and require specific management by the commercial seed grower: (1) their range of activity is relatively short—usually only a few hundred feet around the colony and (2) like most wild bees, their populations are sporadic and undependable.

Alkali bees (*Nomia melanderi*) are highly effective pollinators that are native to many western alfalfa seed production areas. Large colonies become established in bare, moist, salt flats, where the female digs a pencil-sized tunnel 8–10 in. into the soil and hollows out several egg-laying cells. She provides each cell with balls of pollen moistened with nectar on which eggs are laid, and the cell is then plugged with soil. The eggs hatch in a few days, and the larva feed on the pollen and nectar that has been provided. Depending on the time of year, the larva may pupate and emerge as adults in about 30 days or wait until the next season. After emerging, female adults reestablish the colony by laying eggs and starting the life cycle over again. Alkali bees are shown in Figure 9.6.

Alkali bees have become important alfalfa pollinators in certain seed areas where they occur naturally or have been introduced. They range much farther than leaf-cutting bees and usually occur in much greater numbers. Many attempts have been made to introduce alkali bees into new seed areas, but with moderate success. This is done by lifting large cores from well-established nesting sites and inserting them into sites where a new bed is desired. Considerable site preparation is necessary if the establishment is to be successful.

Bumblebees (*Bombus spp.*) and melissodes bees (*Melissodes spp.*) are also effective pollinators; however, they seldom occur in numbers sufficient to increase seed production significantly. Bumblebees are limited in intensive agricultural areas because of destruction of their nesting sites. Melissodes bees have shown remarkable ability to survive under intensively cultivated conditions; however, they are not as gregarious as other wild types and almost never occur in sufficient quantities for pollination of commercial seed fields.

Weather conditions influence the activity of all bee types. Temperatures between 75 and 100°F are most favorable. Bees do not work in the rain or while the flowers are wet. Wind velocities over 5 mph also reduce bee activity appreciably.

The use of insecticides is a potential threat to all bee types, and many documented cases exist where bee populations have been seriously reduced by misuse of insecticides.

Legume Seed Production in Areas of Use

Large volumes of alfalfa and clover seed are still produced in the midwest and mideastern regions; however, it is usually a by-product of hay production and is subject to large fluctuations due to weather conditions and availability of insect pollinators. Although these regions have few of the advantages available for seed producers farther west, considerable seed volumes are harvested annually from Midwest and Mideast

Figure 9.6. *Alkali bees: (A) alkali bee starting her nest excavation, (B) portion of an alkali bee nesting site showing wind-blown entrance mounds among clumps of salt grass and samphire, (C) horizontal section of an alkali bee nest at cell level, (D) an artificial (man-made) alkali bee nesting site, (E) sign along an Idaho road in an alfalfa seed production area, (A–D, Courtesy of William P. Nye; E, Courtesy of Howard Roylance.)*

hay fields. Usually the first hay crop is cut or grazed, and the second is left for seed. Clipping of some types of legumes, however, may be detrimental to seed production. For example, in Michigan, seed production of medium red clover is stimulated by early clipping, while mammoth red clover seed yields are reduced. Harvesting is done by combining directly or by windrowing and allowing the crop to dry a few days before threshing. The probability of late summer rains during the harvest season is a serious hazard to dependable seed production in this area.

Unlike seed production of the northern adapted species, seeds of southern legumes are usually produced almost entirely in their area of adaptation and use. Although a specialized seed production industry has developed, a first cutting of hay may still be taken to stimulate better seed production by the second growth. Seed of southern legumes is harvested in the fall, usually in October. Harvesting methods vary from direct combining, shocking, and threshing to combining from windrows.

HYBRID SEED PRODUCTION

Corn

Discovery of hybrid vigor in field crops after the beginning of the 20th century and development of techniques for producing hybrid corn seed have probably contributed as much to American agriculture as any other single factor. The first announcement of potential advantages of hybrid over open-pollinated varieties was made in 1908. In 1917, a method of double-cross seed production was proposed that would supply hybrid seed corn at prices farmers could afford and in large enough quantities to assure an adequate and constant supply of seed. Since these beginnings, hybrid corn has set an enviable standard for performance and acceptance by farmers. In 1933, only 0.2% of the corn acreage in the United States corn belt was planted with hybrid seed. By 1944, its use had grown to 83%; today virtually all corn is grown with hybrid seed.

A hybrid variety is produced by crossing inbred lines that have been developed by inbreeding and selection for at least five successive generations. Inbreeding results in (1) depression of vigor (plant height, yield, etc.), (2) increasing uniformity (homozygosity), and (3) appearance of undesirable recessive gene effects that can be eliminated from the population. The first-generation progeny of hybrid seed results in yield increases far above either parent, and above that of nonhybrid populations. This yield increase is from heterosis, or hybrid vigor, which is due to an accumulation of a large number of dominant favorable growth factors (genes).

Modern corn production may be from seed representing double-cross, single-cross, or three-way hybrids (see Figure 9.7). Until the early 1960s, double-cross hybrids were the predominant types used for commercial seed production. They offer several advantages:

1. They are more variable than single or three-way crosses; they are not all alike genetically. Thus, the plants may be buffered more against unfavorable conditions that occur during the growing season.
2. A longer pollination period than other crosses may provide more complete filling of the ear with grain and result in higher yields.

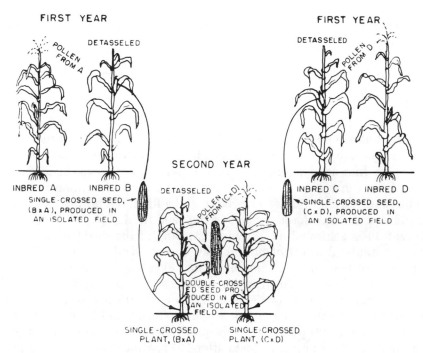

Figure 9.7. *The method by which single-crossed and double-crossed hybrid corn seed is produced. (From Hughes and Metcalf (1972).)*

3. Lower seed cost is an obvious advantage when the yield of double crosses is equal to or better than the best single-cross or three-way hybrids.
4. Double-cross hybrids generally have higher seed quality than single-cross hybrids and may give more nearly optimum stands when adverse conditions occur after planting.

Single-cross corn hybrids have been increasing in popularity in recent years. A corn field planted from single-cross seed is impressive because the plants tend to be uniform. Plant height, ear height, tasseling, silking, and pollen shedding are uniform, giving the field good eye appeal. Also, since only two inbred parents are involved in a single-cross hybrid, a higher level of resistance to diseases, insects, and other unfavorable situations may be incorporated into them during the breeding process. Within a given set of inbred parents, the best single-cross hybrid has a higher yield potential, genetically, than the best double-cross hybrid; however, a particular double cross $[(A \times B) \times (C \times D)]$ may yield better than a particular single cross $(E \times F)$ that uses other inbred parents. The main disadvantages of single-cross hybrids are lower seed yield and relatively lower quality of seed, since seed is produced on inbred parents. To overcome these disadvantages in seed production, some single-cross hybrids are *modified single crosses*. Two closely related sister inbred lines are crossed $(A_1 \times A_2)$, to produce the seed (female) parent. The pollen (male) parent may also be a cross of two different but closely related sister inbreds $(B_1 \times B_2)$ or a full inbred (B). Seed yields are higher since $(A_1 \times A_2)$ plants are somewhat more vigorous and produce better-quality seed than full inbred plants.

Hybrid corn seed is produced throughout the corn belt of the United States by highly specialized producers. Most hybrid corn seed is produced under the control of large seed companies, although a substantial portion is produced by many small companies located throughout the corn belt. Regardless of the control, hybrid corn seed production techniques are quite standardized. Double-cross seed production fields are usually planted in a pattern of six seed (female) parent rows and two pollen (male) parent rows; single-cross production is usually a pattern of two seed parent and one pollen parent, or four seed parent and two pollen parent rows. Seed fields must be well isolated from other corn fields that represent potential contamination from out-crossing. The required isolation can be reduced if additional border rows of the pollen parent are planted around the perimeter of the seed fields.

Hybridization is achieved by allowing crossing of the desired male (pollen) and female (seed) parents. Control of pollen may be achieved by detasseling the female parent or by use of male-sterile female parent inbreds. The first hybrids were all a result of detasseling; however, during the 1960s, use of male sterility seemed likely to completely eliminate the need for detasseling. The situation was dramatically altered in 1971 by outbreaks of a new race "T," of Southern Corn Leaf Blight and the recognition that the Texas source (T-cytoplasm) cytoplasmic male sterility was much more suscepti-ble to the new race "T." For a few years after 1971, almost all seed was produced by detasseling, however, incorporation of genetic sterility and other types of cytoplasmic sterility (other than the Texas source) has heralded another movement away from mechanical detasseling. When male sterility is used for hybrid seed production, restorer mechanisms must be incorporated so the subsequent commercial corn plants will produce pollen as well as seed.

Hybrid corn seed is usually harvested and dried while still on the ear. After drying down to about 12% moisture, it is cleaned and graded into different size and shape classes (e.g., small rounds, large flats) to facilitate planting precision. It is almost always slurry-treated with a fungicide before bagging and marketing.

Sorghum

Like corn, virtually all sorghum produced in the United States is from hybrid seed. However, unlike corn, the sorghum plant is largely self-pollinated, so mechanical removal of pollen-producing structures is not practical. Consequently, hybrid sorghum seed production became practical only after discovery of cytoplasmic and genetic sterility mechanisms. Restorer mechanisms are used for producing hybrid sorghum seed of grain sorghum; when sorghums are to be used as forage and silage, restorer mechanisms are not needed.

Wheat

Commercial hybrid seed production of most self-pollinated crops is considered more difficult than that for cross-pollinated crops. Their flowers and pollen dispersal are not structured for cross-pollination with other plants. Aside from the flower structure and pollen dispersal pattern, male sterility and restorer mechanisms must be incorporated. A few hybrid barley varieties have been released; however, hybrid wheat is the real goal towards which millions of dollars have been invested by both public and private agencies.

The first obstacle to overcome in hybrid wheat was the development of suitable male sterile lines. Now that male sterility has been found, the greatest obstacle to hybrid wheat development is the lack of simple, effective mechanisms for restoration of the fertility of the male line. Genetic restorer systems that can be incorporated into male parents do exist, but they vary widely in effectiveness, are influenced by climate, are genetically complex, and require a long development program.

Another method of producing hybrid wheat involves the use of chemical stamatacides on the female parent. This allows fertilization by pollen from adjacent male rows and assures the production of hybrid seed. Male rows are harvested for grain while female strips are harvested and sold for hybrid seed.

If hybrid wheat proves successful, it could create a huge and specialized seed production industry, since seed must be produced every year for planting the next season's crop. The magnitude of such a program is difficult to conceive of compared to small-grain seed production today. Not only would large amounts be needed, but isolation from commercial wheat fields would be necessary. Whether adequate isolation can be found within the commercial wheat production region is questionable.

Figure 9.8 illustrates one method of hybrid wheat seed production.

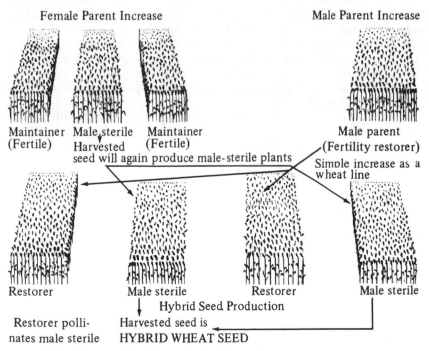

Figure 9.8. *Production of hybrid wheat seed. (From Rogers and Lucken (1973).)*

Hybrid Alfalfa Seed

In 1971, a patent was granted to a commercial seed company for producing hybrid alfalfa seed by a process using cytoplasmic male sterility. This is the first successful use of the hybrid concept in the forage legumes, and the first patent awarded for using the hybrid concept. Unlike hybrids of grain crops, a restorer mechanism is not needed for alfalfa, since the vegetative material rather than the seed is commercially important.

Cytoplasmic male sterile plants are usually discovered by screening large plant populations. These are then increased and maintained by crossing to "maintainer type plants," resulting in a seed population from which male sterile plants can be produced. Commercial hybrid alfalfa seed is produced by growing male sterile plants in rows adjacent to those of male fertile plants, allowing insect cross-pollination. The resulting single-cross hybrid seed is then harvested from the male sterile rows. This seed could be marketed as a single cross but is generally used as a female parent to produce a three-way hybrid. The alternate pollinator rows in hybrid-seed fields are also harvested and marketed as a byproduct of hybrid-seed production.

Questions

1. Why has the small-seeded grass and legume seed industry largely moved to the western United States and Canada?
2. What are the advantages and disadvantages of field burning for grass seed production?
3. How can an optimum balance be maintained between insect control and adequate populations of pollinating insects for cross-pollinated legume seed crops?
4. What are the advantages and disadvantages of contract seed production?
5. How much seed is sold directly to consumers from the farms on which it is produced? How does this practice vary among various crops?
6. What is the future of hybrid seed production for crops other than corn, sorghum and cotton? What about self-pollinating crops?
7. What is genetic drift in a variety? Explain its likelihood in self-pollinated versus cross-pollinated crops?
8. Explain sod binding in grass seed production fields and how it can be avoided.
9. List the relative merits of solid versus row seedings for seed production of grasses and legumes.
10. Explain the relationship between seeding rate and seed yields of grasses and legumes.
11. What do you consider to be the ideal seed moisture content for harvesting seed to avoid mechanical damage? Do you consider mechanical damage to be a serious factor in seed? Which species are most seriously affected?

General References

Bohart, G. E., and T. W. Koerber. 1972. Insects and seed production. In: *Seed Biology*, vol. III, ed. T. T. Kozlowski, pp. 1–54. New York: Academic Press.

Cowan, J. R. 1961. Producing high quality seed. In: *Turfgrass Science*, ed. A. A. Hanson and F. V. Juska, pp. 424–441. Madison, Wisc.: American Society of Agronomy.

Douglas, J. E., ed. 1980. *Successful Seed Programs: A Planning and Management Guide*. Boulder, Colo.: Westview Press.

George, R. A. T. 1985. *Vegetable Seed Production*. New York: Longman Press.

Hebblethwaite, P. D., ed. 1980. *Seed Production*. London and Boston: Butterworth and Company.

Hughes, H. D., and D. S. Metcalfe. 1972. *Crop Production*, 3rd ed., p. 185. New York: MacMillan.

Kelly, A. F. 1988. *Seed Production of Agricultural Crops*. New York: Longman Press.

McDonald, M. B., and L. O. Copeland. 1995. *Principles and Practices of Seed Production*. New York: Chapman & Hall.

Rogers, K. J., and K. A. Lucken. 1973. Hybrid wheat seed production in North Dakota. *North Dakota Agricultural Experiment Station Report No. 806, Farm Research* 30(6):4.

Thompson, J. R. 1979. *An Introduction to Seed Technology*. New York: John Wiley and Sons.

U.S. Department of Agriculture. 1961. *Seeds: The Yearbook of Agriculture*. Washington, D.C.: U.S. Government Printing Office.

1. Airy, J. M., L. A. Tatum, and J. W. Sorenson, Jr. Producing seed of hybrid corn and grain sorghum, pp. 145–153.
2. Bodger, H. The commercial production of seeds of flowers, pp. 216–220.
3. Bohart, G. E., and F. E. Todd, Pollination of seed crops by insects, pp. 240–246.
4. Cochran, L. C., W. C. Cooper, and E. E. Blodgett. Seeds for rootstocks of fruit and nut trees, pp. 233–239.
5. Culbertson, J. O., H. W. Johnson, and L. G. Schoenleber. Producing and harvesting seeds of oilseed crops, pp. 192–199.
6. Graumann, H. O. Our sources of seeds of grasses and legumes, pp. 159–163.
7. Hanson, E. W., E. D. Hansing, and W. T. Schroeder. Seed treatments for control of diseases, pp. 272–280.
8. Harmond, J. E., J. E. Smith, Jr., and J. K. Park. Harvesting the seed of grasses and legumes, pp. 181–188.
9. Hawthorn, L. R. Growing vegetable seeds for sale, pp. 208–215.
10. Hills, C. A., K. E. Gibson, and W. F. Rochow. Insects, viruses, and seed crops, pp. 258–263.
11. Hoekstra, P. E., E. P. Merkel, and H. R. Powers, Jr. Production of seeds of forest trees, pp. 227–232.
12. Kreitlow, K. W., C. L. Lefebre, J. T. Presley, and W. J. Zaumeyer. Diseases that seeds can spread, pp. 265–272.
13. Lieberman, F. V., F. F. Kicke, and O. A. Hills. Some insect pests of important seed crops, pp. 251–258.
14. McMurtey, J. E., Jr. Producing and harvesting tobacco seed, pp. 206–207.
15. Pedersen, M. W., L. G. Jones, and T. H. Rogers. Producing seeds of the legumes, pp. 171–181.
16. Ricker, P. I. The seeds of wild flowers, pp. 288–294.
17. Rogler, G. A., H. H. Rampton, and M. D. Atkins. The production of grass seeds, pp. 163–171.
18. Rudolf, P. O. Collecting and handling seeds of forest trees, pp. 221–226.
19. Shaw, W. C. and L. I. Danielson. The control of weeds in seed crops, pp. 280–287.
20. Stevens, H., and J. R. Goss. Seeds of oats, barley, wheat, and rice, pp. 153–159.
21. Stewawrt, D. New ways with seeds of sugar beets, pp. 199–205.
22. Todd, F. E., and S. E. McGregor. Insecticides and honeybees, pp. 247–250.
23. Waddle, B. M., and R. F. Colwick. Producing seeds of cotton and other fiber crops, pp. 188–192.

Wheeler, W. A., and D. D. Hill. 1957. *Grassland Seeds*. New York: D. Van Nostrand Company.

10

Seed Conditioning and Handling

Seed as it comes from the field is never pure. It usually arrives at the cleaning plant containing large quantities of trash, green leaves, weeds, other crop seeds, and insects. If it contains such materials as green leaves and other high-moisture materials, it cannot be safely stored, efficiently handled, nor accurately cleaned until most of the foreign material has been removed. The process of removing these unwanted materials from a seed lot is known as seed conditioning.

Seed conditioning is a vital part of the total technology involved in making available high-quality seed of improved varieties. It assures farmers of high-quality seed with minimum adulteration. A good seed conditioning job can assure that the previous efforts of plant breeders in developing superior varieties, and of seed producers in growing them, can result in maximum-quality seed. If seed is not conditioned and handled properly, all past efforts in varietal development and seed production may be lost.

Figure 10.1 shows the steps involved in a typical seed conditioning plant.

PRINCIPLES OF SEED CONDITIONING

The seed conditioner has five objectives when cleaning seed: (1) complete separation—removal of all contamination, (2) minimization of seed loss—some good seeds are removed along with contaminants in almost every conditioning operation and this loss must be kept at a minimum, (3) upgrading—improvement of seed quality through removal of decayed, cracked, broken, insect-damaged, or otherwise injured or low-quality seed, (4) efficiency—the highest capacity consistent with effectiveness of separation, and (5) minimization of labor—labor is a direct operating cost and cannot be recovered.

The quality of seed is improved during conditioning in two ways: (1) *separation* of contaminating seeds of other crops, weeds, and inert matter, and (2) *upgrading,* or the elimination of poor-quality seed. The ultimate goal of seed conditioning is to obtain the maximum percentage of pure crop seed with maximum germination potential. This

242

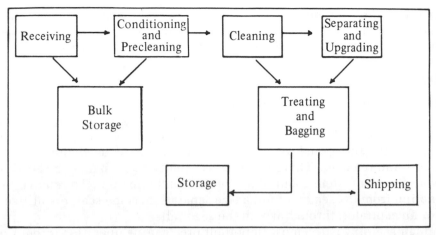

Figure 10.1. *Basic flow diagram showing essential steps in seed conditioning. (Reprinted through the courtesy of the publisher, Vaughan and Delouche (1967).)*

concept is reflected in terms of the *pure live seed* percentage. This is calculated by multiplying the percent purity and the percent germination as in the following example:

Seed Lot X
    Pure seed content—95%
    Germination—93%
    Pure live seed—88.35%   (.95 × .93 × 100)

The pure live seed content provides a more realistic picture of actual quality of a given seed lot than either the purity or germination alone.

Seeds can be separated by mechanical means only if they differ in some physical characteristic that can be detected by some mechanical or electrical process. Thus, the seed conditioner uses differences in physical characteristics of the seed crop being cleaned and those of seeds of other crops and weeds as well as inert matter. Inert matter, such as chaff, stems, and stones, is usually the easiest to remove; other crop seeds and weed seeds may be much more difficult, especially those similar in appearance and physical characteristics. Physical characteristics that are used to separate seeds are: size, length, width, thickness, shape, weight (specific gravity), surface texture, color, affinity for liquids, and electrical properties.

SEED CLEANING EQUIPMENT

The seed conditioner should carefully analyze each seed lot as it comes from the field to determine what the conditioning problem will be and which machine or machines will do the best job. Since usually more than one machine must be used to completely condition a given seed lot to maximum purity and germination, their sequence of use is also important.

Frequently, seed arriving at a conditioning plant contains excessive trash, which makes it difficult to move through the elevators, thus interfering with proper conditioning by lowering efficiency and slowing capacity. When this occurs, the seed may require one or more precleaning operations to improve conditioning efficiency and separation precision, and prevent loss of seed during subsequent conditioning.

Precleaning Equipment

Probably the most commonly used precleaning machine is a scalper. This machine is used to rough-clean various kinds of trash from the seed lot. Although many different scalpers are available, they usually consist of a vibrating or rotating screen (or sieve) through which the small seed pass readily, while the larger seed and inert matter are "scalped off" and removed. This separation is usually accompanied by an air flow that blows away the chaff, stems, and other lighter contamination. To remove the lighter contamination from the seed without a size separation, some scalpers utilize only the air flow of an aspirator, through which the seed falls.

Although scalping is the most important precleaning operation, other operations may also be needed to increase the cleaning efficiency. For example, oats, barley, and many noncereal grass seeds have awns, hairs, or other chaffy appendages that cause them to cling together, and thus make them hard to condition. This is particularly true for the so-called chaffy grasses of the Great Plains region. Seed of these species is often preconditioned by a vigorous hammering, rubbing, or abrading action. Several kinds of machines are available; however the *debearder* perhaps is used most frequently. It has a hammering or flailing action that removes awns, beards, or lint from seed and tends to break up seed clusters of the chaffy grasses, as well as multiple seed units of nonchaffy types, permitting them to be conditioned more efficiently. Another conditioning machine, the *huller-scarifier,* removes hulls or pods from seeds, such as crown vetch, crambe, and lespedeza by an abrading or rubbing action. This machine also reduces the hard seed content of many species, such as sweet clover, crown vetch and alfalfa, whose seed coats are impermeable to water. It will do a good job of scarification, but must be carefully adjusted to prevent seed injury.

Air Screen Machine

The air screen machine (Figure 10.2), sometimes called the fanning mill, is the basic seed-cleaning machine in most conditioning plants. It uses a combination of airflow and perforated metal or wire screens to separate seed on the basis of size, specific gravity, and resistance to airflow. Many sizes of air screen cleaners exist— from the small, two-screen farm models to large industrial cleaners with seven or eight screens and three or four air separations.

The air screen machine works in three different ways. Seed first enters the machine by gravity from a feed hopper. In some machines, it falls directly onto a *scalping* screen, which allows the good seed to fall through, while separating out the larger-seeded species and foreign material. In other machines, the seed first falls through an *aspirating* airstream which blows off the lighter material from the seed mass. The last operation *grades* the seed by allowing the good seed to ride over screen perforations, while the smaller particles drop through. All air screen machines use these three principles, though their sequence may differ.

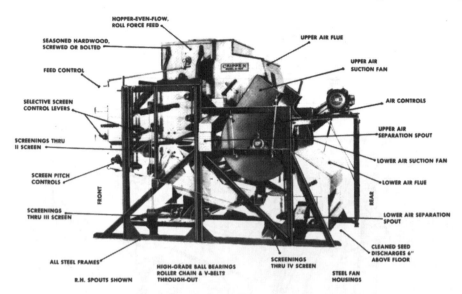

Figure 10.2. The air screen machine, with its parts labeled. (Courtesy of Crippen Manufacturing Company, Alma, Michigan.)

Screens used in an air screen machine are constructed of either perforated sheet metal or woven wire mesh on a wooden frame. Metal screen openings may be either round, oblong, or triangular, while openings in wire mesh screens are either square or rectangular. The perforated metal screens are measured in 64ths of an inch (hole size) and the wire mesh screens are measured in number of openings per inch. Most air screen machines used in conditioning seed use three or four screens. These screens can be easily changed, and almost 100 sizes are available.

The screens are kept clean during operation by tappers, or hammerlike screen knockers (which jar loose material wedged in the perforations), and brushes (which move back and forth underneath the screen to dislodge and brush away material clinging to it). Some of the newer machines have highly resilient rubber balls that bounce between the screens and help dislodge such material.

Gravity Separator

The specific gravity separator (Figure 10.3) is perhaps the second most commonly used piece of seed cleaning equipment. It can be used to separate undesirable seed and inert contaminants that are so similar in size, shape, and seed coat characteristics to the crop seed that they cannot be removed in any other way. In addition to its separation of other crop and weed seed contamination, it is probably the best machine available for upgrading seed quality. For example, deteriorated, moldy, or decayed seed, which are usually similar in size and shape to good seed but have a lower specific gravity, can be removed by this machine. Insect-damaged seed, empty seeds, off-color seeds and any other seeds that have defects that decrease their specific gravity can

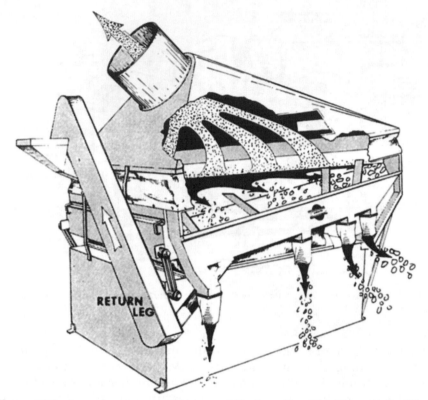

Figure 10.3. *A gravity separator. (Courtesy of Forsbergs Inc., Thief River Falls, Minn.)*

also be removed. Heavy nonseed particles such as mud balls, soil particles, and small stones can also be removed.

Although several types and styles of gravity machines exist, they all have basically the same components and use the same principle of separation. They consist of a base (or frame), one or more fans, a plenum chamber (air chest), a porous vibrating deck, a feed hopper, and a seed discharge system. Seeds are introduced from the feed hopper onto the porous metal or fabric deck, where the combination of shaking and airflow up through the deck causes them to stratify according to specific gravity, producing a layering effect; the heavier particles remain close to the deck surface, while the lighter particles float on a cushion of air above them. The deck can be tilted in two directions to facilitate seed separation and discharge. The slight incline and vibration of the deck causes the heavy particles in close contact with the deck surface to "walk" or be forced toward the top of the deck, where they ultimately fall off and are separated. The lighter particles tend to float on the air cushion above the heavier seeds, following the path of least resistance, and drift to the lower end of the deck where they fall off and are discharged. Seeds and other particles in the medium specific gravity ranges, often called *middlings,* fall from the deck near the middle area; shields, or discharge aprons, can be mounted on the discharge side of the deck to direct the middlings into one spout from which they are returned for another separation on the deck.

The gravity separator is widely used for cleaning and upgrading seed of a great many species. However, it is normally used only after other machines (such as the air screen machine) have eliminated most chaff, stems, and off-sized contamination. It is especially useful for upgrading seed of field beans, alfalfa, grasses, as well as many other crops.

A modification of the specific gravity separator is represented by a machine called the *stoner*. This machine separates seed on the same principle as conventional specific gravity machines; however, it discharges only at each end. The desirable seeds flow to the lower end and are discharged, while stones and heavier concreted earthen material "walk" their way up to the upper end of the deck and are discharged. This machine is useful in conditioning seed of field beans, since it eliminates most of the small stones that commonly occur in bean seed lots.

Dimensional Sizing Equipment

Several types of seed conditioning machines separate or grade seeds on the basis of thickness, length, or width.

Length separators are specifically designed to separate seed differing in length. Two types are available, and both are commonly used for conditioning seed, especially for cleaning smaller-seeded grasses and legumes. These are the *indent disk* and *indent cylinder* separators (Figures 10.4 and 10.5). The indent disk separator uses a series of indented disks that are rotated inside the tilted cylinder through which the seed moves. Disks may be used with indents that will lift out shorter contaminating seeds from the longer crop seeds, or remove smaller crop seed and leave the large-seeded contamination to flow on through the cylinder. The indent cylinder works in much the same way, but the indentations are in the inside cylinder housing, which lifts out undesirable seeds as it rotates on an almost horizontal axis. The cylinder is tilted so that the seeds may flow through it by gravity and discharge from the lower end after a separation is attained. Both the indent cylinder and indent disk separators are best suited to free-flowing heavier-seeded species.

Width and thickness separators are commonly called *graders* or *sizers* by the seed trade and are primarily used for separating hybrid corn seed into different size and shape grades. There are three types of graders used: (1) ribbed horizontal-flat screen types, (2) vertical ribbed screen types, and (3) cylindrical screen types.

Surface Texture Separators

Three types of seed conditioning machines separate on the basis of differences in surface texture. The *roll machine* is probably the best known and most widely used. It is also known as the "velvet roll" or "velvet roll mill," but is most frequently called the "dodder mill" because of its effectiveness in removing dodder from clover and alfalfa seed.

Like gravity separators, roll mills are finishing machines and should be used only

Figure 10.4. *The indent disk separator. The illustration on the left shows how the disks revolve through the seed mass, and the one on the right shows how the shorter seeds are lifted from the seed mass, while the longer ones are rejected by the disk indents. (Reprinted through the courtesy of the Carter-Day Company, Minneapolis, Minn.)*

for seed that has already been conditioned by other machines. This machine is effective in removing rough-textured, irregular, broken, cracked, and immature seed, or inert matter from smooth-textured crop seeds. The machine consists of two slightly inclined velvet-covered cylinders which roll in opposite directions. The seed to be cleaned is fed through rolls at the upper end of the incline, slides down between them, and is discharged from the lower end. Rough-textured seeds are separated by their tendency to cling to the velvet-covered rolls and be lifted out. A metal shield covers each set of rolls and directs the rejected seed and inert matter into the discharge spout. The machine can be adjusted by varying the roll speed, cylinder tilt, and shield clearance. Its capacity may be increased by using multiple sets of stacked rolls. Most commercial units use up to 10 sets of rolls.

The roll mill has done much to aid in separating seed of dodder from western-produced alfalfa seed and is to a large extent responsible for the elimination of this noxious weed as a serious problem in many alfalfa seed consumption areas.

Two other machines that use differences in surface texture are *magnetic separators* and *inclined drapers*. The magnetic separator depends on the affinity of rough-textured, undesirable seed for metal dust. When the seed is pretreated with a combination of

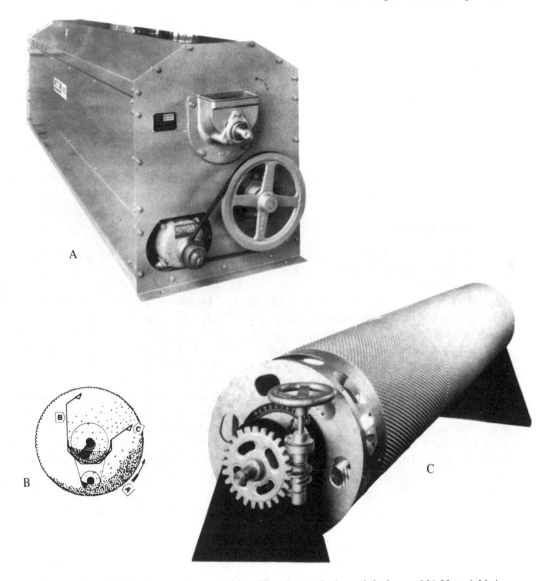

Figure 10.5. *Three views of the indent cylinder machine. The view at the lower left shows: (A) Material being lifted by the cylinder, (B) material being deposited into the conveying trough, and (C) the position of the separating edge. (Reprinted through the courtesy of the Carter-Day Company, Minneapolis, Minn.)*

oil and water and exposed to a fine iron powder, the rough-textured contaminants such as dodder attract the metal dust and they are removed by a magnetized metal drum over which the seed moves. The revolving drum is then brushed or scraped clean on the lower turn, and the seed is eliminated. The inclined draper, a special-purpose finishing machine that does not have wide use, separates seeds on the basis of their different tendency to roll or slide down an inclined surface. Crop seeds containing contamination with different sliding or rolling tendencies can often be cleaned very thoroughly by this method.

Spiral Separator

This device has no moving parts but permits separation of seeds that differ in their ability to roll down an inclined spiral. Round seeds can be easily separated from flat or irregular seeds since they roll much easier. Thus, when seed mixtures are fed into the spiral, the round seeds move at much faster speeds and are thrown by centrifugal force off the top spiral into a second spiral below, where they are discharged separately. The spiral separator is very useful for removing nonround contamination from vetch seed. It is also effective in separating vetch from small grain seeds.

Color Separators

Color separators make it possible to separate seeds that cannot be separated by any other method. This machine requires only that contamination, or undesirable seed, be slightly different in color than that of the good crop seed. This machine should be used only after the seed is cleaned and graded by other processes. If properly used, it can be valuable in upgrading germination and overall seed quality by eliminating off-color, poor-quality crop seeds. For example, it can separate weathered, deteriorated, and off-color diseased seeds from high-quality seed of navy beans. The use of color separators has completely eliminated long hours of tedious and comparatively inefficient hand-sorting belts previously used in the bean industry.

Color separators contain a photoelectric cell that changes its electric characteristics in relation to the amount of light, or radiant energy, it receives. Phototubes can be built that are sensitive to a particular color of light by varying the construction materials, and the light reaching them can be controlled by placing various filters in front of them. Light is then reflected into the cell from a variable background of known color. It is through this beam of light that the seed is dropped by gravity (one at a time). When an off-color seed appears between the preset background and the photocell, a rejection system is triggered which ejects the seed into a discharge area apart from the desirable seed.

Electrostatic Separators

Electrostatic separators separate seed by using differences in their natural or induced electrical properties. Their effectiveness depends on the natural charges of various seeds in the mixture and their relative ability to accept and retain an induced charge.

Seed is passed over a positively charged metal-embedded belt a few inches away from a negatively charged electrode. Those seeds with positive electric charges are attracted toward the electrode, lifted from the metal drum, and fall outside the normal path of seed discharge. Negatively charged seeds are repelled from the electrode and are pinned to the belt and carried underneath the belt where they are brushed off into a separate discharge spout. A third separation effect is achieved by the difference in the ability of seeds to accept and retain an electric charge. Those that accept a negative charge from the electrode (conductors) are attracted to the belt, where they adhere until the charge is lost, causing them to fall at varying distances past the location of normal seed discharge.

Timothy Bumper Mill

The bumper mill is a specialized machine used only for cleaning timothy seed. It utilizes a knocking action to remove weed seeds from timothy and separates on the basis of differences in shape, surface texture, and weight of seeds. The seed to be separated is fed onto a series of slightly inclined, identical, superimposed decks. When the decks are sharply bumped by a knocking action, the timothy seeds tend to roll back a small distance as they move uphill, while the contaminating seeds and inert matter are moved further uphill. By the time the seeds move from the feeder to the discharge area, the timothy seeds are separated far enough from the contamination so that they fall into a separate discharge spout.

Vibrator Separator

This separator has a coated vibrating deck onto which the seed is fed. It separates seeds on the basis of their different reaction to the vibrating deck and materials that coat it. The interaction of some seeds with the deck material creates friction causing the seed to be "walked" upward before discharge at the upper edge of the deck, while others slide or roll downward and are discharged into a separate spout.

Single-vibrator separator decks are sometimes used to separate small samples in seed-testing laboratories. In order to condition commercial seed lots, multiple decks provide greater capacity. The vibrator separator removes curly dock from crimson clover and dog fennel from timothy. As new deck materials are used, it has the potential for many other separations.

Affinity for Liquids

One process uses affinity for liquids to effect a separation. It is used to separate seeds of buckhorn plantain from crop seeds of similar size, shape, and density. Buckhorn seeds are covered with a mucilaginous layer that becomes sticky when moistened. Seed producers mix the seed with water followed by sawdust, which readily adheres to the buckhorn seed coat, increasing its size, and changing its specific gravity, allowing it to be separated by either the air screen machine or gravity separator. Later the sawdust is dried and reused.

SEED TREATMENT

Seed treatment is the process of applying chemical substances to seeds in order to reduce, control, or repel seedborne, soilborne, or airborne organisms. Chemical treatment of seeds is accepted as a sound agronomic practice for seeds of many field and garden crops and is usually a routine part of seed conditioning. In the past 20 years, the treatment of seeds with protective chemicals prior to planting has become a standard and widely accepted practice.

History of Seed Treatment

Although seed treatment has only recently become an important part of seed conditioning, it is by no means a new concept. According to the 1966 DuPont *Seed Treating* manual, a knowledge of the benefits of treating seed dates back to the 17th century, when saltwater was accidentally discovered to reduce bunt and stinking smut infestation of wheat seed. Later, in 1755, Matthieu du Tillet, a French botanist, recommended the use of lye and lime for chemical treatment of wheat seed. Fifty years later, Prevost, a Swiss botanist, introduced the use of copper sulfate seed treatments. However, a new concept of treating seeds began in the 1920s with the introduction of Ceresan and Semesan, the first of the organic mercurial compounds. In recent years, the increasing recognition of the environmental implications of organic mercurials has led to their disrepute and spurred the development of effective alternatives to mercurials for seed treatment.

The Ideal Seed Treatment Chemical

The ideal seed-treatment chemical should be: (1) highly effective against pathogenic organisms, (2) relatively nontoxic to plants, (3) harmless to humans and livestock, even if misused, (4) stable for relatively longer periods of time during seed storage, (5) easy to use, and (6) economically competitive. Unfortunately, none of the chemicals currently available meets all these requirements.

Treatment Formulations and Equipment

The types of pathogenic organisms controlled by seed treatment include: (1) fungi and bacteria—causing seed rots, seedling blights, and smuts. These pests may be seed-borne, soilborne, or airborne; (2) soil insects—such as seed corn maggot and wireworm, (3) storage insects—including weevils, moths, and beetles.

The activity of seed treatment chemicals falls into three principal categories: (1) seed surface disinfestation—chemicals in this group cover the seed and kill or suppress the activity of spores, and other disease agents on the seed surface; (2) seed protection—chemicals in this group protect seed before and during germination from soilborne diseases and insects; and (3) systemic protection—chemicals in this group penetrate the seed and kill or suppress the activity of pests or pathogens. They may also extend into and protect the resulting plant.

Seed treatment chemicals are normally combined with other materials that enhance or maintain their activity. Many formulations contain several inert ingredients in addition to the active ingredients. Inert ingredients act as carriers, binders, wetting agents, sticking agents, emulsifiers, suspending agents, and dyes. These materials do not have to be listed on the label since they are added to the formulation to improve appearance, increase coverage and adherence, prevent dusting off, or make the formulation easily recognizable.

Seed treatment chemicals may be applied as slurry (a mixture of a wettable powder and water), a liquid, or a dust. Other formulations are used, but the slurry method is most common. Special equipment is available for applying each type of formulation. If no commercial application equipment is available, the treatment may be applied by home methods, such as in a revolving metal drum or even in a cement mixer. Special care should be taken to see that the chemical is applied uniformly to the seed.

Both state and federal agencies normally require licensing of all businesses that offer pesticide application to the public, including those that custom-treat seed for a fee. If seed is already treated and then sold, no pesticide applicator license is required. An individual in each licensed facility must normally be certified in the category of seed treatment, and licenses are not issued until one person is certified. Seed treaters who do not offer their services to the public will not need certification unless they use restricted-use pesticides.

Identifying Treated Seed

State and federal laws require that treated seed be identified in two ways: (1) by incorporating a dye into the treatment that will give the seed a contrasting color, and (2) by a statement indicating the seed has been treated and the name of the chemical(s) used (either the common name, chemical (generic) name or abbreviated chemical name). Seed treated with highly toxic pesticides must bear a label with a skull and crossbones and a precautionary statement. The skull and crossbones must be at least twice the size as the type on the label and the precautionary statements must be in red letters on a contrasting background. Seed treated with substances not listed as highly toxic must bear a label with an appropriate precautionary statement. This information may appear on a separate tag or be printed conspicuously on the side or top of the seed container.

SEED HANDLING

Receiving, Elevating, and Conveying Equipment

Every seed conditioning plant must have adequate equipment for receiving seed and conveying it throughout the plant. A well-equipped plant has a pit area where incoming trucks can be unloaded quickly. From the pit area, conveying facilities must be available to move the seed throughout the plant vertically, horizontally, or on an inclined plane as needed. Conveyors can be classified as: (1) bucket elevators, (2) belt conveyors, (3) vibrating conveyors, (4) pneumatic (air) conveyors, (5) screw conveyors, (6) chain conveyors, or (7) lift trucks.

Bucket elevators are normally used to move seeds vertically within the conditioning plant, especially from the pit area to overhead storage bins. They consist of an endless belt or chain equipped with evenly spaced buckets. Several types are available, but the most satisfactory ones have good capacity, are relatively quiet, require little maintenance, and are self-cleaning. The space between the rear of the bucket and the belt should be checked frequently because it is particularly troublesome as a seed hang-up area. Another trouble spot that must be checked and cleaned frequently is the boot area at the lower end of the elevator leg.

Aside from the seldom-used screw and pneumatic conveyors, most other conveyors move the seed horizontally within the plant, especially between storage bins or conditioning operations. Horizontal seed conveyors are suitable for bagged or unbagged seed.

Scales

Seed is weighed at least twice in most seed conditioning operations: (1) when it is received, and (2) when it is bagged. The first weighing requires heavy-duty platform scales for weighing trucks, trailers, or wagons. The second scale needs to weigh up to 100-lb bulks for measuring the amount of seed put into bags. These are usually placed at the bottom of a bagging bin to permit a predetermined amount of seed to flow into the bag before it is stopped by a tripping mechanism, which may be either manually or automatically operated. In addition to these types of scales, most conditioning plants have small portable platform scales for weighing small amounts of seed.

Miscellaneous Equipment

All seed conditioning plants must have considerable miscellaneous equipment to aid in adequately conditioning seed and to prepare it for marketing. This includes sewing machines, heat-sealing devices, air compressors, vacuum cleaners, and tag print equipment.

DESIGN OF SEED CONDITIONING PLANTS

The design and layout of seed conditioning plants should be carefully planned to ensure that: (1) the seed receives the necessary conditioning in the proper sequence, (2) there are no bottlenecks, (3) operating costs are kept to a minimum, (4) the seeds are not injured from excessive handling, (5) facilities are completely cleanable, and (6) the chance of contamination is kept to a minimum.

If possible, the seed should be elevated only once to overhead storage bins, where it can be held until conditioning. From there it should be allowed to flow by gravity between the various pieces of conditioning equipment. In order to accomplish this, an initial elevation of about 40 to 60 ft is needed. Mistakes in design will require two or more vertical elevations and increase the opportunity for seed injury and seed lot contamination, as well as require extra maintenance.

Figure 10.6 shows a suggested design for a seed conditioning layout that can handle most crops, including small-seeded grasses and legumes. Figure 10.7 shows the sequence of conditioning operations for cleaning seed of various species.

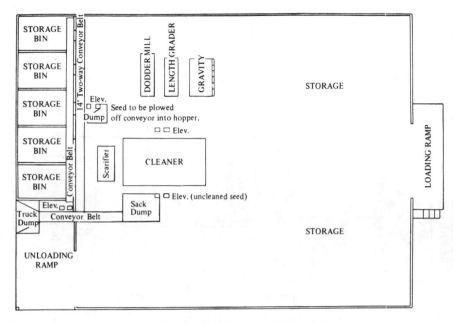

A. — Floor plan for seed conditioning plant.

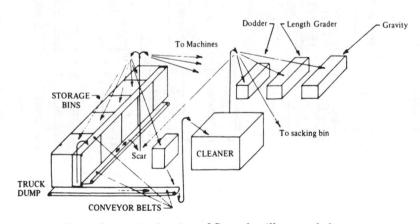

B. — Isometric drawing of flow plan illustrated above.

Figure 10.6. *A suggested layout for a single-story seed conditioning plant. (Reprinted through the courtesy of the publisher, Vaughan et al. (1967).)*

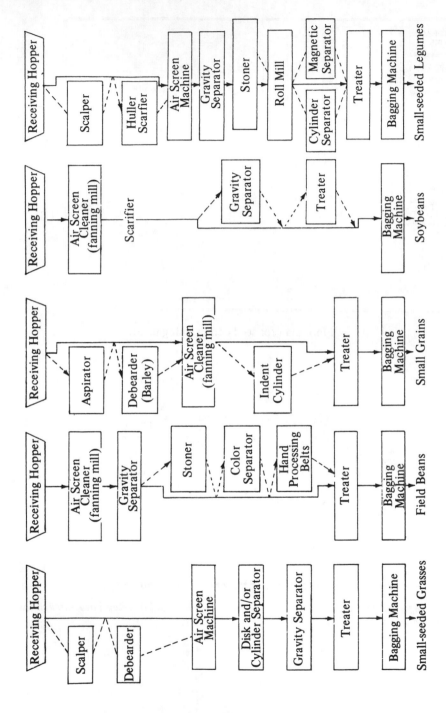

Figure 10.7. *Suggested sequence for conditioning seeds of various kinds. Discontinuous lines indicate optional processes.*

Questions

1. What is the single most important piece of equipment for cleaning seeds?
2. Can you explain the operation principle of each piece of seed cleaning equipment, as well as its advantages and disadvantages?
3. Do you believe that the proper use of seed cleaning equipment could eliminate almost any quality problem in most seed lots? Can you think of poor-quality factors that could not be eliminated by the proper selection and use of equipment?
4. What is the pure live seed concept?
5. Can proper selection and use of equipment compensate for a poor job of seed production?
6. Do you consider seed treatment to be a desirable practice? What do you consider to be the future importance of seed treatment?
7. How does the widespread use of augers for conveying seed affect its quality? Name the places where a seed lot is likely to encounter an auger throughout the production process.
8. Can you suggest principles of adjustment that might apply to almost any piece of seed cleaning equipment? Can you suggest ways that seed damage might be minimized during seed conditioning?
9. Do you believe most seed conditioners only attempt to condition to those quality levels that will meet minimum acceptance levels of the seed trade?

General References

Douglas, J. E., ed. 1980. *Successful Seed Programs: A Planning and Management Guide*. Boulder, Colo.: Westview Press.

Gregg, B. R., A. G. Law, S. S. Virdi, and J. S. Balis. 1970. *Seed Processing*. New Delhi, India: Mississippi State University, National Seeds Corporation, and United States Agency for International Development, New Delhi.

Harmond, J. E., N. R. Brandenberg, and L. M. Klein. 1968. Mechanical Seed Cleaning and Handling. *Agricultural Handbook No. 354*. Washington, D.C.: Agricultural Research Service, U.S. Department of Agriculture in cooperation with Oregon Agricultural Experiment Station.

Klein, L. M., J. Henderson, and A. D. Stoess. 1961. Equipment for cleaning seeds. In: *Seeds: The Yearbook of Agriculture*, ed. Alfred Stefferud, pp. 307–321. Washington, D.C.: U.S. Department of Agriculture.

Thompson, J. R. 1979. *An Introduction to Seed Technology*. New York: John Wiley and Sons.

Vaughan, C. E., B. R. Gregg, and J. C. Delouche, eds. 1967. *Seed Processing and Handling*. Handbook No. 1. State College, Miss.: Seed Technology Laboratory, Mississippi State University.

Wheeler, W. A., and D. D. Hill, 1957. *Grassland Seeds*. New York: D. Van Nostrand.

11
Seed Enhancements

Seeds have evolved over time to respond to a variety of environments. As a result, these adaptations generally produce satisfactory performance in a range of environments. Seed enhancement technology has a central objective to further improve seed performance under very specific regimes and with certain planting equipment. Various techniques have been employed to assure this superior performance and most have found commercial application. This chapter considers three of these seed enhancements: seed hydration, biological seed treatments, and seed coatings. While not a direct manipulation of a zygotic seed, another approach to improving seed performance that will be considered in this chapter has been the propagation of somatic embryos that are then coated and marketed as synthetic seeds.

Seed hydration is a process whereby seeds are hydrated using various protocols and then redried to permit routine handling. This process results in increased germination rate, more uniform emergence, germination under a broader range of environments, and improved seedling vigor and growth. Biological seed treatments are a new approach to adding biological organisms to seeds that effectively control soil and seed pathogens. This technique demonstrates an industry sensitivity to the increasing use of synthetic pesticides which are avoided by biological seed treatments. While seed hydration and biological seed treatments improve the physiological performance of the seed, seed coatings physically add external substances to seeds that further enhance their performance. These range from surrounding the seed with a pellet to improve precision planting to coatings that contain products which protect the seeds against an array of pests or even modify the time that water is absorbed by the seed. One of the prospects emerging from recent innovations in biotechnology and plant tissue culture has been the potential to produce synthetic seeds. These are "seeds" that are essentially embryos produced without the benefit of sexual fertilization. Such seeds are genetically uniform and offer the promise of greater economic yields and superior crop products. All of these seed enhancement technologies are constantly being improved and offer great hope for enhancing seed performance in the future.

SEED HYDRATION

The objective of seed hydration technology is to increase the percentage and rate of germination, expand the range of temperatures over which the seed will germinate, and increase the uniformity of stand establishment. To accomplish these objectives, seeds must be hydrated in some way at a moisture level sufficient to initiate the early events of germination (Phase II of imbibition) but not sufficient to permit radicle protrusion (Phase III) (Akers and Holley 1986). Three general approaches to hydration have been developed: prehydration, priming, and solid matrix priming.

Prehydration

Soaking seeds in water prior to planting enhances germination and seedling growth by controlling the imbibition conditions and reducing the vagaries of adverse weather and soil conditions (Bradford 1986). While the soaking process can involve water, seeds are often pregerminated in gels at 20°C and then planted; a process known as fluid drilling (Gray 1981). The gel both protects the germinated seed and maintains seed moisture. Examples of gels used in fluid drilling include hydroxyethyl cellulose, magnesium silicate, and polyacrylamide (Taylor and Harman 1990). In some cases, the gel may contain activated carbon to detoxify soil herbicides (Taylor and Warlrolic 1987). In other cases, it may contain added nutrients and pesticides to further improve seedling performance (Finch-Savage 1984). Prehydration can also occur followed by redrying if the seeds have not visibly germinated (Phase III) to facilitate subsequent handling, storage, and planting using traditional agricultural equipment. In such cases, the seeds still germinate under a broader range of temperatures than seeds which have not been prehydrated (Gelmond 1965; Guedes and Cantliffe 1980; Gerber and Caplan 1989). Because the amount of water absorbed by the seed is precisely controlled during prehydration to ensure that germination does not occur, the physiological mechanism that results in greater seedling performance is considered the same as that for priming or solid matrix priming (Karssen and Weges 1987).

Priming

The terms *priming* (Heydecker and Coolbear 1977) and *osmoconditioning* (Khan et al. 1978) have been used to describe the soaking of seeds in aerated low water potential osmotica. Examples of such compounds include polyethylene glycol (PEG), KNO_3, K_3PO_4, KH_2PO_4, $MgSO_4$, NaCl, glycerol, and mannitol. The benefit of such salts is to supply the seed with nitrogen and other nutrients essential for protein synthesis during germination. Their disadvantage is their occasional toxicity to the germinating seedling. Even so, given the correct concentration and time, carrot and tomato seeds perform better than untreated seeds when salts are used as the osmoticum (Haigh and Barlow 1987). Today, the most preferred osmoticum is PEG. It is a high molecular weight (from 6,000 to 8,000 daltons) inert compound whose large molecular size precludes it from entering the seed and creating toxic side effects associated with the use of salts (Michel and Kaufmann 1973). A major disadvantage of PEG is that oxygen solubility is inversely related to its concentration (Mexal et al. 1975). As a result, when PEG is used as the osmoticum, the solution is often aerated to ensure an adequate supply of oxygen to the seed (Akers 1990). In addition, it is difficult to commercially treat large quantities of seeds using this technique. Generally, seeds are soaked at 15°C (Bradford 1986) in osmotica that possess a water potential of −0.8 to −1.6 MPa (Khan 1992) for several hours (Guedes and Cantliffe 1980) to several weeks (Khan et al. 1980/1981) for optimum performance.

Priming has been successful for such crops as tomato, carrot, onion (Haigh and Barlow 1987; Alvarado et al. 1987; Dahal et al. 1990), pepper (Bradford et al. 1990), celery (Brocklehurst and Dearman 1983), parsley (Pill 1986), wheat, barley, sorghum (Bodsworth and Bewley 1981), and ryegrass (Danneberger et al. 1992). In other cases, priming has been able to overcome thermodormancy in crops such as lettuce by expanding the range of temperatures at which the seed will germinate (Cantliffe et al. 1984; Valdes and Bradford 1987). Based on these studies, priming is consistently successful with small-seeded crops. The technique, however, has been less successful with large-seeded crops such as soybean (Helsel et al. 1986) and sweet corn (Bennett and Waters 1987). Armstrong and McDonald (1992) showed that priming of soybean seeds without an intervening air-dry treatment increased plumule and radicle length and weight. However, when these seeds were air-dried, seed performance was decreased due to excess leakage of electrolytes from cracked cotyledons.

After priming, the seeds are dried back to enable normal handling, storage, and planting. The drying treatment slightly depresses the germination advantages gained during soaking (Brocklehurst et al. 1984). In addition, use of rapid drying rates or excessive temperatures can cause seed injury, therefore, priming must be conducted under carefully controlled conditions. Primed seeds can be stored successfully for short periods without losing the benefits gained from the treatment (Thanos et al. 1989). However, long storage periods cause faster loss of vigor and viability compared to nontreated seeds (Alvarado and Bradford 1988; Argerich et al. 1989).

Solid Matrix Priming

Another approach to controlled seed hydration is the use of solid carriers with low matric potentials (Kubik et al. 1988; Taylor et al. 1988); a process called *solid matrix priming* or *matriconditioning*. Some of the ideal characteristics of the carrier are: (1) a low matric potential, (2) negligible water solubility, (3) high water holding capacity, (4) a high surface area, (5) nontoxicity to the seed, and (6) the ability to adhere to the seed surface (Khan 1992). Natural substances with these characteristics are vermiculite, peat moss, and sand. Commercially available substances include Celite and Micro-Cel, which are diatomaceous silica products, and Zonolite, which is a vermiculite (Khan 1992). Other substances include a Leonardite shale and bituminous soft coal that also have an appreciable osmotic effect (Taylor et al. 1988) and a calcined clay (Kubik et al. 1988).

To accomplish solid matrix priming, seeds are generally placed into the carrier at matric potentials ranging from −0.4 to −1.5 MPa at 15°C for 7 to 14 days and brought to moisture equilibrium. After that period, the solid material is sieved away from the seeds. The technique has proven successful in enhancing seed performance of a number of crops (Table 11.1). The technique has also been applied to large-seeded crops such as sweet corn (Bennett and Waters 1987; Harman et al. 1989).

Physiological Mechanisms Responsible for Priming

Seed priming reduces the imbibitional damage associated with planting seeds in cold soils (Bennett and Waters 1987). It also results in less secondary dormancy caused

Table 11.1. The Effect of Matriconditioning in Micro-Cel E on the Performance of Vegetable Seeds at 20°/10°C (Khan 1992)

Seed cv.	Treatment[1]	Total emergence (%)	$T_{50}(d)$[2]	Top fresh wt.[3]
Red beet	Matriconditioned	155a[4]	3.8c	1.29a (13)
Cardinal	Matriconditioned + dried	156a	3.9c	1.25a (13)
	−1.2 MPa PEG 8000	140b	5.5b	1.03b (13)
	Untreated	131b	7.5a	0.81c (13)
Sugar beet	Matriconditioned	88ab	2.3c	1.46a (13)
E-4	Matriconditioned + dried	95a	3.4b	1.38a (13)
	−1.2 MPa PEG 8000	88ab	3.6b	0.94c (13)
	Untreated	82b	4.9a	0.72d (13)
Onion	Matriconditioned	98a	3.9c	0.62a (15)
Texas early grano	Matriconditioned + dried	97a	4.0c	0.61a (15)
	−1.2 MPa PEG 8000	92b	6.8b	0.48b (15)
	Untreated	93b	7.9a	0.36c (15)
Tomato	Matriconditioned	95a	4.3d	1.32a (15)
FM jackpot	Matriconditioned + dried	94a	5.3c	1.12b (15)
	−1.2 MPa PEG 8000	88a	8.2b	0.80c (15)
	Untreated	89a	11.8a	0.69d (15)
Pepper	Matriconditioned	96a	7.4b	2.26a (21)
Rino	Untreated	82b	14.1a	1.18b (21)
Carrot	Matriconditioned	88a	5.0b	0.58a (18)
Nantes	Matriconditioned + dried	74b	8.5a	0.42b (18)
	−1.2 MPa PEG 8000	78b	8.4a	0.36b (18)
	Untreated	89a	9.3a	0.31b (18)
Celery	Matriconditioned	92a	6.9c	0.19a (17)
FM 1218	Matriconditioned + dried	68c	9.4b	0.10b (17)
	−1.2 MPa PEG 8000	86a	9.3b	0.06bc (17)
	Untreated	78b	13.8a	0.03c (17)

[1]Matriconditioning was conducted at 15°C in light in a mixture of seed:carrier:water as shown in Table 4.4. After matriconditioning seeds were planted with or without air-drying in Cornell Peat-Lite mix. Seeds conditioned in PEG 8000 were washed and wipe-dried before planting. Emergence recorded at 12-hour day, 20°/10°C, night-temperature regime.
[2]Time to 50% emergence.
[3]Fresh weight of 15 tops. Data in parentheses are days after planting.
[4]Means in columns for each seed type separated by DMRT at 5% level.

by planting seeds such as lettuce in excessively warm soils (Valdes and Bradford 1987). Seeds which have been primed also leak less metabolites (Styer and Cantliffe 1983). Morphological changes have been observed in seeds following priming. For example, a portion of the lettuce seed endosperm is hydrolyzed during priming that permits faster embryo growth (Dahal et al. 1990). Increases in cell wall elasticity have also been noted (Karssen et al. 1989). It is also believed that hydration of the seed to Phase II during priming permits early DNA replication (Bray et al. 1989), increased RNA and protein synthesis (Fu et al. 1988; Ibrahim et al. 1983), and more ATP availability (Mazor et al. 1984). In addition, studies have suggested that an increase in seed vigor may also occur following priming. These have shown that repair of deteriorated seed parts occurs during the hydration phase of the process (Karssen et al. 1989; Saha et al. 1990). Importantly, these changes are irreversible and are tolerant of the subsequent desiccation that follows priming.

Factors Affecting Seed Priming

The factors which affect seed priming include (1) ambient conditions during hydration (temperature, light), (2) type(s) of osmotica, (3) oxygen availability, (4) duration of the treatment, (5) control of microbial contamination, and (6) drying.

Generally, lower temperatures require a decreased PEG concentration to obtain the desired water potential as at a higher temperature, which leads to increased oxygen concentration. Thus, priming is usually more successful when conducted at a lower temperature. Light may be necessary in some priming treatments if the species requires light for germination. The type of osmoticum employed also affects the degree of priming success. For example, the use of a solid matrix such as vermiculite or a finely ground shale allows more air to get to the seed than when an osmoticum is used. This is because the solid matrix carrier is not completely saturated to achieve the desired water potentials and the seed is therefore in an aerobic environment. Compounds used in priming also may change water potentials since the seeds are actively absorbing the water from the solution which lowers the osmotic water potential further. This problem is of greater concern when osmoconditioning larger seeds or where there is a high seed-to-solution ratio. Evaporation of the water from the osmoticum should also be considered, particularly when germination temperatures are high. Controlling microbial contamination during priming is also an important quality control step. In the case of solid matrix priming, seeds are often surface sterilized or a slurry fungicide treatment added. When priming, fungicides can be added directly to the osmoticum to prevent microbial growth. Finally, the method of seed drying can influence the degree of priming. Drying can be performed by ambient air, forced air, or vacuum drying. Ambient air drying occurs by placing the primed seed on a flat uniform surface and allowing it to dry slowly. This provides a uniform drying rate which, in some instances, may be considered too slow. In addition, this approach requires considerable space. Forced air drying is another option which reduces the requirement for space and is more rapid. However, careful attention must be given to ensure that mechanical seed damage does not occur as the seed dries (Armstrong and McDonald 1992). Vacuum drying provides the most uniform drying environment for primed seed but its cost makes this option less attractive.

BIOLOGICAL SEED TREATMENTS

Biological seed treatments are those treatments that use fungi or bacteria to control soil and seed pathogens instead of a synthetic chemical seed treatment. These are gaining increasing popularity because of safety concerns for humans and the environment as well as phytotoxicity problems associated with excess use of pesticides. In addition, biological seed treatments offer the potential for protecting the plant throughout its entire life cycle rather than just during the seed/seedling stage.

In contrast to most synthetic pesticides which kill a wide range of microorganisms, biological seed treatments typically have a narrow range of specificity, often killing just one specific pest. As a result, they disrupt ecosystems to a smaller degree than broad-spectrum pesticides (Cook and Baker 1983). This trend, however, has been reversed in recent years. Biologicals such as *Trichoderma* can control a number of pathogens such as *Pythium, Rhizoctonia,* and *Fusarium* species. Some biological seed treatments are able to colonize plant roots and contribute to long-term pest control effectiveness (Harman et al. 1989) as long as the soil pathogen does not affect the population of the biological seed treatment on the roots (Mazzola and Cook 1991). Biological seed treatments have been reported to induce growth increases that are persistent and ultimately lead to increased yield enhancements as a consequence of long-term disease control (Ahmad and Baker 1987; Schippers et al. 1987).

Biological seed treatments markedly enhance plant performance. Fungi such as *Trichoderma harzianum* (Smith and Wehner 1987), *Pythium oligandrum* (Martin and Hancock 1987), and *Chaetomium globosum* (Walther and Gingrat 1988) control damping off and other diseases. Bacteria such as *Rhizobia* species have long been applied to seed coatings to enhance root nodulation and nitrogen fixation. These beneficial bacteria (Jawson et al. 1989), and others such as *Bacillus subtilis* (Tschen 1987) and pseudomonads (Digat 1989), are being examined for their incorporation into seeds using seed enhancement technologies. Not all biological seed treatments, however, have proven successful under a wide range of soil conditions. Certain *Trichoderma* species are ineffective in lower-iron soils, presumably due to the more effective competition for iron from pseudomonads (Hubbard et al. 1983). In other cases, the biological seed treatments are not persistent in the soil, seed, or the plant under natural conditions. For example, a genetically engineered biocontrol *Pseudomonas fluorescens* was added to corn seeds that were planted in nutrient solution and in the field. The biological control was found 43 days after planting in the seed remnant and plant roots of nutrient-grown plants but decreased over time for those planted in soil (Fisher et al. 1993).

The application of biocontrol agents to seeds is influenced by the efficacy of the biocontrol, the number of propagules added to the seed, the application of the biocontrol, and control of other microbes in the application process. Biological seed treatments that are effective in protecting a germinating seed will not require as high a population on the seed as one which is less effective. The number of propagules per seed must be determined to ensure adequate control of the pest and this number often must be adjusted since repeated subculturing of biocontrol agents results in diminishing effectiveness (Callan et al. 1990). Other amendments that enhance biological seed treatments include food bases for the microorganisms, pH of the seed coating, and the rapidity with which the microorganism establishes itself.

The handling and application of biological seed treatments, in contrast to traditional seed treating equipment may require sterile conditions to prevent contamination of pathogenic organisms with the beneficial microbes. Equipment which has been used to apply several biological control agents should be sterilized to avoid contaminating the desired microbe with less efficacious ones. If conventional seed treatment equipment is used in the application of biocontrol microbes, it should be thoroughly cleaned since remaining fungicide residues may be harmful to the biocontrol agent. Conventional seed treatment equipment may also create physical damage to the biological organism during application to the seed that make it less active. The application of biological seed treatments to seeds to achieve a uniform coating may present a problem of seed agglutination (clumping) during drying. Capper and Higgins (1993) developed microgranule formulations of *Pseudomonas fluorescens* that were applied in furrow at planting instead of direct application to the seed. This same concept was applied to a fungal biological seed treatment, *Trichoderma harzianum,* where granular formulations were made from hyphal segments, chlamydospores, and conidia mixed with sodium alginate and polyethylene glycol (Dandurand and Knudsen 1993). Finally, it is important to surface sterilize seeds with sodium hypochlorite and a few drops of an adjuvant to ensure complete seed coat coverage prior to application of the biological seed treatment. This rids the seed coat of any other pathogens that may compete with the biological control agent.

SEED COATINGS

One of the most useful areas of seed enhancements is seed coatings. Seed placement and performance can be greatly enhanced by altering the shape of the seed or placing chemicals on the seed coat which regulate and improve germination. Two types of seed coatings are in commercial use: seed pelleting and seed coating (Figure 11.1).

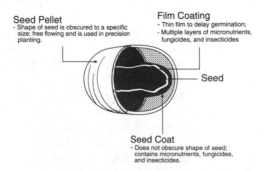

Figure 11.1. Diagrammatic illustration of various types of seed coatings.

Seed Pelleting

A seed pellet is applied to a seed to improve its plantability and performance. Many seeds, particularly vegetable seeds, are not uniformly round, which hinders precision planting for optimum crop yields. In other cases, seeds are so small and light that their accurate placement in/on the soil is uncertain (Smith and Miller 1987). To facilitate the free flow of these seeds in planters, many seed companies provide seeds with coatings of materials that change the shape and size of the seed so that it becomes heavier and rounder. A seed pellet is characterized by its ability to totally obscure the shape of the encased seed.

Seeds are introduced into a coating drum or pan that resembles a cement mixer. An amalgam of fillers (clays, limestone, calcium carbonate, talc, vermiculite) and cementing additives (gum arabic, gelatin, methylcellulose, polyvinyl alcohol, polyoxylethylene glycol-based waxes) are used to form the pellet and other compounds such as innoculants, fungicides, etc. may be added to enhance seed performance (Taylor and Harman 1990). A unique application of pellets is with the addition of calcium oxide and peroxides to the pelleting materials. These are believed to release oxygen to the seed under flooded field conditions thus minimizing anaerobic damage (Ollerenshaw 1985; Langan et al. 1986). An example of pelleting formulations for clover seeds is provided in Table 11.2. As the drum rotates, the seeds are first sprayed with water followed by the addition of the filler with binder. The wet seed attracts and becomes coated with the dry filler and the pellet gradually increases in size with each turn of the coating drum. Longer rotation times with greater amounts of filler lead to greater pellet size and roundness. At the end of the pelleting process, a binder is added to harden the outer layer of the pellet. This also reduces the amount of dust produced during handling, shipping, and planting. After pelleting is complete, the pelleted seeds are dried and handled in the same way as unpelleted seeds.

Certain technical problems must be monitored during the pelleting process. The pelleting material must be compatible with the seed so that seed quality is maintained

Table 11.2. Composition of Pellets, Containing Clover Seed, Used in This Investigation

Component	Formulation								
	1	2	3	4	5	6	7	8	9
Kaolin clay	59.0	57.5	57.5	57.5	75.0	51.0	58.5	52.0	57.5
Seed	2.0	2.0	2.0	2.0	2.0	2.0	2.0	2.0	2.0
Inoculum	1.0	1.0	1.0	1.0	1.0	1.0	1.0	1.0	1.0
PVA 205S		1.0					0.5	1.0	
PVA 523S			1.0						1.0
PVA 540S				1.0					
Sodium silicate					0.8				
Calcium sulfate						8.0			
DAP[1]								15.0	SOP[2]
Water	38.0	38.5	38.0	38.0	21.2	38.0	38.0	39.0	38.0

[1]DAP: diammonium phosphate.
[2]SOP: sprayed dried pellet with saturated aqueous solution of DAP (100 ml solution/200 g pellets).
From Smith and Miller (1987).

and germination is not hindered. For example, pelleting materials are "wet" during pelleting so that inadvertent seed hydration occurs which leads to increased respiration and possibly reduced seed quality. In other instances, if the pelleting material is too hard after drying, it may be difficult for radicle emergence through the pelleted material. While pellet strength is an important asset which ensures that the pellet remains intact during routine handling, if the pellet is too strong, germinability will be reduced. Successful pelleting also requires that a pellet form around an individual seed. If this does not occur, precision planting success is compromised. While this objective is often achieved, this is not always the case. For instance, some pellets may form around multiple seeds. In other cases, particularly when the seed lot is not thoroughly cleaned, pellets may form around inert matter such as sand or undesired plant material. The size of the pellet can also be an issue to seed purchasers. Therefore, pellets are routinely sized during and after the pelleting process. Seed analysts also have a greater responsibility when conducting a purity test with pelleted seeds. They must ensure that each pellet contains a seed and, furthermore, they must also identify the purity of the pelleted seed lot which will contain a high percentage of inert matter contributed by the filler present in the pellet. Table 11.3 illustrates some of the variability in seed performance and pellet attributes following seed pelleting using differing formulations.

Despite these technical problems, pelleting of seed is recognized as an important addition to the precision planting in many vegetables such as onion, lettuce, carrot, and various flowers. The addition of other beneficial chemicals such as plant hormones,

Table 11.3. Percentage Germination of Ladino Clover Seed Pellets at 21 Days After Planting, Pellet Strength, and Pellet Destructability. Pellets Were Planted Immediately After Drying (Planting 1) and After Six Months Storage at Room Temperature (Planting 2).

Formulation	Germination Planting 1	2	Seedlings pellet^{-1} Planting 1	2	Pellet strength Longitudinal	Diametrical	Pellet destructability[1]
	——(%)——		——(no.)——		————(g)————		——(%)——
1	85	86	1.6	2.3	270	293	3.3
2	82	89	2.0	2.5	3573	2363	0.8
3	89	91	1.9	2.6	6882	7061	0.4
4	82	80	1.9	2.1	6772	6200	0.4
5	74[2]	62	2.0	1.8	795	776	1.2
6	58	52	1.7	1.5	362	442	1.2
7	82	89	1.8	2.2	2520	2974	1.0
8	1	1	0.1	0.1	1846	1515	1.8
9	48	16	1.4	0.7	4879	4988	0.8
LSD	10	10	0.3	0.5	751	825	0.8

[1]Pellet destructability was determined by weight lost when tumbled for 1 hour.
[2]Means for plantings 1 and 2 are significantly different (P = .05) for each formulation as determined by student's T test.
From Smith and Miller (1987).

micronutrients, microbes, and fungicides to the pellet further improves seed performance in the field.

Seed Coating

Seed coating is one of the most economical approaches to improving seed performance. A seed coating is a substance that is applied to the seed but does not obscure its shape. Often, the purpose of the coating is to apply substances such as fungicides, insecticides, safeners, micronutrients, and other compounds directly to the seed. This enables the seed company to precisely tailor the seed to avoid specific stresses anticipated in certain planting environments. The ideal traits of a seed coating polymer are that it should (Rushing 1988) (1) be a water-based polymer, (2) have a low viscosity range, (3) have a high concentration of solids, (4) have an adjustable hydrophilic-hydrophobic balance, and (5) form a hard film upon drying. These traits should lead to excellent plantability, contain no "dust off" of additives, and provide for excellent germination under all environmental conditions.

The application of seed coatings is very similar to slurry seed treatments and similar equipment is used. One of the major benefits of such seed enhancements is that they are placed directly on the seed and in the immediate vicinity of the germinating seedling. This means that less chemical is required compared to broadcast or furrow applications with far less cost, while avoiding environmental damage from excess pesticide use.

Another approach to seed coatings is film coating of seeds. This is the application of additives dissolved in a dyed solution of a "sticky" polymer. The addition of the coating is very different from pelleting since it only represents an increase of 1–10% of the seed weight and the shape of the seed is still retained. The seeds are dipped or sprayed with the dissolved polymer and then immediately dried. An interesting advantage of this approach is that the minimal increase in seed weight allows the formulation to be changed several times during the spraying and drying process so that the seeds can contain a multilayer film coat. Other advantages of film coatings are the absence of "dusting off" problems and improvement of seed flow in planting equipment compared to other coating processes.

Many of the seed coatings described to this point have emphasized the use of chemical or biological controls to improve seed performance. A new area of seed coating research has emphasized the application of water-impermeable plastic film coatings to delay germination until a specified time (Figure 11.1). This approach permits planting of the seed in the fall and degradation of the seed coating during the winter, with resultant germination in the spring. This avoids problems often encountered in wet fields in the spring which consequently delay planting and reduce yields. Another application of water-impermeable plastic seed coatings is the delay in germination of incompatible male or female lines in hybrid seed production so that synchronous flowering or nicking is achieved. While the potential for such seed coatings is exciting, the mechanisms which control seed coat degradation are still difficult to precisely predict. Further research in these areas is needed.

Integration of Seed Priming and Other Seed Enhancements

Seed priming treatments are often integrated with other seed treatments to enhance performance. For example, pregerminating primed seeds further reduces emergence time and increases shoot weight (Pill 1986). Seed coating and pelleting technologies have been developed to improve plantability of flat seeds and also to permit the addition of bioactive chemicals, nutrients, and beneficial microbes (Halmer 1988). Pelleting of primed seeds has been shown to produce greater performance than either the pelleting or priming treatment alone (Valdes et al. 1985; Bennett 1988). Another integrative approach is to include growth regulators and/or pesticides in the osmoticum or solid matrix carrier soak water. The addition of GA_3 to the PEG solution has reduced the time to germination, improved the rate of germination, and prevented the induction of secondary dormancy in lettuce (Khan 1992). Fungicides (Giammichele and Pill 1984; Osburn and Schroth 1989) and plant nutrients (Finch-Savage and Cox 1982) also have been added to seeds during priming.

SYNTHETIC SEEDS

In some cases, the sexual reproduction of seeds is undesirable because it assures that seeds are not alike genetically following meiotic recombination. In fact, many crops yield best when their seed is produced by hybridization of male and female lines. This hybrid production is expensive, particularly when conducted manually, and yields of hybrid seed are often low and hybridization success not always assured. In addition, many crops are vegetatively propagated when planting of seed might be preferred. For these reasons, the clonal or vegetative production of synthetic seeds has been explored and is a rapidly expanding area of activity. A review of progress on this subject has been presented by Redenbaugh (1993).

Synthetic seeds may have value for a number of traditional crops (Table 11.4). For example, synthetic cultivars of alfalfa and orchardgrass are commonly used in which seeds are nonuniform and each plant is a distinct genotype. Synthetic seeds would allow the incorporation of specific new genes into single outstanding hybrids that could then be produced asexually by somatic embryos. Synthetic seed production might also benefit the planting efficiency of crops that are vegetatively propagated such as fruit and nut trees because of self-incompatibility and long breeding cycles. Similarly, forest conifers such as pine and spruce trees, which are currently planted only by seed, would benefit from synthetic seed production because the long conifer life cycle before the trees are capable of bearing seeds would be avoided. In certain self-pollinated crops such as cotton and soybean, the production of hybrid seeds has been hindered because of the closed flowers at the time of pollination and the requirement for hand pollination. Somatic embryogenesis would circumvent this labor-intensive process. Other crops, such as tomato and seedless watermelon, demand a high value for hybrid seeds, which makes the production of synthetic seeds for these particular markets more competitive (Redenbaugh et al. 1991).

Table 11.4. Potential Applications of Synthetic Seed Technology (SST) for Selected Crop Species.

Crop	Somatic Embryo Quality[a]	Relative Seed Cost[b]	Application[c]	Relative Need for SST[d]
Alfalfa	h	l	s	m
Corn	p	m	i	m
Cotton	p	m	h	m
Grape	h	na	s,g	m
Loblolly pine	p	h	c	h
Norway spruce	h	–	c	h
Orchardgrass	h	l	s	m
Soybean	p	m	h	m
Hybrid Tomato	n	v	d	h
Seedless Watermelon	n	v	d	h

[a]Relative somatic embryo quality: h—highly developed embryos; p—poorly developed embryos; n—somatic embryos not obtained.

[b]Relative cost of seed: v—seed cost limits planting; h—seed is costly; m—moderate; l—relatively inexpensive; na—seed is not used.

[c]Application for synthetic seed: c—circumvent long breeding cycles; d—decrease hybrid seed cost; g—germplasm conservation; h—mass production of hybrids; i—eliminate need for inbreds; s—circumvent self-incompatibility.

[d]Relative need: h—highly useful if implemented; m—existing methods are effective but implementation should yield improvements.

Principle

Synthetic seeds rely on the production of somatic embryos instead of zygotic embryos formed from sexual fertilization of male and female parents (Figure 11.2). Somatic embryos are produced asexually from vegetative tissue that has been given specific nutrient and environmental regimes causing the undifferentiated vegetative cells to form into the parts of an embryo. As a result, somatic embryos are clones that possess identical genotypes. Both somatic and zygotic embryos show typical embryonic development characteristic of monocots (Gray and Conger 1985) and dicots (Becwar et al. 1989). However, they differ because multiple somatic embryos are produced from a single callus while a single zygotic embryo is produced from each fertilization.

Challenges

Multiple somatic embryos are often found on a single callus, in which multiple stages of embryo development are also observed. This causes the nonuniform embryos to be subjected to changing nutrient conditions since the nutrients are depleted by the developing tissues and then replenished. Consequently, many somatic embryos have organs developing at differing rates which contribute to asynchronous embryo development (Conger et al. 1989). In some cases, this leads to precocious germination, while in others the prevailing nutrient environment may be conducive to shoot or root devel-

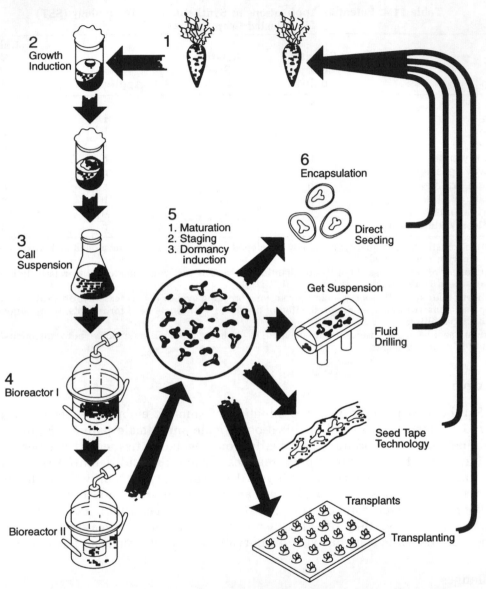

Figure 11.2. *Schemes for vegetative propagation via somatic embryogenesis were published as one of the frontpieces for Volume 2 of the* Handbook of Plant Cell Culture, *and includes the growth of embryos for transplants, fluid drilling, encapsulation, and seed tape technology. It is not clear if these projected systems were researched. (From Sharp, W. R., Evans, A., Ammirato, P. V., and Yamada, Y.,* Handbook of Plant Cell Culture, Vol. 2, Crop Species, *Macmillan, New York, 1984. With permission.)*

opment but not both. In still other cases, somatic embryos often develop extra cotyledons or poorly developed apical meristems (Ammirato 1987). This asynchronous embryo development makes harvest of uniformly mature somatic embryos difficult in synthetic seed production. Attempts to better synchronize somatic embryo development have included physical separation of proembryonal cultures to assure uniform callus size and physiological synchronization by adding abscisic acid to the culture medium. Abscisic acid appears to cause cell water (turgor) content to decrease, thereby slowing embryo growth which inhibits germination of embryos that would tend to germinate precociously (Gray 1989).

Somatic embryos also do not typically exhibit a quiet resting or dormancy (quiescence) phase characteristic of orthodox seeds (Gray 1987). Usually, somatic embryos continue to grow into seedlings or they revert back into disorganized callus tissue. This inability to produce a resting phase where all embryos are at the same arrested physiological and morphological state also is a challenge to synthetic seed development. Without this arrested growth stage, synthetic seeds cannot be successfully stored or treated using traditional seed technology practices. Attempts to introduce dormancy into somatic embryos are being explored using orthodox seeds as a developmental model. Since orthodox seeds express dormancy after the seeds begin to dehydrate on the parent plant, studies have been initiated to determine whether dry down of somatic embryos might induce this important trait and there is evidence of success (McKersie et al. 1988; Carman 1988; Seneratna et al. 1989).

Encapsulations

Assuming that satisfactory techniques are developed for inducing dormancy into somatic embryos, the dry, delicate embryos would benefit from a seed coating to protect against injury incurred from traditional seed handling and planting. The protective encapsulations can be designed to provide physical protection as well as essential nutrients, antibiotics, and fungicides to assist the embryo during germination. Two approaches, hydrated and dry encapsulation, have been studied. The hydrated encapsulation is most effective for nonquiescent somatic embryos that can be planted directly without storage (Figure 11.3). Types of hydrated encapsulations are composed primarily of calcium alginate (Redenbaugh et al. 1987) although the success of seedling establishment in the field using this approach has been limited (Fujii et al. 1987). The preferred approach is the dry, hardened synthetic seed coating surrounding somatic embryos which would permit conventional seed storage and handling. To date, only a water-soluble plastic resin surrounding carrot embryos has been examined and this did not result in satisfactory seedling establishment (Kitto and Janick 1985).

SUMMARY

The concept of direct genetic manipulation of the embryo through somatic embryogenesis offers immense potential for the production of superior performing seeds of a number of crops. However, numerous technical hurdles associated with dormancy imposition and satisfactory encapsulation techniques to permit normal seed handling need to be better developed. Overriding these technical concerns are the additional

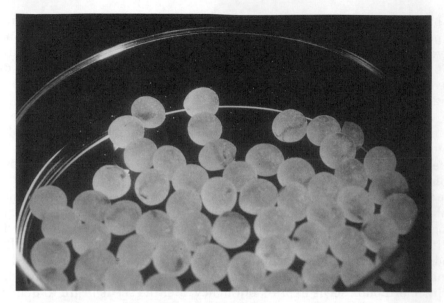

Figure 11.3. *Synthetic celery seeds coated in a calcium alginate gel. Note small somatic embryos inside gel coating. Photograph courtesy of K. Redenbaugh.*

costs incurred for synthetic seed production. It has been estimated that the cost of one alfalfa synthetic seed would be 3.3 cents (Redenbaugh et al. 1987), which is far in excess of conventional seed costs. Until the cost of synthetic seeds becomes competitive with seeds that are naturally produced, this promising technology will require further research to refine and automate synthetic seed production systems.

Questions

1. Identify the central objective of seed enhancement technology.
2. What are the practical consequences of seed hydration? What seed moisture level is necessary to obtain the benefits of seed hydration?
3. Name three general approaches to seed hydration and describe how each is accomplished commercially.
4. Describe fluid drilling.
5. What is polyethylene glycol and how is it used as a seed enhancement technology? What beneficial effects does it have on seed performance?
6. Outline six ideal characteristics of a solid matrix carrier and describe why each is important.
7. Name six factors that affect seed priming success and explain why they are important.
8. What are biological seed treatments, why is the seed industry interested in continuing their development, and how are they effective?
9. Distinguish between seed coatings, seed pelletings, and seed film coatings.
10. What are some of the commercial advantages of synthetic seed production? How are synthetic seeds produced? Is this seed enhancement technology practical from a seed industry perspective?

General References

Ahmad, J. S., and R. Baker. 1987. Competitive saprophytic ability and cellulolytic activity of rhizosphere-competent mutants of *Trichoderma harzianum*. *Canadian Journal of Microbiology* 34:229–234.

Akers, S. W. 1990. Seed response to priming in aerated solutions. *Search* 19:8–17.

Akers, S. W., and K. E. Holley. 1986. SPS: A system for priming seeds using aerated polyethylene glycol or salt solutions. *Horticultural Science* 21:529–531.

Alvarado, A. D., and K. J. Bradford. 1988. Priming and storage of tomato (*Lycopersicon lycopersicum*) seeds. I. Effects of storage temperature on germination rate and viability. *Seed Science and Technology* 16:601–612.

Alvarado, A. D., K. J. Bradford, and J. D. Hewitt. 1987. Osmotic priming of tomato seeds: Effects on germination, field emergence, seedling growth and fruit yield. *Journal of American Society Horticultural Science* 112:427–432.

Ammirato, P. V. 1987. Organizational events during somatic embryogenesis. In: *Plant Tissue and Cell Culture*. New York: Alan Liss.

Argerich, C. A., K. J. Bradford, and A. M. Tarquis. 1989. The effects of priming and aging on resistance to deterioration of tomato seeds. *Journal of Experimental Botany* 40:593–598.

Armstrong, H., and M. B. McDonald. 1992. Effects of osmoconditioning on water uptake and electrical conductivity in soybean seeds. *Seed Science and Technology* 20:391–400.

Becwar, M. R., T. L. Noland, and J. L. Wycoff. 1989. Maturation, germination, and conversion of Norway spruce (*Picea abies* L.) somatic embryos to plants. *In Vitro Cellular and Developmental Biology* 25:575–580.

Bennett, M. A. 1988. Evaluation of seed coating and priming treatments for stand establishment of processing tomatoes. *Proceedings of International Conference on Stand Establishment Horticultural Crops*. American Society of Horticultural Sciences, Lancaster, PA.

Bennett, M. A., and L. Waters. 1987. Seed hydration treatments for improved sweet corn germination and stand establishment. *Journal of American Society of Horticultural Sciences* 112:45–49.

Bodsworth, S., and J. D. Bewley. 1981. Osmotic priming of seeds of crop species with polyethylene glycol as a means of enhancing early and synchronous germination at cool temperatures. *Canadian Journal of Botany* 59:672–676.

Bradford, K. J. 1986. Manipulation of seed water relations via osmotic priming to improve germination under stress conditions. *Horticultural Science* 21:1105–1112.

Bradford, K. J., J. J. Steiner, and S. E. Trawatha. 1990. Seed priming influence on germination and emergence of pepper seed lots. *Crop Science* 30:718–721.

Bray, C. M., P. A. Davison, M. Ashraf, and R. M. Taylor. 1989. Biochemical events during osmopriming of leek seeds. *Annals of Applied Biology* 102:185–193.

Brocklehurst, P. A., and J. Dearman. 1983. Interactions between seed priming treatments and nine seed lots of carrot, celery, and onion. II. Seedling emergence and plant growth. *Annals of Applied Biology* 102:585–593.

Brocklehurst, P. A., J. Dearman, and R. K. L. Drew. 1984. Effects of osmotic priming on seed germination and seedling growth in leek. *Scientia Horticulturae* 24:201–210.

Callan, N. W., D. E. Mathre, and J. B. Miller. 1990. Biopriming seed treatment for biological control of *Pythium ultimum* premergence damping-off in sh2 sweet corn. *Plant Disease Reporter* 74:368–372.

Cantliffe, D. J., J. M. Fischer, and T. A. Nell. 1984. Mechanism of seed-priming in circumventing thermodormancy in lettuce. *Plant Physiology* 75:290–294.

Capper, A. L., and K. P. Higgins. 1993. Application of *Pseudomonas fluorescens* isolates to wheat as potential biological control agents against take-all. *Plant Pathology* 42:560–567.

Carman, J. G. 1988. Improved somatic embryogenesis in wheat by partial simulation of the in-ovulo oxygen, growth regulator and desiccation environments. *Planta* 175:417–422.

Conger, B. V., J. C. Hovanesian, R. N. Trigiano, and D. J. Gray. 1989. Somatic embryo ontogeny in suspension cultures of orchardgrass. *Crop Science* 29:448–452.

Cook, R. J., and K. F. Baker. 1983. *The Nature and Practice of Biological Control of Plant Pathogens.* St. Paul, Minn.: American Phytopathological Society.

Dahal, P., K. J. Bradford, and R. A. Jones. 1990. Effects of priming and endosperm integrity on seed germination rates of tomato genotypes. II. Germination at reduced water potential. *Journal of Experimental Botany* 41:1441–1453.

Dandurand, L. M., and G. R. Knudsen. 1993. Influence of *Pseudomonas fluorescens* on hyphal growth and biocontrol activity of *Trichoderma harzianum* in the spermosphere and rhizosphere of pea. *Phytopathology* 83:265–270.

Danneberger, T. K., M. B. McDonald, C. A. Geron, and P. Kumari. 1992. Rate of germination and seedling growth of perennial ryegrass seed following osmoconditioning. *Horticultural Science* 27:28–30.

Digat, B. 1989. Strategies for seed bacterization. *Acta Horticulturae* 253:121–130.

Finch-Savage, W. E. 1984. A comparison of seedling emergence from dry-sown and fluid-drilled carrot seeds. *Journal of Horticultural Science* 59:403–410.

Finch-Savage, W. E., and C. J. Cox. 1982. Effect of adding plant nutrients to the gel used for fluid drilling early carrots. *Journal of Agricultural Science* 99:295–303.

Fisher, P. J., S. A. Broad, C. D. Clegg, and H. M. Lappin Scott. 1993. Retention and spread of a genetically engineered pseudomonad in seeds and plants of *Zea mays* L.—A preliminary study. *New Phytology* 124:101–106.

Fu, J. R., S. H. Lu, R. Z. Chen, B. Z. Zhang, Z. S. Liu, Z. S. Li, and D. Y. Cai. 1988. Osmoconditioning of peanut (*Arachis hypogaea* L.) seeds with PEG to improve vigour and some biochemical activities. *Seed Science and Technology* 16:197–212.

Fujii, J. A., D. T. Slade, K. Redenbaugh, and K. A. Walker. 1987. Artificial seeds for plant propagation. *Trends in Biotechnology* 5:335–340.

Gelmond, H. 1965. Pretreatment of leek seeds as a means of overcoming superoptimal temperatures of germination. *Proceedings of International Seed Testing Association* 30:737–742.

Gerber, J. M., and L. A. Caplan. 1989. Priming sh2 sweet corn seed for improved emergence. *Horticultural Science* 24:854.

Gray, D. 1981. Fluid drilling of vegetable seeds. *Horticulture Review* 3:1–27.

Giammichele, L. A., and W. G. Pill. 1984. Protection of fluid-drilled tomato seedlings against damping-off by fungicide incorporation in a gel carrier. *Horticultural Science* 19:877–879.

Gray, D. J. 1987. Effects of dehydration and other environmental factors on dormancy in grape somatic embryos. *Horticultural Science* 23:786–791.

Gray, D. J. 1989. Synthetic seed for clonal production of crop plants. In: *Recent Advances in the Development and Germination of Seeds,* ed. R. B. Taylorson, pp. 29–45. New York: Plenum Press.

Gray, D. J., and B. V. Conger. 1985. Somatic embryo ontogeny in tissue cultures of orchardgrass. In: *Tissue Culture in Forestry and Agriculture,* ed. R. R. Henke, K. W. Hughes, M. J. Constantin, and A. Hollaender. New York: Plenum.

Guedes, A. C., and D. J. Cantliffe. 1980. Germination of lettuce (*Lactuca sativa*) at high temperature after seed priming. *Journal of American Society of Horticultural Science* 105:777–781.

Haigh, A. M., and E. W. R. Barlow. 1987. Germination and priming of tomato, carrot, onion, and sorghum seeds in a range of osmotica. *Journal of American Society of Horticultural Science* 112:202–208.

Halmer, P. 1988. *Technical and Commercial Aspects of Seed Pelleting and Film Coating,* pp. 191–204. Thornton Heath, U.K.: British Crop Protection Council.

Harman, G. E., A. G. Taylor, and T. E. Stasz. 1989. Combining effective strains of *Trichoderma harzianum* and solid matrix priming to improve biological seed treatments. *Plant Disease Reporter* 73:631–637.

Helsel, D. G., D. R. Helsel, and H. C. Minor. 1986. Field studies on osmoconditioning soybeans, *Glycine max*. *Field Crop Research* 14:291–298.

Heydecker, W., and P. Coolbear. 1977. Seed treatments for improved performance-survey, an attempted prognosis. *Seed Science and Technology* 5:353–425.

Hubbard, J. P., G. E. Harman, and Y. Hadar. 1983. Effect of soilborne *Pseudomonas* spp. on the biological control agent, *Trichoderma hamatum* on pea seeds. *Phytopathology* 73:655–659.

Ibrahim, A. E., E. H. Roberts, and A. H. Murdoch. 1983. Viability of lettuce seeds. II. Survival and oxygen uptake in osmotically controlled storage. *Journal of Experimental Botany* 34:631–640.

Jawson, M. D., A. J. Franzluebbers, and R. K. Berg. 1989. *Bradyrhizobium japonicum* survival in and soybean inoculation with fluid gels. *Applied Environmental Microbiology* 55:617–622.

Karssen, C. M., A. Haigh, P. van der Toorn, and R. Weges. 1989. Physiological mechanisms involved in seed priming. In: *Recent Advances in the Development and Germination of Seeds,* pp. 269–280. New York: Plenum Press.

Karssen, C. M., and R. Weges. 1987. Osmoconditioning of lettuce seeds and induction of secondary dormancy. *Acta Horticulturae* 215:165–171.

Khan, A. A. 1992. Preplant physiological seed conditioning. *Horticulture Review* 13:131–181.

Khan, A. A., N. H. Peck, and C. Sammimy. 1980/1981. Seed osmoconditioning: Physiological and biochemical changes. *Israel Journal of Botany* 29:133–144.

Khan, A. A., K. L. Tao, J. S. Knypl, B. Borkowska, and L. E. Powell. 1978. Osmotic conditioning of seeds: Physiological and biochemical changes. *Acta Horticulturae* 83:267–278.

Kitto, S. L., and J. Janick. 1985. Hardening treatments increase survival of synthetically coated asexual embryos of carrot. *Journal of American Society of Horticultural Science* 110:282–287.

Kubik, K. K., J. A. Eastin, J. D. Eastin, and K. M. Eskridge. 1988. Solid matrix priming of tomato and pepper. *Proceedings of International Conference* Stand Est. Hortic. Crops. American Society of Horticultural Sciences, Lancaster, PA, pp. 86–96.

Langan, T. D., J. W. Pendleton, and E. S. Oplinger. 1986. Peroxide coated seed emergence in water-saturated soil. *Agronomy Journal* 78:769–772.

Martin, F. N., and J. G. Hancock. 1987. The use of *Pythium oligandrum* for biological control of preemergence damping-off caused by *Pythium ultimum*. *Phytopathology* 77:1013–1020.

Mazor, L., M. Perl, and M. Negbi. 1984. Changes in some ATP-dependent activities in seeds during treatment with polyethylene glycol and during the redrying process. *Journal of Experimental Botany* 35:1119–1127.

Mazzola, M., and R. J. Cook. 1991. Effects of fungal root pathogens on the population dynamics of biocontrol strains of fluorescent pseudomonads in the wheat rhizosphere. *Applied Environmental Microbiology* 57:2171–2178.

McKersie, B. D., S. R. Bowley, T. Senaratna, D. C. W. Brown, and J. D. Bewley. 1988. Application of artificial seed technology in the production of alfalfa (*Medicago sativa* L.). *In Vitro Cellular and Developmental Biology* 24:71–76.

Mexal, J., J. T. Fisher, J. Osteryoung, and C. P. Reid. 1975. Oxygen availability in polyethylene glycol solutions and its implications in plant-water relations. *Plant Physiology* 55:20–24.

Michel, B. E., and M. R. Kaufmann. 1973. The osmotic potential of polyethylene glycol 6000. *Plant Physiology* 51:914–916.

Ollerenshaw, J. H. 1985. Influence of waterlogging on the emergence and growth of *Lolium perenne* L. shoots from seed coated with calcium peroxide. *Plant and Soil* 85:131–141.

Osburn, R. M., and M. N. Schroth. 1989. Effect of osmopriming sugar beet seed on germination rate and incidence of *Pythium ultimum* damping-off. *Plant Disease Reporter* 73:21–24.

Pill, W. G. 1986. Parsley emergence and seedling growth from raw, osmoconditioned and pregerminated seeds. *Horticultural Science* 21:1134–1136.

Redenbaugh, K., ed. 1993. *Synseeds: Synthetic Seeds to Crop Improvement.* Boca Raton, Fla.: CRC Press.

Redenbaugh, K., J. A. Fujii, and D. Slade. 1991. Synthetic seed technology. In: *Cell Culture and Somatic Cell Genetics of Plants,* Vol. 8, ed. I. K. Vasil, pp. 35–74. New York: Academic Press.

Redenbaugh, K., D. Salde, P. Viss, and J. A. Fujii. 1987. Encapsulation of somatic embryos in synthetic seed coats. In: *Proc. Symp. Synthetic Seed Technology for the Mass Cloning of Crop Plants: Problems and Perspectives. Horticultural Science* 22:803–808.

Rushing, K. W. 1988. Additives to enhance seed quality and field performance. *Proceedings of Mississippi State University, Mississippi State, MS State Seed Short Course* Vol. 30.

Saha, R., A. K. Mandal, and R. N. Basu. 1990. Physiology of seed invigoration treatments in soybean (*Glycine max* L.). *Seed Science and Technology* 18:269–276.

Schippers, B., B. Lutenberg, and P. J. Weisbeek. 1987. Plant growth control by fluorescent pseudomonads. In: *Innovative Approaches to Plant Disease Control,* ed. I. Chet, pp. 19–39. New York: John Wiley & Sons.

Seneratna, T., B. D. McKersie, and D. C. W. Brown. 1989. Artificial seeds: Desiccated somatic embryos. *In Vitro Cellular and Developmental Biology* 25:39–45.

Sharp, W. R., A. Evans, P. V. Ammirato, and Y. Yamada. 1984. *Handbook of Plant Cell Culture, Vol. 2, Crop Species.* New York: Macmillan, NY.

Smith, A. E., and R. Miller. 1987. Seed pellets for improved seed distribution of small seeded forage crops. *Journal of Seed Technology* 11:42–51.

Smith, E. M., and F. C. Wehner. 1987. Biological and chemical measures integrated with deep soil cultivation against crator disease of wheat. *Phytopathology* 19: 87–90.

Styer, R. C., and D. J. Cantliffe. 1983. Evidence for repair processes in onion seeds during storage at high seed moisture contents. *Journal of Experimental Botany* 34:277–282.

Taylor, A. G., and G. E. Harman. 1990. Concepts and technologies of selected seed treatments. *Annual Review of Phytopathology* 28:321–339.

Taylor, A. G., D. E. Klein, and T. H. Whitlow. 1988. SMP: Solid matrix priming of seeds. *Scientific Horticulture* 37:1–11.

Taylor, A. G., and D. T. Warholic. 1987. Protecting fluid drilled lettuce from herbicides by incorporating activated carbon into gels. *Journal of Horticultural Science* 62:31–37.

Thanos, C. A., K. Georghiou, and H. C. Pasam. 1989. Osmoconditioning and aging of pepper seeds during storage. *Annals of Botany* 63:65–69.

Tschen, J. S. M. 1987. Control of *Rhizoctonia solani* by *Bacillus subtilis. Transactions of the Mycological Society of Japan* 28:483–494.

Valdes, V. M., and K. J. Bradford. 1987. Effects of seed coating and osmotic priming on the germination of lettuce seeds. *Journal of American Society of Horticultural Science* 112:153–156.

Valdes, V. M., K. J. Bradford, and K. S. Mayberry. 1985. Alleviation of thermodormancy in coated lettuce *Lactuca sativa* cultivar 'Empire' by seed priming. *Horticultural Science* 20:1112–1114.

Walther, D., and D. Gingrat. 1988. Biological control of damping-off of sugar beet and cotton with *Chaetomium globosum* or a fluorescent *Pseudomonas* sp. *Canadian Journal of Microbiology* 34:631–637.

12
Seed Certification

Seed certification is a program to maintain and make available to the public high-quality seeds and propagating materials of genetically distinct crop varieties. Under this program, certified seed is produced by outstanding farmers and seed producers using careful quality control, pedigreed planting stock, field inspections during the growing season, and seed inspections following harvest. Certification is an officially recognized method for maintaining varietal identity of seed on the open market. Consequently, it has become especially important for field crops (except hybrid corn) because most varieties of field crops traditionally have been publicly released and their seed sold on the open market. It is of lesser importance for other kinds of crops, whose varieties are frequently privately released and seed production controlled by private companies.

Certification is also widely used for seed destined for international sales. In the United States alone, over 111,000 tons of seed were produced during 1991/1992 for export under the OECD certification program.

HISTORY

Seed certification in the United States and Canada dates back to the early 1900s when the first new varieties appeared from state land grant colleges and government experiment stations. Prior to this, most field crops originated from plant materials introduced from other countries. When new varieties became available, they were distributed to farmers on a haphazard, inefficient, and often inequitable basis. Frequently such varieties were contaminated, lost, or of poor physical quality.

During the period of 1900 to 1920, organizations were set up in various states through which seeds of new "college-bred" varieties were distributed to farmers. These organizations were often outgrowths of state experiment associations and soon became known as crop improvement associations, or seed certification agencies. These agencies were frequently administered by experiment station or extension service staff of the land-grant institutions where they were located.

During the 20's, 30's and 40's, under the guidance and influence of the universities, seed certification became an established institution for increasing and making available to the public high-quality seed for improved varieties—varieties that were almost without exception products of the university or government agency breeding programs.

CERTIFICATION TODAY

Seed certification in the United States is the responsibility of each individual state, and within each state there is an agency designated to certify seed. Regardless of the agency responsible, the basic authority for certification is derived from the seed law of the individual state. Several states have certification programs administered by state departments of agriculture. In a few states, certification is administered by the Cooperative Extension Service. Most states have certification programs administered by grower-controlled crop improvement associations, although university personnel frequently act as their secretary-managers. In other states, the secretary-manager is hired by the board of directors of the crop improvement associations. Regardless of how the project is organized, certification programs in the United States and Canada are generally non-profit programs, but must generate funds to cover salaries, overhead, and operating expenses. Ordinarily, even these state agencies maintain close association with university personnel, who may serve on their boards of directors.

In Canada, the seed certification program is administered by the Canadian Seed Growers Association, representing pedigreed (certified) seed growers from all Canadian provinces.

THE ASSOCIATION OF OFFICIAL SEED CERTIFYING AGENCIES

The Association of Official Seed Certifying Agencies (AOSCA) is an organization of certification agencies in the United States, Canada, and New Zealand. However, official agencies from several other countries have indicated an interest in joining. Its purposes are: (1) to establish minimum standards for genetic purity and recommend minimum standards for the classes of certified seed, (2) to standardize seed certification regulations and procedures, (3) to encourage cooperation with all individuals, agencies, groups, and organizations to accomplish these purposes, and (4) to assist its member agencies in seed promotion, production, and distribution.

The history of the AOSCA dates back to 1919, when representatives from Michigan, Minnesota, North Dakota, South Dakota, Wisconsin, and the Canadian Seed Growers Association met in St. Paul, Minnesota, to explore the possibilities of developing some type of organization that would be helpful in solving mutual problems. After considerable discussion, a Seed Improvement Federation was proposed, but the group felt that more states should be represented. Another meeting was set for December 1919, at the International Grain and Hay Show in Chicago. At that meeting, 13 states and Canada were represented, and the International Crop Improvement Association (ICIA) was formed. The objectives of the ICIA were to promote the agricultural interests of the various states as well as the provinces of Canada, emphasizing especially the improvement of field crops in general and seed improvement in particular. These objectives were to be attained by:

1. encouraging the breeding and improvement of field crops and seeds;
2. husbanding, propagating, and disseminating elite, registered, certified, and improved seeds;
3. creating a more active interest in better seeds through circulars, reports, and other publicity as well as by encouraging local, state, national and international shows;
4. assisting in the standardization of seed improvement and certification work being done by member agencies.

From the beginning, the ICIA had a major influence on certification throughout the United States and Canada. It has been instrumental in enunciating the fundamental concepts of certification, establishing field and laboratory inspection standards, and encouraging uniformity in certification procedures among its member agencies. Membership in the organization was voluntary, as were its standards and policies; however, almost all certification agencies in both the United States and Canada were members of the ICIA and were greatly influenced by it. In 1968 the name of the ICIA was changed to the Association of Official Seed Certifying Agencies.

THE GENERATION SCHEME OF CERTIFICATION

Inherent in the certification concept is a generation system whereby the pedigree of superior crop varieties is maintained through subsequent seed production. A four-generation scheme (Figure 12.1) has been devised to do this, and seed of each generation is identified by a special color labeling tag.

1. **Breeder seed** is produced under the direct supervision of the plant breeder and represents the true pedigree of the variety.
2. **Foundation seed** is the first generation seed from breeder seed and is ordinarily produced under contract by a foundation seed organization. Foundation seed is also labeled with white certification tags.
3. **Registered seed** is the seed from foundation seed and is intended for the purpose of increasing seed another generation before the production of certified seed. Registered seed is not intended to be a commercial class of seed. It is designated by purple seed tags. In two states (Michigan and Wisconsin), all certified seed is the progeny of foundation seed, and no registered class is used. Most states still maintain the registered generation. However, this class is eliminated in many cross-pollinated crops, particularly with species where seed is produced outside the area of adaptation.
4. **Certified seed** is produced from foundation or registered seed and represents the final product of the certification program. It is labeled with the familiar blue tag, which has become associated with the public image of certification.

Although the four-generation scheme for seed certification has been an integral part of certification since the very beginning, it has not always been applied in the strictest sense. It was not until the mid-1960s that the four-generation concept was adopted in practice by all certifying agencies. However, many variety developers have

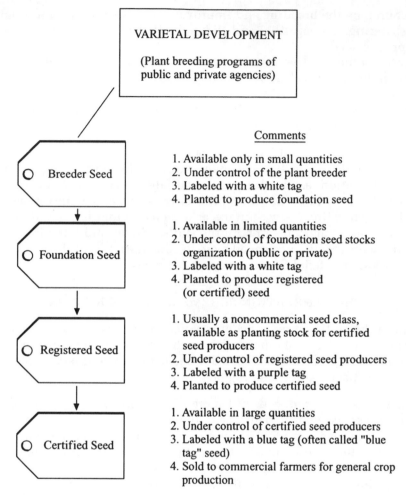

Figure 12.1. *Diagrammatic scheme of the overall limited generation seed certification program, from the development of a new variety to its availability to commercial farmers.*

elected to restrict some or all of their varieties to a three-class (Breeder, Foundation, Certified) system in order to better maintain varietal purity.

The Canadian generation system is the same as that in the United States except that there is a *select* class between the breeder and foundation classes for wheat, oats, barley, rye, flax, triticale, buckwheat, field peas, lentils, field beans, faba beans, and soybeans. Breeder seed of these crops is allocated to members of the Canadian Seed Growers Association (CSGA) who have become established as select seed growers after serving a three-year probationary plot-production program. For select status, growers may grow no more than 2.5 acres of one variety of select seed or no more than 5 acres of all crop varieties. The plots must have no more than 1 off-type plant per 20,000 crop plants. Seed from each plot is postcontrolled for varietal purity by Agriculture

Canada. Select seed may produce select seed for five multiplications before the grower is required to obtain new breeder seed. Select seed is used to produce foundation seed crops which are also postcontrolled for varietal purity. A Breeder Seed Crop Certificate is issued by the CSGA for the initial increase of breeder seed and each increase made thereafter by the originating plant breeder. Crops of all classes, including breeder, are field inspected for varietal purity by Agriculture Canada for the CSGA.

FOUNDATION SEED PRODUCTION

Foundation seed is sometimes called the "vital link" between breeder seed produced under the control of the plant breeder and certified seed produced by the certified seed grower. It is the seed stock from which registered and certified seed are produced. It is produced by a foundation seed organization, which may be a private association of seed growers, a special project within a university experiment station, or an independent private business.

Foundation seed agencies receive breeder seed of new crop varieties as they are released and increase them to foundation seed. After the initial release of breeder seed, it must be maintained and made available every year. This is done in one of several ways. First, a small portion of foundation seed fields may be designated as eligible for breeder seed production. This area is carefully inspected, rogued of off-types, and classified as breeder seed in cooperation with the experiment station or releasing agency. Thus, foundation seed of annual crops can be maintained on a permanent basis. An alternate, but less frequently used method is for the releasing institution to grow small lots of breeder seed each year for annual release to the foundation seed association. For perennial crops, maintenance of breeder seed is much simpler, since small blocks of each variety can be maintained to provide breeder seed for transfer to the foundation seed agency.

If the foundation seed organization does not have adequate production and processing facilities of its own, it may arrange for contract production, which is then made available to registered or certified seed growers after the seed has been cleaned, conditioned, and bagged to the proper standards.

Several factors should be considered when producing foundation seed. First, only the best seed growers with the right combination of experience, land, facilities, and ability are accepted as foundation seed growers. Second, the supply of foundation seed should not exceed the demand. To anticipate demand and provide for foundation seed production calls for advance planning. If excess foundation seed is produced, it must be carried over at extra expense, downgraded to certified seed, sold as commercial grain, or destroyed at a considerable loss. Third, the foundation seed should be available to all certified seed growers on an equitable basis at a reasonable price. When the supply of foundation seed is limited, consideration is given to the growers' production history, their facilities, and their ability to produce high-quality seed.

Although most foundation seed programs operate on a nonprofit basis they must be self-sustaining. Costs such as labor, overhead, buildings, and conditioning facilities must be met from the sale of foundation seed-requiring operation on sound business principles. Responsibility for this rests with the manager, who is employed by a board of directors. The board establishes overall policy and sees that the aims and function of the organization are met.

Foundation seed organizations usually have a formal agreement with the state agricultural experiment station which defines the terms of breeder seed release of new crop varieties. Exceptions may occur when foundation seed organizations are private businesses not associated with a university experiment station, or when the foundation seed project is administered by an experiment station staff.

HOW VARIETIES BECOME ELIGIBLE FOR CERTIFICATION

Varietal Release

Most agronomic varieties have been released from university experiment stations; however, private seed companies also release new crop varieties. Privately released varieties have long been available for vegetable crops, corn, sorghum, and cotton, and are now common for other field crops.

Regardless of the releasing agency, a procedure must be available for evaluating potential varieties and recommending their release. Most of our experience comes from release by university experiment stations. When plant breeders have a candidate for release, they submit to the appropriate review board a description of their variety, its identifying characteristics, and performance data. Release procedures may consist of one or two steps; however, it is quite common first to submit it for consideration by a commodity committee, comprised of persons closely concerned and familiar with the crop. In experiment stations, this usually involves the plant breeder, a plant pathologist, an entomologist, extension agronomists, and other trained personnel. After favorable action, it often goes to a larger, more formal committee composed of experiment station personnel from disciplines who are responsible for release of new varieties of all crops. The first group provides closer knowledge and involvement with specific crop areas, while the second committee observes uniformity of release procedures and provides perhaps a more objective evaluation of candidates for release. The second committee also advises on specific release procedures and other seed increase matters.

Definition of a Variety

To be eligible for certification a variety must be properly released, named, and described. In the past, the term *variety* has been simply defined as "a specific close subdivision of a kind with definite distinguishing genetic characteristics that can be maintained or inherited when the plants are reproduced over a period of years." Because of the different kinds of crops and germ plasm available for certification, it has often been unclear whether many candidates for certification actually qualified as varieties. To help clarify this situation, an ad hoc committee, representing the United States Department of Agriculture, the Association of Official Seed Certifying Agencies, the American Society of Agronomy, and the American Seed Trade Association has developed a comprehensive consensus definition of different kinds of varieties. These definitions are published and made available to all concerned organizations.

Varietal Review Boards

Individual certification agencies are aided in determining the eligibility of varieties for certification by national variety review boards which have been established by the AOSCA. Four review boards have been set up, representing alfalfa, grasses, soybeans, and small grains. Each board is composed of six members, representing the: (1) American Seed Trade Association, (2) Association of Official Seed Certifying Agencies, (3) Crop Science Society of America, (4) National Council of Commercial Plant Breeders, (5) United States Department of Agriculture, and (6) Agricultural Research Service, USDA.

The functions of the boards are to review and evaluate information presented by breeders of the respective crops in industry and in public agencies who request certification of new varieties and to advise the AOSCA on their acceptability as bona fide varieties that have been properly released and described. The goal of this arrangement is the acceptance by any member agency for certification of all varieties given favorable action by these boards. However, in actual practice, many individual agencies still require varieties to meet adaptability and performance standards established for their particular state.

CERTIFICATION PROCEDURE

Planting Stock

The proper planting stock is essential for certified seed production. It provides a pedigree, which is the basis for the certification concept. Certified seed is usually produced from registered seed, although it may be produced from foundation or breeder seed. Likewise, other classes of seed may be produced from earlier seed generations.

Application

An application for certification must be submitted to the appropriate state or country certifying agency requesting certification as foundation, registered, or certified seed. The application must be accompanied by at least one official tag substantiating the class of seed, planted-breeder, foundation, or registered. Each certification agency has its own application procedures that must be followed.

Field Inspections

Inspections are performed on all fields for which applications are received. These are timed so that varietal off-types and other crop and weed contamination are most easily detected. In some crops (e.g., clover and alfalfa), a seedling inspection may be required to check for volunteer crop plants. These are normally made a few weeks following seeding. Small grain inspections are normally made after the chaff color has changed, and during the hard dough stage when different chaff color of off-types is most easily distinguished. However, oats are often inspected while the plants are green and the seed is in the soft dough stage. Grass and legume seed fields are usually inspected during pollination, when off-types and weeds are most easily seen and when isolation from adjacent cross-fertile fields is apparent. Table 12.1 shows the AOSCA genetic standards for several crops. Note that in this table the cross-pollinated crops must be separated (isolated) specified minimum distances from other fields with which they could potentially outcross. No isolation is required between self-pollinating crops, except for short distances (e.g., 5–10 m) to help prevent mechanical mixing.

Harvesting

A crop is harvested for certified seed in the same way as for other purposes except that more care is given to moisture content, purity, and prevention of mechanical damage. Seed harvested with excessive moisture will not maintain its quality during storage, while excessively dry seed is more susceptible to mechanical injury. Particular attention is directed to thorough cleaning of combine and handling equipment before starting to harvest a field to avoid contamination.

Conditioning

All seed must be cleaned thoroughly (to remove other crop and weed seed, chaff, straw, and other inert matter) before it will meet the purity standards for certification in most states. The amount and kind of cleaning depends on the kind of seed and its composition. It is absolutely essential that the conditioning facilities be thoroughly cleaned between seed lots. Although this is a difficult and exacting task, its importance can hardly be overemphasized. Utmost care should be given to avoid mechanical damage during conditioning. Seed of certain species are especially fragile and must be handled with extreme care. Seed must be elevated at least once to reach the top of the conditioning flow; however, if possible, it should be elevated only once, and gravity should be used to move it through the different steps in cleaning and bagging. The use of augers to move seed should be avoided whenever possible, because they tend to damage the seed.

Sampling

The sample that is tested for determining seed quality and acceptance for certification is taken after the last conditioning operation. It may be taken by an automatic sampling device, but is more commonly taken from the bagged or bulk seed by hand sampling methods by an official of the certifying agency. The sample should be taken at a time and by a procedure to represent accurately the seed to be marketed. If the seed is treated with a fungicide, the submitted sample should be drawn from the treated seed.

Seed Inspection

A sample of the conditioned seed to be certified is tested for purity, germination, and noxious weed seed content. Sometimes phytosanitary tests are also required by the certification agency for disease assessment. Whether the analysis is performed by the certification agency or by a state seed laboratory, it is used to determine whether

Table 12.1. AOSCA Minimum Genetic Standards (Used as an Example)

Crop Kind	Foundation				Registered				Certified			
	Land¹	Isolation²	Field³	Seed⁴	Land¹	Isolation²	Field³	Seed⁴	Land¹	Isolation²	Field³	Seed⁴
Alfalfa	4	600	1000	0.1	3	300	400	0.25	1	165	100	1.0
Barley	1	0	3000	0.05	1	0	2000	0.1	1	0	1000	0.2
Hybrid	1	660	3000	0.05	1	660	2000	0.1	1	330	1000	0.2
Bird's-foot Trefoil	5	600	1000	0.1	3	300	400	0.25	2	165	100	1.0
Clover (all kinds)	5	600	1000	0.1	3	300	400	0.25	2	165	100	1.0
Corn Inbred lines	0	660	1000	0.1	—	—	—	—	—	—	—	—
Foundation Single Cross	0	660	1000	0.1	—	—	—	—	—	—	—	—
Hybrid	—	—	—	—	—	—	—	—	0	660	—	0.5
Open-Pollinated	—	—	—	—	—	—	—	—	0	660	200	0.5
Sweet	—	—	—	—	—	—	—	—	0	660	—	0.5
Cotton	0	0	0	0	0	0	35000	0.01	0	0	7000	0.1
Cowpeas	1	10	1000	0.1	1	10	500	0.2	1	10	200	0.5
Crambe	1	660	2000	0.05	1	660	1000	0.1	1	660	500	0.25
Crown vetch	5	600	1000	0.1	3	300	400	0.25	2	165	100	1.0
Field & Garden beans	1	0	2000	0.05	1	0	1000	0.10	1	0	500	0.20

¹Number of years that must elapse between destruction of a stand of a variety and establishment of a stand of a specific class of a variety of the same crop kind.
²Distance in feet from any contaminating sources.
³Minimum number of plants or heads in which 1 plant or head of another variety or off-type is permitted.
⁴Maximum percentage of seed of other varieties or off-types permitted.
Data from Certification Handbook, AOSCA, p. 9.

Table 12.2. Suggested AOSCA Nongenetic Seed Standards for Alfalfa

Factor	Standards for each class		
	Foundation	Registered	Certified
Pure Seed (minimum)	99.00%	99.00%	99.00%
Inert Matter (maximum)	1.00%	1.00%	1.00%
Weed Seeds (maximum)	0.10%	0.20%	0.50%
*Objectionable or Noxious Weed			
Seeds (maximum)	None	None	None
Total Other Crop Seeds (maximum)	0.20%	0.35%	1.00%
Other Varieties (maximum)	0.10%	0.25%	1.00%
†Other Kinds (maximum)	0.10%	0.10%	0.50%
Germination & Hard Seed (minimum)	80.00%	80.00%	80.00%

*Objectionable or noxious weed seeds shall include the following: bindweed (*Convolvulus arvensis*), Canada thistle (*Cirsium arvense*), dodder (*Cuscuta* spp.), dogbane (*Apocynum cannabinum*), Johnson grass (*Sorghum halepense*), leafy spurge (*Euphorbia esula*), perennial sow thistle (*Sonchus arvensis*), quackgrass (*Agropyron repens*), Russian knapweed (*Centaurea repens*), and white top (*Lepidium draba,* L. *repens, Hymenophysa, pubescens*).

†Sweet clover seed shall not exceed 9 per lb for foundation seed; 90 per lb for registered seed; and 180 per lb for certified seed.

Data from *Certification Handbook,* AOSCA, p. 20.

the lot qualifies for certification. Table 12.2 shows the suggested nongenetic AOSCA seed quality standards for alfalfa. Unlike genetic standards, they are only suggested standards, and are not required for certification, except when required by the state certification agency.

Seed Tagging

A few certification agencies have a *one-tag* system in which the analysis information (purity, germination, etc.) is printed on the certification tag. However, most agencies have adopted a *two-tag* system, in which the analysis tag and certification tag are different. Some agencies maintain laboratories that determine only if seed meets certification requirements; labeling information is obtained from tests performed in a state or commercial laboratory. With the two-tag system, the seed quality information can be changed or updated without removing the official blue certification tag.

Regardless of the system used, the tags should be attached in a way that will reveal evidence of any opening, reclosing, or other tampering with the contents of the container. It should be impossible to open the container without breaking or defacing the tag. This is easily accomplished by sewing the tag into the seam or by attaching the tag to the stitching with a metal seal. Metal seals were very common in the earlier days of certification but are seldom used today.

Marketing

The marketing of certified seed is the responsibility of the producer. It requires both promotion and a reputation for delivering quality, and most experienced growers have established regular customers through which most of their seed is marketed. Many certification agencies have advertising programs to help promote the image of certified seed and to carry out educational programs in seed improvement.

Many growers avoid marketing problems by growing seed on contract for elevators, seed dealers, or other larger seed growers. The contractor may submit the application, pay all certification charges, and even condition, bag, and tag the seed. (Normally, the grower only plants the seed, harvests the crop, and delivers it to the contractor, although various other arrangements may be used.) Contract seed production is becoming more important as smaller growers find it increasingly difficult to compete with larger seed dealers. It is common in the western grass and legume seed production area, which is outside the major area of utilization and requires extensive transportation and marketing arrangements. Contract seed production offers the security of a fair and stable price for the individual seed grower and provides the seed dealer the security of having adequate supplies of seed at a competitive price.

Certified seed may be marketed cooperatively with that of other growers. Successful marketing cooperatives usually have either a unique product, a common geographical location, or some other factor that promotes the common interest of the members and allows them to compete with other seed sources. Furthermore, by group action they can handle larger volumes, promote their product, offer it at a stable price, and generally be more competitive in the seed market.

Crop Improvement Associations in North America commonly assess additional fees to provide revenues to support programs to help promote certified seed. Such programs generate funds to promote certified seed in the broadcast and print media, as well as other kinds of promotional methods. Such promotion may focus on general advantages of certified seed or be used to promote particular certified varieties. Some are done in cooperation with individual certified seed growers or groups of growers.

Benefits of Certification

Benefits for the farmer. There are several benefits for the average farmer in planting certified seed. First, it provides access to seed of excellent varieties with good assurance of high genetic purity. Thus, it helps avoid unnecessary losses in yield from planting seed of unknown or contaminated varieties. Such off-types are likely to yield plants of different maturity, susceptibility to diseases and insects, or be less productive. Similarly, certified seed which is high in mechanical purity provides assurance to the user against the introduction of weeds, diseases or other crop seeds. Contamination by undesirable plants of any kind can reduce productivity and lower crop quality.

Benefits for Certified Seed Producers. Historically, only outstanding farmers in each state have produced certified seed on a sustained basis. There are several reasons for this in addition to the increased income potential from seed production. They tend to be generally more willing to accept greater effort and timely management required for success. They also recognize the inherent advantage of early access to new varieties. Finally, they demand the highest quality seed possible for their farming operation and take pride in meeting these demands.

INTERAGENCY CERTIFICATION

Interagency certification is the participation of two or more certifying agencies in performing the services required to complete certification on an individual lot of seed. It is frequently useful when a demand for seed exists in a state other than that where the seed is produced. The plan was first proposed in 1943 to facilitate the interstate movement of hybrid corn seed, where the production and field inspection would be done in one state, and the seed inspection and final certification completed in another. Under this scheme, seed growers in Michigan have historically produced several thousand acres of certified oat seed annually, which have been field inspected and then shipped into New York and Ohio for final certification.

OECD—AN INTERNATIONAL CERTIFICATION PROGRAM

The Organization for Economic Cooperation and Development (OECD) provides a scheme for the varietal certification of herbage seed moving in the international market and is the nearest existing program to a completely international seed certification organization. Instituted in 1960, the OECD is an outgrowth of the Organization for European Economic Cooperation (OEEC), which includes membership from several European countries, Japan, and North America. The entire program involves trade agreements, economic expansion, financial stability, and overall economic well-being of the members. This program is funded in the United States and Canada by an additional assessment on each unit of seed certified.

Certification under the OECD scheme is on a basis of varietal purity only, and standards have been established to help maintain varietal purity. Requirements include: (1) authentication of the proper planting stock, (2) documentation of previous cropping history, (3) minimum isolation between adjacent seed fields, (4) the number of harvest years during which a field can produce certified seed, and (5) field inspection criteria. As long as the varietal purity is intact, no fields or seed lots are rejected because of seed quality factors such as germination and pure seed content.

The OECD has an agreement with the U.S. Agricultural Research Service (ARS) to conduct the program in the United States. State agencies that negotiate a memorandum of agreement with the ARS may operate the program by certifying seed according to OECD standards and attaching official OECD certification tags. Ordinarily, this procedure is followed when the seed is to be shipped to another OECD member country. Under this program, thousands of acres of grass and legume seed crops are certified each year in the United States and are sold (usually under contract) by U.S. seed companies to companies in other countries. This is possible because the OECD tag assures that the seed has met internationally recognized standards for genetic purity. The program has greatly aided international trade in seed. Table 12.3 shows the tonnage of total OECD certified seed produced in 1991/92 and the relative production in different countries. The United States is by far the largest participant in this program and its production is about equally divided between domestic and foreign cultivars.

The OECD certification standards are printed in OECD Publication *Documentation in Food and Agriculture*.

Table 12.3. Weight (Tons) of Seed Certified Under OECD Seed Schemes (1991/92).

Country	Home/Foreign Cultivars	Herbage and Oil	Cereal	Maize	Beet	Total All Schemes	Country Totals	Percent
Argentina	H	334		1,500		1,834		
	F	88		3,627		3,715	5,549	2.0
Australia	H	5,932	2,188			8,120		
	F	837				837	8,957	3.0
Austria	H		37	104	21	162		
	F	653	71	6,134	244	7,102	7,264	2.0
Belgium	H	291			132	423		
	F	364	10	1,191	7	1,572	1,995	1.0
Canada	H	14,312		51		14,363		
	F	16,284		2,784		19,068	33,431	11.0
Czech Republic	H	7,790	1,872			9,662		
	F						9,662	3.0
Denmark	H	81			150	231		
	F	258	275		0.7	534	765	0.2
France	H	2,595	297			2,892		
	F						2,892	1.0
Germany	H	1,818	503	156	256	2,733		
	F	158	25	73	50	306	3,039	1.0
Hungary	H	1,666	3,074	2,457		7,197		
	F	27,161	582	35,502	205	63,450	70,647	23.0
Ireland	H	50				50		
	F		21			21	71	0.2
Israel	H	2,719				2,719		
	F						2,719	1.0
Italy	H	27	0.0	35	50	112		
	F	339			101	440	552	0.2
Japan	H	11				11		
	F						11	0.004
Netherlands	H	1,344			67	1,411		
	F	73	15			88	1,499	0.5
Norway	H	0.9	1.1			2		
	F						2	0.006
Poland	H	9,334	1,788		10	11,132		
	F	3,076				3,076	14,208	5.0
Romania	H	261				261		
	F	863		12,143		13,006	13,267	4.0
South Africa	H					25		
	F	25					25	0.008
Spain	H	0.9				0.9		
	F	1,040	54	1,938	332	3,364	3,365	1.0
Sweden	H	338	84			422		
	F	9			0.8	10	432	0.1
Tunisia	H		16,099			16,099		
	F						16,099	5.0
Turkey	H							
	F	442		1,902	25	2,369	2,269	1.0
United States	H	47,371	14,491	38,286		100,148		
	F	8,817		2,151		10,968	111,116	36.0
Total Domestic	179,983							
Total Foreign	129,926						309,909	100.0

Subterranean Clover Seed Scheme
Australia 4,083

CHANGING CONCEPTS AND SERVICES

From its origin in the early 1900s up to the late 1950s, seed certification was built around three primary concepts: superior varieties, genetic purity, and high seed quality standards. These concepts were seldom criticized, and over the years they became almost synonymous with certification. However, in the late 1950s, a re-evaluation of this philosophy by the certification agencies greatly changed both the concept and practice of certification.

Performance and Recommendation Criteria

Traditionally, crop varieties became eligible for certification only after they were recommended to growers within the jurisdiction of a given certification agency. Such a requirement evolved quite naturally, since only superior varieties were eligible for release by state experiment stations. As plant breeding programs grew and new varieties were developed, older varieties were removed from the recommended lists and new ones added. This policy was strengthened by strong crops extension programs that complemented the varietal release programs. It became customary for state universities to publish annual lists of variety recommendations, and the appearance of a variety on these lists normally meant eligibility for certification.

In the late 1950s, recommendation and performance came under question as valid criteria for certification. Two factors were primarily responsible for this: (1) the volume of seed produced away from the area of consumption, and (2) the appearance of private field crop varieties. Today, evidence of varietal performance (merit) is usually not a factor in certification. Most agencies will certify any variety that has been properly identified and described by any public or private agency and has met the criteria described on page 288.

Varietal Purity Only (VPO) Certification

Perhaps the most controversial issue to confront certification in the United States is the concept of varietal purity only certification. Under this concept, seed is certified if the field and seed inspections show the crop to meet minimum standards of varietal purity; seed lots are rejected only for excessive contamination by off-types, inadequate field isolation, or other genetic purity factors. The occurrence of weeds or other crops, disease infestation, or even low germination does not constitute cause for rejection of any field or seed lot. Under the practice of VPO certification, consumers are assured of variety purity by the certification agency and choose seed lots that meet their own seed quality (purity, germination, etc.) criteria on the basis of information on the seed tag. Several state certification agencies have adopted VPO certification, though most still require certified seed to meet minimum seed quality standards.

Certification of Blends

Varietal blends of seed are eligible for certification by several certification agencies. Where this is permitted, all components of the blend must represent certified seed and must be blended in specific, predetermined, and commercially acceptable proportions. The components of the blend are confidential between the producer and the certification agency. Some agencies are reluctant to certify blends on the basis that they do not represent a pure variety.

Sod Certification

Some certification agencies have been certifying turf sod for many years. Certification is usually performed on the basis of elite varietal and mechanical purity of seed-stocks, and verification of vegetative varietal composition, freedom from diseases and insects, and absence of weeds and other crop plants. In some states, sod certification is principally for phytosanitary condition.

Tree Seed Certification

Progress in the breeding and improvement of trees has caused an interest in the seed certification of improved tree varieties. Seed orchards of new tree varieties have been established from which certified seed and seedlings are harvested and sold to help establish improved tree stands. Certification agencies have responded to forestry industry requests for help in developing procedures by which customers can be assured of varietal purity and high seedling quality. The AOSCA standards have now been modified to include trees, shrubs and native plant species.

Some agencies also certify the source or origin of forest tree seeds and seedlings. This type of certification is performed in the absence of, or as a supplement to, seed from established seed orchards. Certification of the origin of seed is important when it is desirable to obtain seed from locations of climate, elevation, and exposure similar to the sites where the seed is intended for planting. This assures that the resulting plants will be ecologically suited for the planting site and also ensures their survival and performance.

Phytosanitary Certification

Phytosanitary certification does not qualify as seed certification in the usual meaning of the term. It certifies only that the seed and the field from which it came is free of specified diseases. Phytosanitary certification is normally performed by pathologists from the official government agency, and a tag is attached to specify what has been done. Usually a phytosanitary certificate is issued for the seed. This enables seed suppliers to provide the certificate to their customers and to officials of the state or country into which the seed is shipped. When seed is shipped internationally, phytosanitary certification is usually required by the receiving country. Most phytosanitary certificates expire within 14 days after shipment of the seed or plant materials.

The United States (USA) Certification Concept

A diversity of certification programs in the United States and different standards and procedures remain in spite of efforts toward standardization. This lack of uniformity has frequently caused problems for certified seed produced in the United States in spite of progress made under OECD certification which sets minimum genetic standards for all seed produced under the OECD program. This has caused interest in the formation of a "USA certification" program in the United States. The goal of such an organization would be to certify seed by standardized procedures that would make the seed acceptable in international commerce and also appeal to the private seed industry in the United States who normally does not support certification or who would like to see more uniformity in certification procedures among U.S. certification agencies. Although at this point (1994), the USA seed certification concept is still in its infancy, it appears to have good potential for the future.

Ancillary Programs

Because of financial difficulties, many certification agencies have had to consider other ways to maintain profitability in addition to providing traditional certification services. Consequently, they have explored the possibility of providing ancillary programs to help utilize their talents and to help make ends meet. Two of these programs are discussed below.

Identity Preserved Programs. The identification and maintenance of genetic purity has been the strength and focus of seed certification programs since their inception. This has enabled improved varieties to be made available to farmers both quickly and efficiently, and has contributed enormously to crop production in the United States and around the world. In recent years it has become evident that this same expertise could also be extended beyond the farmer level to the consumer and end-user of agricultural grain products. For example, millers of certain types of soybean prefer high oil soybean varieties and are willing to pay a premium for soybeans that are documented to be the preferred type. This provides a way to avoid the necessity of the mixing of different varieties or quality levels of grains, oilseeds, or other farm products during storage or marketing. The further development of this concept could provide a valuable service to agriculture and food-related industries which require high levels of product quality and uniformity.

Identity preserved programs for grain products have been established by seed certification agencies in several states. For the most part, these programs utilize the same procedures and practices of conventional certification, including careful record-keeping, field (in some cases), and post-harvest inspections for genetic purity and other aspects of quality, including uniformity.

Quality Assurance. Many certification agencies offer quality assurance programs to the seed industry. These provide field inspection and evaluation services when there is no interest in completing the certification process. This kind of service is used by seed companies that desire the expertise of the certification agency in providing field or laboratory inspections and advice on quality control. In many ways, it is not unlike a pest scouting or crop consulting service by an agency with a particular expertise in quality seed programs. In such cases, the seed may be labeled with a specially developed quality assurance tag which indicates the kind of services performed.

THE FUTURE OF SEED CERTIFICATION

Past Contributions

There is little doubt of the contribution of seed certification to the development of North American agriculture. It has provided a rapid and highly efficient way for seed increase and distribution of superior varieties developed and released by state experiment stations. This has had great impact on both the seed industry and North American agriculture, and has provided a model for similar development in countries around the world. However, many changes have recently occurred in the seed industry that give cause for concern about the changing form of certification and perhaps its very survival.

The Changing Seed Industry

The nature of the North American seed industry has developed and changed dramatically in the last 100 years. Today, it is well supplied with a wide range of varieties of all major crops. For the most part, it has sophisticated production, quality control, and marketing programs. Furthermore, those in the industry are well aware that farmers have become more discriminating in their needs and demands. This has made the seed industry highly competitive, efficient, and responsible to seed customers.

Today's seed industry is showing greater interest and ability to provide its own quality assurance programs, with decreasing dependence on third-party umpire services such as certification to evaluate the quality of their product. The present interest in the International Standards Organization (ISO-9000) program is an indication that the seed industry feels that it can provide quality assurance programs for itself. Such programs have rigid guidelines and requirements for quality assurance regarding facilities, qualification and training of inspectors, precision and instrumentation, as well as continuous monitoring to assure that standards are met and quality is maintained. While large seed companies are better able to afford such self-policing quality assurance programs without the need for third-party certification, smaller companies are more likely to use the quality control and referee services of crop improvement agencies.

The passage of the Plant Variety Protection Act (PVPA) in 1970 and the varietal explosion that followed has had a profound effect on seed certification in the United States. Seed buyers today have a great many choices of both private and public varieties compared to the relatively few, mainly public varieties of a half century ago. Varietal protection has allowed commercial companies to invest in genetic improvement programs and protect their varieties against infringement by other parties. Thus, many seed companies have invested heavily in variety development and have strong production and sales programs involving hundreds of varieties across most field crops.

The professionalism of the modern seed industry, along with its arsenal of high quality, productive varieties has allowed the private seed industry to thrive and prosper in competition with certified seed producers, who for the most part produce only public varieties. Although the private seed industry may also produce seed of public varieties, private varieties are more profitable because of the marketing advantage provided by exclusivity and promotional programs. This off-sets many of the traditional advantages that certification has offered for public varieties. Consequently, there may be little incentive for commercial companies to use the certification process. This is especially true for large companies with well-established research and quality control programs. Furthermore, private companies promote and market their seed aggressively and use their resources and reputation to stand behind their products. The use of brands and special marketing programs is also an important part of their programs. These factors have led to the perception that private varieties or brands are superior to public varieties and have helped to erode into the market that certified seed has had in the past. Perhaps the crop that reflects this trend more than any other is soybean where seed of private varieties command as much as 30% higher price than seed of comparable certified varieties.

Competition and Survival

Certified seed producers have competed with the private seed industry in various ways. One is by aggressive marketing programs that promote the advantages that certified seed has always enjoyed, i.e., excellent varieties, third person quality verification, and cooperative promotional efforts. Two examples of this are Public Varieties of Indiana (PVI) and Public Varieties of Minnesota (PVM). These programs, composed of certified seed producers, assess themselves additional fees that are used to collectively promote certified public varieties. Many other agencies have similar programs.

Certification can exist only when it provides a real or perceived quality advantage or a service to the seed industry. In the past, its major advantage has been that it represented access to superior germplasm as new public varieties were released from public research institutions and made available through certification. Certification represents an unbiased and official third-party assessment of a high level of genetic and mechanical quality. It is ideally suited for seed of public varieties which is produced and made available by many different seed producers in competition with each other and with uncertified commercial or bin-run seed of the same or different varieties. It has been less successful for seed of private varieties.

Certification is very important for almost all kinds of field seeds moving in international commerce, since most countries require seed to be certified in order to be imported. Consequently, even hybrid corn seed moving in international commerce is certified. Otherwise, little, if any, hybrid corn seed sold domestically is certified, since it offers no particular advantage over uncertified hybrid corn seed. Forage and turf seed certified under the OECD program also move easily in international commerce, although, unlike hybrid corn, this seed is also certified for domestic use.

New Dimensions; New Horizons

As the 20th Century ends the future of the role of certified seed in a modern, sophisticated seed industry appears somewhat uncertain. Although certification continues for publicly released field crop varieties, the increasing importance of private varieties and the ability of the private seed industry to provide many of the benefits of certification has led to an uncertain future for certification programs. As a result, many U.S. certification agencies are broadening their role by offering other services that provide a strong financial base in order to secure their future. These include (1) quality assurance programs, (2) identity preserved programs, (3) pest scouting services, and (4) seed sampling and seed testing services. Though none of the programs are necessarily identical to past certification roles, all can make valuable contributions to the seed industry and to agriculture, especially for smaller seed companies.

While it is evident that certification will continue to evolve and adapt to a changing seed industry, it also seems clear that it can continue to have an important role in the seed industry. First, it should be a vehicle in the seed increase and availability of public varieties. Second, it should continue its role in the collection of research assessments or royalties on public varieties and thus continue to fund variety development. This partnership with public research institutions should be mutually beneficial while providing a valuable service to the seed industry.

Questions

1. What basic authority defines the responsible certification agency in any given state? Can you name three types of agencies that are responsible for certification in various states? Why do these differ? What agency certifies seed in your state?
2. What is meant by OECD certification?
3. What are the advantages and disadvantages of certification on the basis of varietal purity only? Do you approve of CVPO? Why?
4. What is the importance of certification for vegetable and hybrid corn seed?
5. Do you believe certification will eventually outlive its usefulness? Why or why not?
6. If you were a farmer, why would you buy certified seed?

General References

Armstrong, J. 1994. The Role of Certified Seed in the Twenty-First Century. Presented at the 1994 Annual Meeting of the Association of Official Seed Certifying Agencies, Fort Mitchell, Kentucky.

Clapp, A. L. 1970. The Kansas Seed Grower: A History of Seed Certification in Kansas. Manhattan, Kansas: The Kansas Crop Improvement Association.

Cowan, J. R. 1972. Seed Certification. In *Seed Biology*, vol. 3. ed. T. T. Kozlowski, pp. 371–398. New York: Academic Press.

Douglas, J. E., ed. 1980. *Successful Seed Programs: A Planning and Management Guide*. Boulder, Colo.: Westview Press. 302 pp.

Hackleman, J. C. *History of the International Crop Improvement Association, 1919–1961*. International Crop Improvement Association.

McDonald, M. B. and W. D. Pardee (eds.). 1985. The Role of Seed Certification in the Seed Industry. CSSA Spec. Publ. No. 10, Madison, WI.

Thompson, J. R. 1979. *An Introduction to Seed Technology*, Toronto: John Wiley and Sons. 252 pp.

U.S. Department of Agriculture. 1961. *Seeds: The Yearbook of Agriculture*. Washington, D.C.: U.S. Department of Agriculture.

Wheeler, W. A., and D. D. Hill. 1957. *Grassland Seeds*. New York: D. Van Nostrand Company, 734 pp.

13

Seed Testing

Seed testing is the science of evaluating seed quality for agricultural purposes. Although initially developed for evaluating the planting quality of field crop and vegetable seeds, it is also valuable for determining the quality of lawn, flower, and tree seeds.

Even though humanity's use of seeds dates back to prehistoric times, the art and science of seed testing have only developed in the last century. Until about 300 years ago, our knowledge concerning seed morphology and physiology was limited. As a result, it was possible for unscrupulous vendors to market seed of such crops as alfalfa with sweet clover seeds or other contaminants. Such practices became so widespread and of such serious concern that laws were passed and seed testing procedures established. Berne, Switzerland was the first city to enact seed legislation prohibiting the sale of adulterated clover seed in 1816. The first seed testing laboratory was established in Saxony, Germany, in 1869, and the first one in America was set up at the Connecticut Agricultural Experiment Station in 1876. Today official seed testing laboratories are found in almost all 50 states and in nearly every country in the world, and many privately operated commercial laboratories are located in North America.

Other important milestones in seed testing include the publication of the first "Rules and Apparatus for Seed Testing" by the United States Department of Agriculture in 1897 and the first drawings of seeds in the publication "The Viability and Germination of Seeds" by F. A. Hillman in 1902. In 1915, E. G. Boerner developed the first divider for separating grain, seed and other material; in 1916, H. D. Hughes developed the first seed counter for preparing germination tests; and in 1917, G. N. Collins developed the first seed blower.

The expression *seed quality* is used loosely to reflect the overall value of seed for its intended purpose. Seed quality is usually a composite of several factors, all of which contribute to the desirability, or planting value, of the seed. The key question is "Why do we test seeds?" There are several reasons. First, and most obvious, is that the dry seed's potential to establish a seedling cannot be determined until the seed has been germinated. However, we also test seeds to determine the genetic (variety) and mechanical (weed/other crop) components of the seed lot.

Seed testing values provide important information to both the seed producer and purchaser. The seed producer wants to ensure that only a quality product is marketed so that consumers will return for their further seed needs. Seed purchasers need assurance that associated expenses such as field tillage, pesticide applications, and other costs are not lost due to stand failure as a result of planting poor-quality seeds. Finally, both buyers and sellers recognize that seed laws have been enacted to aid in the orderly marketing of seed based on the principle of truth in labeling. In some cases, differences in reported values on the seed label exist and are litigated in court. The seed testing information and how it was acquired form the foundation of these legal cases and emphasize the need for proper conduct and interpretation of seed tests.

Because of the universal importance and value of seed, many organizations rely on the results of routine seed testing. These organizations vary from local or state agencies to national and international agencies. As a result, it is important that test procedures be standardized and that results be widely reproducible. This means that the tests must be conducted under the same conditions with uniform interpretations. The process toward standardization in the United States began in 1905 when the Annual Appropriations Act was passed which gave the U.S. Department of Agriculture authority to purchase seeds on the open market, to test and publish the results, and to identify individuals or organizations found to market mislabeled seed. The standardized testing of seeds required that a compendium of test procedures be developed. The Rules for Testing Seeds of the Association of Official Seed Analysts were developed to meet this objective. The following considerations discuss some of the specifications put forth in those rules. Individuals interested in a greater knowledge of the procedures for conducting a seed test should consult the AOSA Seed Analyst Training Manual (McDonald et al. 1992).

OBTAINING THE SAMPLE

No seed analysis, regardless of how carefully or accurately accomplished, is any better than the sample on which it is performed. The importance of a representative sample cannot be overemphasized; that is, the sample must truthfully represent the quality of the seed lot from which it is drawn. It is generally assumed that a seed lot is homogenous. If this were the case, it would be satisfactory to extract a portion (sample) from the seed lot and presume that it represents the bulk of the seed. However, this seldom occurs. Seed lots are almost never completely homogenous for at least four reasons. First, heavy and light seeds segregate within the bulk or bag due to gravity, with heavier seeds being found predominantly at the bottom of the container. Second, harvesting of the crop in the field combines seed from differing locations, thus altering the composition of the seed as a result of variations in maturity, lodging, disease, or the occurrence of weeds. Third, failure to adequately blend two or more lots from differing locations at the time of bagging can result in seed lot heterogeneity. Fourth, lack of uniformity in harvesting, storage, and conditioning results in seed lot heterogeneity.

As a result of this heterogeneity in most seed lots, a seed lot must be sampled and the sample must be representative of the seed lot. Sampling is usually done in two steps. First, the sample to be submitted to a seed laboratory is drawn from the bulk seed lot and sent to the laboratory for analysis. This is known as the *submitted sample*. Second, when it reaches the laboratory, it must be divided further to a size that can be analyzed. This latter sample is used for the actual analysis and is called the *working sample*.

The Submitted Sample

The sample may be drawn at any time during seed conditioning or after the seed is offered for sale. If drawn during conditioning, it may be taken by hand or small container at frequent intervals during the conditioning operation or automatically drawn at specified intervals by a *mechanical sampler*. Either technique, if done properly, will give a representative sample.

Seed is usually sampled for testing while still in storage or as it is offered for sale (Figure 13.1). Because of the variety of ways in which seeds are stored and offered for sale, they may be found in various types of containers, from small vegetable and flower seed packets, to boxes and cans of grass seed, to large bulk lots of cereal grain seed. Regardless of the container, the seed lot must be properly sampled so the sample is representative. Rules and procedures for sampling under various conditions have been established by the Association of Official Seed Analysts and the International Seed Testing Association. These rules provide for sampling by mechanical samplers, by use of standard sampling *probes* or *triers* (Figure 13.2), by hand, or by taking the entire container as the submitted sample.

The Sampling Process

Bulk Seed. A trier, or probe, is recommended for sampling bulk seed, although hand sampling may also be performed if handsful are taken from well-distributed points throughout the bulk. Hand sampling is limited by the difficulty of reaching all portions of large bulk lots, whereas large probes up to 72 in. in length can be used to sample hard-to-reach locations within the seed lot.

Seed in Bags. When a seed lot consists of six bags or less, each bag should be sampled from well-distributed points throughout the bags. When lots consist of more than six bags, samples should be taken from five bags plus 10% of the remaining bags. Regardless of the lot size, however, it is not necessary to sample from more than 30 bags. Here are some examples:

No. of bags in lot	5	7	10	23	50	100	200	300	400
No. of bags to sample	5	6	6	7	10	15	25	30	30

Seed in Small Containers. Seed in small containers should be sampled by taking at random an entire unopened container from the supply in order to obtain the minimum amount required for the working sample.

A

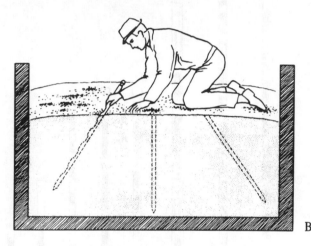

B

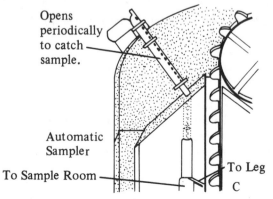

Opens periodically to catch sample.

Automatic Sampler

To Sample Room

To Leg

C

Figure 13.1. *Seed sampling techniques: (A) bag sampling; (B) bulk sampling; and (C) mechanical sampling. ("A"—Courtesy of Bob Neumann.)*

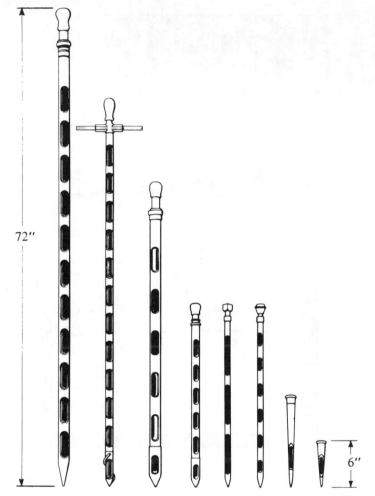

72"

6"

Figure 13.2. *Examples of sampling probes. On the far right is a "thief probe."*

Subdividing the Sample

The sample drawn by any of the various techniques may be too large for the submitted sample and should be divided further before submitting it to the laboratory. Further subdivision should be done by a mechanical halving device, such as the Boerner divider. Absolute care should be taken at this point to guard against introducing bias into the sample to be submitted. During the subdividing process, there may be a tendency to unconsciously remove stones, stems, damaged seeds, or even noxious weed seeds. Such deviations from the correct sampling procedure make all subsequent testing results meaningless.

Mailing the Sample

After the properly sized sample is obtained, it should be carefully labeled and placed in a container suitable for mailing to the seed laboratory. Cloth, plastic, or paper bags are acceptable; however, these should be placed inside a sturdy cardboard box that can be labeled and can withstand the rigors of mailing. Each sample should be labeled as follows: (1) name and address of owner, (2) crop kind and variety, (3) tests requested, and (4) lot number and number and weight of containers (bags) in the lot.

SUBSAMPLING

When the submitted sample arrives at the seed laboratory, it is entered in the official logbook, assigned a number, and the accompanying information is recorded. The sample then goes to the subsampling area of the laboratory, where it is divided into working samples for the various tests that will be performed. Here the working samples for the purity examination, noxious weed examination, germination, and other special tests are obtained. The remaining portion of the sample is retained as an official sample in case future tests are desired. The weight of the working sample for purity analysis is determined by the weight of seed required to comprise a minimum of 2500 seeds and will vary greatly from small- to large-seeded species.

The Importance of Subsampling

Dividing procedures must be absolutely precise and unbiased if the test results are to be meaningful. The working sample must accurately represent the sample submitted to the laboratory, which in turn represents the seed lot only if sampling procedures were properly followed. In contrast to sampling, sample dividing procedures are generally dependable, because this operation is performed in the laboratory by professional, trained personnel, while sampling from bulk seed lots is often done by persons who may not realize the importance of a representative sample.

Subsampling Techniques

The Rules for Testing Seeds state only that the working sample shall be taken from the submitted sample in such a manner that it will be representative. The actual procedure may be either by manual or mechanical methods. Several hand methods are used. One very simple method, often called the *pie method*, consists of spreading the sample on a clean, flat surface, and dividing it into sections as if cutting a pie. Any of the sections, if randomly selected, may be used alone or in combination with other sections as a working sample. Another hand technique, called the *random cup method*, consists of placing a number of uniformly sized thimbles or cups on a clean, flat surface and slowly pouring the sample so the seed is distributed evenly over the flat surface filling the cups as the seed is distributed. The working sample may then be obtained by randomly selecting several of the cups until sufficient seed is obtained.

Mechanical halving devices (Figure 13.3) are most often used for subdividing and are dependable for providing a representative sample. These are devices that divide the sample into two equal portions, both in size and content. The working sample is obtained by dividing the submitted sample one to several times until the proper weight of the working sample is obtained. Any of the divided portions may be combined and redivided to yield the proper-sized working sample.

Three mechanical dividers are commonly used for subsampling. Any of these dividers will yield a representative sample; however, each has certain advantages over the others. The Boerner divider is probably the most common; however, some chaffy grasses and other non-free-flowing seeds will not flow through it. A Boerner divider consists of a hopper, inverted cone, and a series of baffles directing the seed into two spouts. The baffles form alternate channels and spaces of equal width. They are arranged in a circle at their summit and are directed inward and downward, the channels leading to one spout and the spaces to an opposite spout. A valve or gate at the base of the hopper retains the seed. When the valve is opened, the seed falls through the spouts into the seed pans. The Gamet Precision divider requires electrical power to operate and is more suitable for certain seeds that do not flow through the Boerner divider. A

Figure 13.3. *Subsampling dividers: on the left is a Gamet Precision divider and on the right is a Boerner divider. (Courtesy of Bob Neumann.)*

Gamet Precision divider makes use of centrifugal force to mix and scatter the seeds over the dividing surface. In this divider, the seed flows downward through a hopper onto a shallow rubber cup. Upon rotation of the rubber cup by an electric motor, the seeds are thrown out by centrifugal force and fall downward. The circle or area where the seed falls is equally divided into two parts by a stationary baffle so that one-half the seeds fall in one spout and one-half in the other spout. For non-free-flowing chaffy grasses, such as gramma grass and needle grass, the Hay-Bates or a similar divider should be used. For seeds of cotton and certain other species, hand-dividing methods of subsampling may be necessary; however, extreme caution must be taken to obtain a representative sample.

PURITY TESTING

When individuals purchase seed, one of the primary decisions they make is what kind of seed is needed. Their purchase is made with the understanding that the species and variety are the principal constituents of the seed lot. Yet, we know that harvesting and cleaning of seed are not exact sciences. Other types of seed and materials are almost always present. The type and level of contamination of these other components can significantly influence the value of the seed. The purity test, therefore, is designed to identify what these contaminants are and how much of them is present.

Seed purity denotes the composition of a particular seed lot. It is based on a physical determination of the components present and includes percentages by weight of: (1) pure seed, (2) other crop seed, (3) weed seed, and (4) inert matter. *Pure seed* is the portion of the working sample represented by the crop species for which the lot is being tested; in actual practice, it includes the percentage of each crop species present in levels of 5% or more. *Other crop seed* is the percentage of crop seeds, other than the species being tested, present in concentrations of less than 5%. *Weed seed* indicates the percentage of seeds present from plants considered weeds. Sometimes this designation may be arbitrary, since a plant may be considered a crop in one state or country but termed a weed elsewhere. For any particular region, however, the analyst uses well-accepted guidelines for classifying seed as crops or weeds. *Inert matter* denotes the portion of the sample that is not seed. It usually consists of chaff, stems, and small stones, but may include pieces of broken, damaged, or immature crop or weed seeds that do not qualify as entire seeds. The criteria for this distinction are explicit and defined in the Rules for Testing Seeds (AOSA 1991).

The size of the sample on which the purity examination is performed is given in the Rules for Testing Seeds. The sample size (weight) is determined by the size of seed being tested; approximately 2500 seeds are required to yield a statistically acceptable purity test result. The size of the working sample specified in the Rules is larger for large-seeded crops than for small-seeded crops; however, the actual number of seeds in the test is not greatly different.

Philosophy of Purity Testing

The philosophy of purity testing is to avoid judging whether seeds are capable of germinating when performing the test. Consequently, shriveled, immature, frosted, or

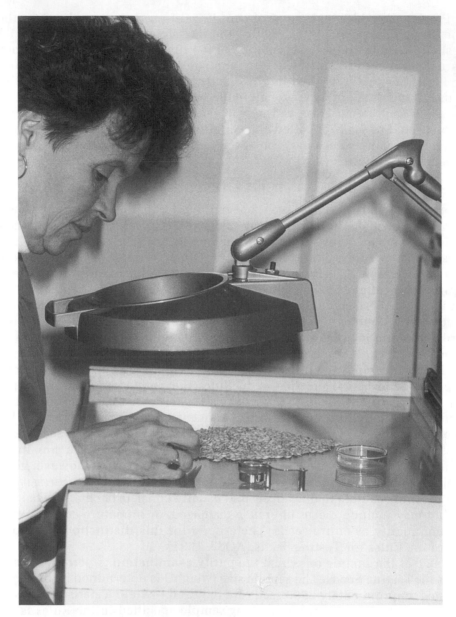

Figure 13.4. *A purity testing station. (Courtesy of Bob Neumann.)*

otherwise damaged seeds are considered as pure seed. This may be called the *quick method*, in contrast to the *strong method* used earlier by European seed analysts, who classified crop seed as pure seed only if it appeared to be capable of germination. The quick method left the determination of germination capability to the germination analyst, but is no longer used.

Procedures for Purity Separations

The purity test is perhaps the most complex and exacting of all tests for seed quality. A seed analyst must have a comprehensive knowledge of seed structure and function and must be able to identify a wide array of differing species. For this reason, it is not uncommon to find that many seed analysts have their own working seed herbaria to assist in the identification of unknown samples. Seed herbaria are useful because published descriptions of seeds are rarely as comprehensive as those for plants and viewing actual specimens can be very helpful. An average seed herbarium contains 2,000–3,000 specimens, but varies in size dependent on the range of seed materials typically encountered in the seed testing laboratory. The herbarium can be arranged alphabetically or by phylogeny. Phylogenetic arrangements are by plant families and then according to species within the family. This approach assures that specimens which are closely allied evolutionarily are placed together. Its disadvantage is that many seed analysts are not familiar with taxonomic relationships which vary according to the authority used to classify them. As a result, some analysts simply file specimens alphabetically according to family and then species. This approach offers the advantage of rapid retrieval of specimens but does not afford the opportunity for direct comparisons with related groups.

Purity separations are usually made by hand, although mechanical aids may be used to speed up the analysis or make it less tiring for the analyst. Purity analysis equipment usually includes a work board, adequate light, forceps, a hand lens, and a stereoscopic microscope for identifying small seeds.

Various mechanical devices are frequently used to aid in the purity analysis. The use of seed blowers (Figure 13.5) has contributed to speeding up and improving uniformity of purity analyses, especially for grass seeds, for use in separating empty florets, bits of stems and leaves, chaff, and other inert material. The Rules for Testing Seeds provide uniform blowing procedures for the purity determinations of small-seeded grass species in lieu of the more time-consuming hand separation procedures. The blowing procedure not only speeds up the test but reduces variability among seed laboratories.

Another difficult area for purity testing is caused by the increased use of coated and pelleted seeds (see Chapter 11 on Seed Enhancements). The Rules for Testing Seeds specify coated seeds as a seed unit covered with any substance which changes the size, shape or weight of the original seed (seeds coated with ingredients such as, but not limited to, rhizobia, dyes and pesticides are excluded). This process alters the shape of flat seeds that are difficult to mechanically plant, for example, and makes them round by adding clay fillers so that the pelleted seed can easily roll into planters. While seed coatings are extremely valuable from a practical perspective, they make the purity analysis more difficult. The analyst must first separate the coating material (often 90% of the dry weight of the seed sample) to determine what is actually present in the clay pellet. This separated portion is weighed as other components of a purity test and recorded as "coating material."

To aid in purity testing, seed scientists at Oregon State University have developed a microscopic inspection station, as well as a vibrator-separator to help purity analysts

Figure 13.5. *Seed blowers: (A) the General seed blower, (B) the Ottawa blower, and (C) the South Dakota blower. (Courtesy of Bob Neumann.)*

make their separations. This type of equipment is used in several North American seed laboratories.

NOXIOUS WEED SEED EXAMINATIONS

Each state has established an official list of noxious weed seeds. In general, the plants from these seeds are particularly troublesome and objectionable. Such lists are a part of the state seed law and are usually defined in two categories, *primary* (or *prohibited*) and *secondary* (or *restricted*) noxious weed seeds. Sale of seed lots containing primary (or prohibited) noxious weed seeds is usually prohibited, while the sale of seed lots containing secondary noxious weed seeds is permitted, but the number of weed seeds per pound of crop seed is limited. Since each state has its own seed law, the weed seeds listed as noxious are not necessarily the same from state to state.

The noxious weed seed examination is an attempt to provide special information about noxious weed seed content of seed lots. Because of the limited occurrence of noxious weed seeds, a large working sample is used for the noxious weed seed examination—often as much as 500 grams of large-seed types down to 2.5 grams for small-seed types. The occurrence of noxious weed seeds is reported in number of seeds per pound or ounce. Since state seed laws and noxious weed seed lists vary among states, the results of this examination may be reported for only one state, or it may be reported for two or more states. Crop seeds transported in interstate commerce should have an "all-state" noxious weed seed examination.

GERMINATION TESTING

Probably the single most convincing and acceptable index of seed quality is the ability to germinate. Seeds are tested for germination because a seed lot is composed of a population of individual seed units; each possessing its own distinct capability to grow and produce a mature plant. A seed germination test is an analytical procedure to evaluate seed viability and germination under standardized (favorable) conditions. It enables a seed vendor to determine and compare the quality of a seed lot *before* it is marketed to the consumer. Furthermore, the percent germination can be used to determine the planting value of a seed lot, its storage potential, and labeling information required to provide for standardized marketing of seed lots. Thus, germination testing is perhaps the most important function of a seed testing laboratory. Since the process of seed germination is covered in Chapter 4, this discussion will cover only the laboratory techniques used for performing the analysis.

Procedures for Germination Testing

The germination test is ordinarily performed on the pure seed of the crop kinds that constitute 5% (or more) of the sample after all inert matter and other crop and weed seeds are removed. Each pure seed kind is germinated and reported separately.

A minimum sample of 400 seeds is recommended for a statistically dependable germination test. These are usually planted in four replicates of 100 seeds each, although various other arrangements are sometimes used (Figure 13.6). Each replicate is evaluated separately, but the official germination report is an average of all replicates.

The exact procedures and regimes under which different kinds of seeds are germinated have been developed over 100 years of experience in germination testing and have been augmented during the last 40 years by a systematic program of referee testing involving interchange of samples and results among laboratories. The testing instructions given in the Rules for Testing Seeds include the germination media (substrata), the temperature required, the duration of the test period, and additional suggestions for optimal results (Figure 13.7).

The time required for germination tests varies among species. Some seeds require less than seven days, while others may require a month or longer. The seeds of some trees and woody shrubs are notorious for their long germination requirements. The Rules for Testing Seeds also specify germination requirements for tree, vegetable, woody shrub, and flower species. However, knowledge of germination requirements, especially for wild and exotic species, is not complete.

Evaluation of Germination

At the end of the prescribed germination period, the tests are evaluated; however, it is sometimes desirable to make preliminary evaluations, called *first counts*. Seedlings

Figure 13.6. *Using a vacuum head for preparing 100-seed replicates for germination testing. (Courtesy of Bob Neumann.)*

Figure 13.7. A light-equipped seed germinator for testing seeds that require light for best germination. (Courtesy of Bob Neumann.)

that have germinated and are normal are counted and removed from the substrate at the time of the first count. This procedure helps subsequent counts, because early-germinating seedlings often tend to grow profusely, causing difficulty in evaluating later-germinating seedlings. Seeds that remain ungerminated at the end of the prescribed period are considered dead or dormant (refer to the discussion on dormant seed in Chapter 6).

The "Normal Seedling." The seed analyst has a somewhat different concept of seed germination than the layperson, to whom germination implies the rupture of the seed coat and the emergence of the root and shoot apexes. The Rules of the AOSA, which most seed analysts in North America follow, prescribe the following definition of germination, embodying the normal seedling concept: "the emergence and development from the seed embryo of those essential structures which, for the kind of seed in question, are indicative of the ability to produce a normal plant under favorable conditions." Not only does this concept include the layperson's definition of germination, it also reflects the agronomic value of the seed (i.e., capacity to produce normal plants under favorable conditions).

Abnormal Seedlings. Any seedling that is not classified as a normal seedling is considered abnormal. The germination analyst may classify seedlings as abnormal for various reasons, for example, the absence of certain essential structures (such as radicle, epicotyl, cotyledon), twisted or otherwise abnormal shape, to greatly reduced growth or seedling vigor.

The ability of a seed analyst to discriminate between and classify normal and abnormal seedlings is one of the most subjective aspects of seed testing. Therefore, constant education and training are required to assure uniformity in interpretations. To help provide uniformity, the AOSA developed a Seedling Evaluation Handbook in 1992 which is now recognized as a formal component of the Rules for Testing Seeds. The Handbook provides line drawings depicting differences between normal and abnormal seedlings to help analysts in discriminating among questionable seedlings (Figure 13.8).

Firm, Ungerminated (Dormant) Seeds.   Seeds other than hard seeds that remain firm (nondecayed) and ungerminated at the end of the prescribed germination period are called firm, ungerminated seeds. This is a type of dormancy commonly found in certain grasses, and should be treated appropriately to stimulate germination.

Hard Seeds.   Hard seeds are those that do not imbibe water and therefore remain hard at the end of the prescribed germination period. Hard-seededness is a type of dormancy that prevents germination of viable seeds because they cannot absorb water through their impermeable seed coat. The percentage of hard seeds is reported as part of the total percentage germination.

Laboratory Methods of Breaking Dormancy

Any time a seed fails to germinate in the time specified in the Rules for Testing Seeds, the analyst must determine whether the seed was ungerminable due to lack of viability or dormancy. If the seed does not appear diseased, it is probably dormant. Once it has been recognized that dormancy exists, the challenge to the seed analyst is to determine approaches that can break the dormancy. Since seeds have evolved many unique ways to maintain dormancy, the analyst must employ various approaches to break the dormancy-imposing mechanism(s). In some cases, a single treatment may be effective. In others, a combination of techniques may be necessary. The Rules for Testing Seeds specify appropriate dormancy-breaking techniques for species where dormancy commonly occurs. These usually are either by prechilling or the use of KNO_3.

Prechilling.   Viable seeds other than hard seeds can often be stimulated to germinate by a cold treatment of the water-imbibed seeds, commonly called *prechilling*, or *stratification*. This is accomplished by placing the seed on or in moist substrata at relatively low temperatures (about 5°C) for a specified period of time—usually about five days; longer durations may be necessary for the seeds of some species (e.g., woody species). The experienced seed analyst recognizes those species in which dormancy is likely to occur and routinely prechills them as a standard part of the laboratory procedure.

Potassium Nitrate (KNO_3).   Seed germination in many species, such as turf grasses, can be stimulated by using a dilute solution (0.1% to 1.0%) of potassium nitrate as moisture for the germination test. Like prechilling, the use of KNO_3 is a valuable aid in germination of those species benefited by it and has become a routine procedure in the germination testing of many species.

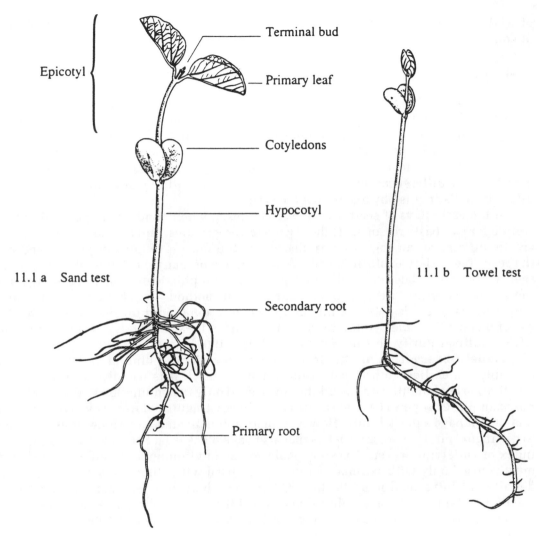

Figure 13.8. *Soybean seven-day seedlings, sand test and towel test. (From AOSA* Seedling Evaluation Handbook. *1992. Contribution No. 35 to the Handbook on Seed Testing. Association of Official Seed Analysts. 101 pp.).*

SPECIAL TESTS FOR SEED QUALITY

Although purity, germination, and noxious weed evaluations are routinely performed on almost every seed sample submitted to the laboratory, many additional tests also reflect seed quality. Such special tests are usually performed only when requested; however, they may be done routinely for certain species or for law enforcement or certification samples. These special tests have been developed as by-products of routine testing procedures in the seed technicians' attempts to learn more about the quality of seed lots. Today most modern, well-equipped seed laboratories have the capability of conducting such tests.

Variety Tests

It is notably difficult to distinguish between crop varieties because of their similarity in morphological characteristics. Generally this is a reflection of their close genetic kinship. Often a small genetic difference may be the basis for a new variety if this difference contributes to resistance to disease or insects, winter hardiness, higher protein content, yield, or some other desirable trait that warrants release of the variety. Small genetic differences may be hard to detect in the mature plant and are even more difficult to distinguish by examining the seed.

In the early days of seed testing, varietal tests, when conducted, were relatively simple for two basic reasons: (1) there were fewer varieties, and (2) there were usually greater differences among varieties. Because of the success of modern plant breeding, the resulting variety explosion, and the appearance of many closely related varieties, seed analysts have been obliged to find newer, more sophisticated ways of distinguishing among varieties in the seed laboratory. Their methods are changing from visual observations of seeds and seedling morphology to detailed grow-out tests, or to the use of biochemical and cytological methods. Most of these methods are detailed in the AOSA Cultivar Purity Testing Handbook (AOSA 1991).

Visual Observation of the Seed. This is the simplest kind of variety test, and probably the earliest to be used. Although still useful, it is usually not reliable for positive varietal identification and should be used only in conjunction with other tests. For example, Chippewa 64 soybean can easily be distinguished from soybean varieties that do not have a black hilum. However, more sophisticated techniques must be used to distinguish it from other varieties that also have black hila. Seed size is also a useful index of variety; however, it is so variable and so environmentally influenced that it must be used only with extreme caution. When used with other characteristics, it is a definite aid in varietal identification. Other seed characteristics that are sometimes used for certain species are: color; shape of rachilla; lemma and palea characteristics; presence, absence, or shape of the awn; and presence or absence of pubescence on the seed.

Visual Observations of Seedlings. Many of the more useful varietal identification tests have been performed on seedlings. Such tests are useful because they may yield more information than do observations of the ungerminated seed and do not require as much time as field grow-out tests. Some seedling characteristics are: (1) leaf coloration patterns, (2) color patterns of the lower crown area, (3) vernation (folded or rolled) pattern of leaf in the sheath, (4) length of internodes, (5) pubescence, and (6) leaf shape or blade width.

Greenhouse, or Field Grow-Out Tests. When time is not a crucial factor in varietal identification, a greenhouse or field grow-out test is usually more dependable than seed or seedling tests. Greenhouse tests are usually performed in conjunction with seed and seedling tests to substantiate decisions made earlier. In grow-out tests, observations on flowering date, flower color, characteristics of seed produced, and vegetative characteristics, such as presence of pubescence on the leaf margins or at the tip of the auricles, may help distinguish one variety from another.

Modern crop varieties have become more closely related than varieties available 20 to 30 years ago. With recent varietal protection (see Chapter 16), and the added importance of accurate distinctions, more refinement must be made in the use of grow-out tests. European plant scientists are apparently far ahead of their North American counterparts in this method of varietal distinction, perhaps because of their history of varietal registration, varietal protection statutes, and strong seed control philosophy.

Ultraviolet Light Tests. Response under ultraviolet light has been used for both seed and seedling variety tests with varying success. The lemma, palea, and glumes of certain oat varieties contain substances that fluoresce when exposed to ultraviolet light. The ultraviolet light test, however, has limited usefulness, because many oat varieties show the same response—either fluorescence or nonfluorescence; therefore, the test is useful only when two varieties with opposite responses are being compared.

For many years the fluorescence test of ryegrass has been used with great success to help distinguish between two species—annual or Italian (*Lolium multiflorum*) and perennial or English (*Lolium perenne*) ryegrass. The past usefulness of this test was due to the fact that seedling roots of all known annual ryegrass varieties exhibited a positive response, while perennial varieties were nonfluorescent. It may be otherwise impossible to distinguish between annual and perennial ryegrass seeds, although the presence of an awn usually indicates annual ryegrass. However, this characteristic is not dependable, since the awn is fragile and may be detached by handling. The accuracy of the fluorescence test has been good enough to use as a measurement of the percentage of the two species when found together in mixtures, and the results are used for seed labeling purposes.

Many new ryegrass varieties have recently been developed, having both annual and perennial parentage; the fluorescence levels among these varieties vary between 0 and 100% and sometimes have little relationship to their annual or perennial habit. Tests have shown that these varieties may be characterized by their inherent fluorescence level and that this characteristic can be used to help in varietal identification.

Unlike the fluorescence test of oats, which uses the response of the seed coat under ultraviolet light, the test for ryegrass is conducted on the roots of five- to ten-day-old seedlings growing on white filter paper. The fluorescing substance has been isolated and designated as *annuoline*. The substance appears as an exudate from the ryegrass roots that are in contact with the paper medium.

Chemical Tests. The ideal variety test would be the exposure of a seed to some chemical that would clearly reveal its varietal identity in comparison with related varieties. Unfortunately, such a test does not exist. The nearest realization may be the *phenol test* (Figure 13.9), which has been used for varietal distinctions of both bluegrass and wheat. The test is performed by placing the seeds on a paper medium moistened with approximately 1% carbolic acid (phenol solution) for about four hours. Tests are evaluated according to the darkness of staining that occurs; the seeds of some bluegrasses stain in the embryo area and can be distinguished from nonstaining varieties,

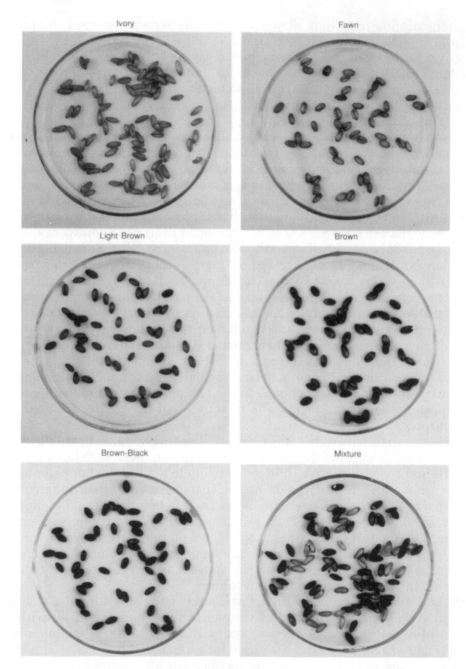

Figure 13.9. *The phenol test for wheat: Examples of the five different color categories (From AOSA Cultivar Purity Testing Handbook, 1991).*

while in wheat the entire pericarp is observed for the degree of staining. Wheat varieties can be categorized according to whether they stain very dark, medium dark, very light, or remain unstained.

Another very useful chemical test for distinguishing soybean varieties is the peroxidase test. Some soybean varieties contain the peroxidase enzyme in the seed coat while others do not. The test is conducted by removing the seed coat from the seed and placing it into a test tube to which is added 10 drops of a 0.5% guaiacol solution for 10 minutes. After that period, one drop of a 0.1% H_2O_2 is added to the test tube. A positive test is indicated by the immediate formation of a reddish-brown solution while a negative test will be indicated by a colorless solution. Other useful chemical tests include the hydrochloric acid test for oats, sodium hydroxide test for wheat, and potassium hydroxide test for red rice and sorghum. The procedures for all of these tests are provided in the AOSA Cultivar Purity Testing Handbook (AOSA 1991).

Chromosome Counts. During the 1960s, many new tetraploid grass and legume varieties were released in the United States. The doubled chromosome complement of these varieties as compared to their diploid counterparts has provided seed analysts with a built-in varietal distinction method by merely counting the number of chromosomes in seedling root tips (see Figure 13.10). Like many other variety tests, chromosome counts cannot be used to distinguish between different varieties with the same chromosome number. However, they are useful in detecting contamination, especially diploid contamination of tetraploid varieties. The tests are a valuable aid in monitoring the varietal quality of certain tetraploid certified grasses and legumes. Their use is more limited for higher polyploid species, such as wheat and bluegrasses, because of the difficulty associated with counting the chromosomes and the complexity of their polyploidy.

Electrophoresis. An exciting test that may have much potential for varietal distinction is the electrophoretic analysis for proteins, or isoenzymes—a test that can be used on seeds as well as on other plant parts. It has been used to distinguish among

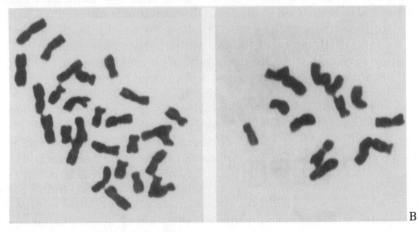

A   B

Figure 13.10. *Metaphase chromosomes of ryegrass: (A) diploid ryegrass, with 14 chromosomes, and (B) tetraploid ryegrass, with 28 chromosomes. (From Will et al. (1967).)*

ryegrass varieties and has been investigated by many workers as an aid in varietal identification. Electrophoresis has generated much interest among certification and seed control officials, particularly because of the increased need for varietal identification with the advent of the Plant Protection Act (see Chapter 16).

The electrophoresis test produces a separation of seed proteins, or isoenzymes, in an agar medium by establishing an electric field and permitting the proteins to arrange themselves within the media according to their polar (positive or negative electrical charge) nature; those with a more positive nature will align themselves near the negative pole, and those with negative tendencies will align near the positive pole. After the protein pattern has been established, it can be photographed and compared with the patterns of known varieties.

Electrophoresis of seed proteins and enzymes has proven to be highly versatile in varietal identification of seeds. As a result, this technique has been incorporated into the ISTA Rules for Seed Testing (ISTA 1993) and is described by the AOSA Cultivar Purity Testing Handbook (AOSA 1991). Still, this technique fails to differentiate a number of varieties in some crops and newer techniques to differentiate varieties continue to be examined. Different enzymes of course are synthesized according to the base sequence of the DNA molecule. A more powerful technique to discriminate varieties therefore would be one which can reveal the base sequences of the DNA molecule that produce these and other enzymes. The powerful technology of molecular genetics has produced a technique called random amplified polymorphic DNA or RAPDs for short. This technique can be used to examine the base sequence of the DNA molecule with techniques that are very similar to electrophoresis of seed proteins. Studies indicate that it is a very versatile approach to varietal identification that may be quite reasonable in cost (McDonald et al. 1994).

Chromatographic Methods. Both thin-layer and paper chromatography have been used with varying success in varietal distinction. Readily observable differences have been reported in thin-layer chromatographic bands among both ryegrass and soybean varieties, and, to a lesser extent, in oat varieties. Thin-layer chromatographic methods have also been used to distinguish soft white from durum wheat and between closely related members of the genus *Trifolium*. Paper chromatography has been used to aid in varietal distinctions of *Brassica* and other species (Payne 1986).

In routine varietal testing, the usefulness of chromatographic techniques is limited, primarily because it is difficult to use, a long time is required for the test, and also because different varieties do not always have observable differences in chromatographic bands. This latter problem can sometimes be overcome by refinement of the technique, for example, by use of different absorbents (thin layer), developers, and ultraviolet light to help distinguish differences on the chromatograph.

Other Cytological Methods. Although other cytological methods have not been used extensively in varietal distinction, they offer considerable potential. Cytological testing methods should become more valuable as new ways are found to introduce and direct chromosomal aberrations for creating new plant varieties having new predetermined characteristics (for example, disease resistance). In such instances, the presence of known chromosomal aberrations, such as deficiencies, duplications, inversions, and translocations, might be used for positive varietal identification.

Disease Resistance for Varietal Identification. A pathogen inoculation test can easily be used to distinguish between disease-resistant and disease-susceptible crop varieties that otherwise appear similar. This test has been successfully used to distinguish between phytophthora-resistant and susceptible soybean varieties (Figure 13.11) and also in corn hybrids with different responses to southern corn leaf blight—for example, Texas male sterile varieties versus normal cytoplasm corn hybrids.

Determining the Effectiveness of Seed Treatments

This test is used to determine the effectiveness of seed treatment with chemical pesticides. Although state and federal laws require treated seed to be dyed a contrasting color (see Chapter 16), the effectiveness of the coloration does not indicate the completeness and effectiveness of the treatment. The common method of testing the effectiveness of fungicide treatment is to plate the treated seed on agar media and apply a covering of *Gibberella, Glomerella*, or *Cingulata* spores over the entire media surface. If seeds are ineffectively treated, the spores should germinate and grow around the seeds. A clear zone soon appears around each effectively treated seed where spore germination is prevented (Figure 13.12).

Figure 13.11. *A test for resistance to pyththora rot in two soybean varieties. The one on the left is susceptible. (Courtesy of A.F. Schmitthenner).*

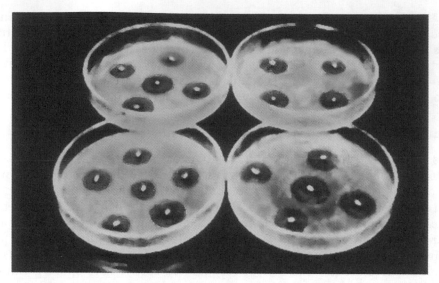

Figure 13.12. Inhibition zones around seed of an agar plating test indicate effective seed treatment.

Effectiveness of Inoculation of Legume Seed

Tests for effectiveness of legume seed inoculation can be performed by the grow-out of seeds in the greenhouse or in growth chambers and their comparison to well-inoculated control samples. This test has been routinely performed in Indiana, and the information obtained used in the enforcement of that state's seed law pertaining to preinoculated seed (Figure 13.13).

Seed Moisture Test

Evaluation of seed moisture content is an extremely important determination in seed testing. Knowledge of the seed moisture content is useful because it provides information regarding the potential for harvesting, cleaning, and planting injury(ies) as well as the likelihood for successful long-term storage. A seed moisture test is conducted by first weighing the seeds to determine their "wet" weight. Then, the seeds are placed into an oven set for 100°C for grass, legume, and cereal seeds or 85°C for tree seeds for 24 hours except for large or thick-coated pine seeds and oily seeds such as *Brassica* species in which 48 hours at 85°C is required. After the drying period, the seeds are removed from the oven, placed in a desiccator for 15 minutes to cool, and the "dry" weight of the seeds determined. Two methods are used to express seed moisture content and are calculated in the following way:

Wet Weight

$$\frac{\text{Weight before drying} - \text{Weight after drying}}{\text{Weight before drying}} \times 100 = \% \text{ Moisture content}$$

(Wet weight basis)

Figure 13.13. *Left: plants from an effective humus inoculum applied to red clover seed at time of planting. Right: ineffective preinoculated seed. (Courtesy of L. C. Shenberger).*

Dry Weight

$$\frac{\text{Weight before drying} - \text{Weight after drying}}{\text{Weight after drying}} \times 100 = \% \text{ Moisture content}$$

$$(\text{Dry weight basis})$$

In the seed trade, seed moisture is described on a fresh weight basis. The value will never exceed 100%. Research scientists often determine this value on a dry weight basis and the value can often exceed 100%. Thus, it is important that the analyst be aware of the specific procedure followed when percent seed moisture content is given, since the calculations can lead to divergent results and interpretations.

SEED TESTING TOLERANCES

The importance of having the working sample be **representative** of the entire seed lot can hardly be overemphasized since a nonrepresentative sample would provide erroneous results of the quality of the seed lot. Moreover, each time the working sample is reduced, determinations of the seed lot quality tend to become less accurate. In addition, a seed sample is composed of individual biological units with their own inherent performance characteristic, so it is not surprising that variability in test results between two working samples obtained from the same submitted sample would be obtained. Thus, variation in test results from the same seed lot is expected and normal. But, how much variation in test results is acceptable or tolerable? That is the purpose of seed testing tolerance.

Tolerances are used to define statistically acceptable limits within which different test results may be expected to vary. Tolerances have been established for the more common tests performed on seeds. They usually provide for the variability expected from random sampling error and some account for variability caused by interpretational errors or seed lot heterogeneity. Most studies of variation have shown that actual variability among different test results often exceeds that accounted for by existing tolerances. Because of this, most seed law enforcement agencies sometimes allow "administrative tolerances" when determining if seed is improperly labeled. When properly applied, tolerances specify when results are "out of tolerance" or if a retest is necessary. Tolerances are based on the fact that the values reported have a probability of error of 5%. Tolerance is defined as the difference permitted between a labeled percentage (the first analysis) and the test results obtained by a laboratory when checking the accuracy of the labeled information.

Tolerances for purity, germination, fluorescence, and noxious weed seed examinations are included in the Rules for Testing Seeds. The following example illustrates the use of germination tolerances. A state seed inspector picked up a Merion bluegrass seed sample from a lawn and garden store which was labeled as germinating 95%; an official germination test in the seed laboratory showed the sample to germinate 87%. To determine if the sample is mislabeled, both tests are given equal chance of being correct; thus they averaged ($\frac{95/87}{2}$) to give a weighted mean of 91%. The tolerance for the weighted mean of 91% is 6. The difference between the labeled germination and the second test is 8; consequently, the sample is out of tolerance and the seed lot is considered to be mislabeled.

SEED TESTING ORGANIZATIONS

The importance of seeds as an agricultural commodity is mirrored in the complexity of international, national, and state organizations concerned with assessing its quality (Figure 13.14). Many of these organizations are interested in the orderly movement of seeds from one country or state where the seeds are produced to the next where they are used. Of particular interest to seed analysts are those organizations devoted exclusively to seed testing. These include ISTA, AOSA, the Society of Commercial Seed Technologists (SCST), and the Commercial Seed Analysts Association of Canada (CSAAC).

International Seed Testing Association

The International Seed Testing Association (ISTA) is the only worldwide organization dedicated to seed testing on an international scope. Its goals include: (1) development of rules for seed testing, (2) standardization of testing techniques, (3) seed research, and (4) cooperation with other international agencies for seed improvement.

The ISTA had its beginnings in the early 1900s, when seed technicians from several European laboratories felt the need for more exchange of seed testing information and communication among seed laboratories in different countries. During this period, the international seed trade was becoming established, creating the need for standardization of seed quality concepts across national borders.

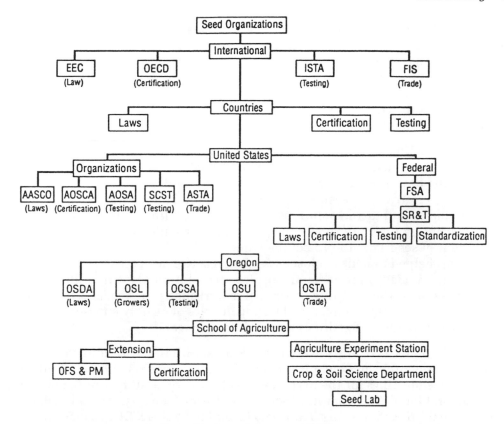

SELECTED SEED ORGANIZATIONS

AASCO - American Association of Seed Control Officials (Uniform Laws)
AOSA - Association of Official Seed Analysts (Standardize Testing)
AOSCA - Association of Official Seed Certifying Agencies (Standardize Certification)
ASTA - American Seed Trade Association
EEC - European Economic Community (Marketing)
FIS - Federation of International Seedsmen
FSA - Federal Seed Act (Regulates Interstate Seed Movement)
ISTA - Interrernational Seed Testing Association (Standardize Testing)
OECD - Organization for Economic Cooperation and Development (International Certification)
RUSSL - Recommended Uniform State Seed Law (Standardize State Seed Laws)
SCST - Society of Commercial Seed Technologists
SR&T - Seed Regulatory and Testing

Figure 13.14. *A typical flowchart illustrating important seed organizations at the international, national, and state levels. (Courtesy of Oregon State University Seed Lab).*

This need was first put into action at the 1905 Botanical Congress in Vienna, during which several people met informally to plan a European seed testing association. Plans were made for a Seed Testing Congress in Hamburg, Germany, in 1906. Another Seed Testing Congress was held in 1910. Due to conditions in Europe, another meeting was not held until Professor K. Dorph Peterson, of Copenhagen, called a Third Testing

Congress in Copenhagen in 1921, where the European Seed Testing Association was formed. Under the auspices of this group, the Fourth International Seed Testing Congress was held in Cambridge, England, in 1924. At this meeting the name was officially changed to the International Seed Testing Association.

Since its beginning, the ISTA has had great growth and accomplishments. It has become truly worldwide in both scope and representation. Membership in ISTA now includes 117 laboratories from 53 countries. Some of its notable accomplishments are:

1. In promoting uniformity of seed testing results among laboratories, it has facilitated movement of seed across international boundaries and helped farmers get the best possible seed regardless of the country of origin.
2. It has arranged for seed scientists and technicians to meet and discuss their problems and to find solutions for them. By drafting seed testing rules and by discussing their interpretations, they have provided a sound basis for enactment of seed laws to protect the farmer.
3. It has helped to achieve closer association between test results and field performance, assisting farmers to recognize seed of high planting value.
4. It has organized training courses and workshops in Africa, Asia, and South America to help promote seed testing in areas of rapidly emerging agriculture.
5. It has provided a focal point of seed knowledge.

The ISTA holds a congress every three years at different locations throughout the world to hear scientific and technical papers from its members and to provide forums and committee meetings for the exchange of information and the finding of solutions to mutual problems. The complete activities at each congress are published in its official journal—*Seed Science and Technology* (prior to 1972—*ISTA Proceedings*).

Association of Official Seed Analysts

The Association of Official Seed Analysts (AOSA) is an organization composed of analysts from official state, federal, and university seed laboratories throughout the United States and Canada. Its contribution in bringing seed testing to a respected and highly sophisticated level in these two countries has been enormous. Perhaps its greatest contribution has been the development of rules and procedures for seed testing, and the standardization of their interpretation. It also has had great influence on seed legislation in every state, as well as at the federal level. The Referee Committee of AOSA distributes problem seed samples to different laboratories for testing. Such activity helps attain standardization in procedures and interpretation among different laboratories.

The AOSA was formally organized in Washington, D.C., in 1908, with 16 states represented. Since its early days, the AOSA has held annual meetings almost every year. The minutes of its annual meetings and the papers presented are published in the *Journal of Seed Technology* (formerly *AOSA Proceedings*). It also publishes a newsletter three times a year, which includes articles on seed testing topics. The Association has published many special publications, among which are a series of handbooks on selected topics.

Society of Commercial Seed Technologists

The Society of Commercial Seed Technologists (SCST) is a society of seed analysts from private or commercial seed laboratories throughout the United States and Canada. This includes self-employed seed analysts who test seed on a custom-fee basis and analysts from seed companies, who ordinarily are salaried or on a commission, and who test seed handled in the company's business.

The SCST originated in the early 1920s, largely as a liaison between the AOSA and the American Seed Trade Association (ASTA), because of their mutual need for better acquaintance and communication. The AOSA had regarded the ASTA suspiciously because some seed producers had flagrantly violated seed labeling laws and occasionally used fraudulent merchandising schemes. ASTA members regarded the AOSA as a well-meaning, but highly technical and regulatory-minded organization that promoted complex and often conflicting seed legislation. By 1922 some of the larger seed companies had their own seed testing laboratories and analysts. At the combined AOSA and ASTA meeting in Chicago in 1922, 13 commercial seed analysts met to form what was first called the American Society of Commercial Analysts. From the time SCST was first organized, there was good cooperation between the SCST and AOSA. These two organizations held their annual meetings at a common place, presented papers, exchanged ideas, and participated in referee testing together. The AOSA welcomed the new organization because it created a new bond of communication with the ASTA on a more technical and professional level. The respect for the SCST was strengthened by the high standards it established for society membership. In 1947, membership standards were further strengthened by the establishment of a comprehensive examination for membership. A minimum score of 80 was established for passing the test, which included: (1) seed identification, (2) purity and germination techniques, (3) evaluation of normal and abnormal seedlings, (4) knowledge of botany, (5) Canadian and United States federal seed laws, (6) official rules for seed testing and tolerances. In addition, a combination of minimum college credit in the biological sciences and experience in seed analysis was established as a requirement. Analysts who pass all requirements and are accepted by a two-thirds vote by SCST members have the right to use the Society Seal and Insignia. The official seal is proof that the SCST member is a *Registered Seed Technologist*, and this becomes part of the analyst's credentials. Thereafter, the seal accompanies the results of any test performed under his or her supervision.

Commercial Seed Analysts Association of Canada

In 1944, six commercial seed analysts, formerly with the Toronto Seed Laboratory of the Canada Department of Agriculture, met in Toronto and formed the Ontario Commercial Seed Analysts Association. The purposes of this organization were (1) to keep abreast of new methods of seed testing, and (2) to assist analysts in overcoming any problems that might arise in their work. Analysts from other Canadian provinces quickly showed an interest in this association, and at the second meeting in 1945, the name was changed to the Commercial Seed Analysts Association of Canada (CSAAC).

By 1967, the association had grown to 34 members, with 11 each in Ontario and Alberta, seven in Manitoba, one in Quebec, three in the United States, and one in England.

The Association holds its annual meetings in Toronto, and proceedings of this meeting are published in the *Maple Leaf*, the official publication of CSAAC.

Most members of the CSAAC are also members of the Society of Commercial Seed Technologists, so close communication is maintained between these two organizations. Members of CSAAC also attend meetings of the Association of Official Seed Analysts.

Questions

1. Do you consider seed testing to be a science, a skill, or an art?
2. What four components are considered to be part of the purity separation?
3. Define a noxious weed. How is the incidence of noxious weed seeds recorded in the purity test results? Why is a larger seed sample examined for noxious weed seeds than for the purity test?
4. Explain the normal seedling concept in interpreting laboratory germination tests.
5. What is the difference, if any, between hard seed and firm ungerminated seeds?
6. List several ways of stimulating faster germination in the seed-testing laboratory.
7. Describe several ways of distinguishing between varieties in a seed-testing laboratory. Which do you consider to be the most practical in routine seed testing? The least practical?
8. Do you feel there should be more emphasis on pathological testing in the United States and Canada? Can more emphasis be justified in view of its costs and results? If so, which kind of pathological tests should be used? which are presently in use? (See Chapter 14.)
9. Do you think most seed samples submitted by seed growers for testing are properly drawn? Do you believe a grower would purposely misrepresent a sample to make a seed lot appear better? How many bags should be sampled from a seed lot containing 250 bags of seed?
10. What is the difference between sampling and subsampling?
11. What are the purposes of tolerances? Do you know how they are applied for the various test results?
12. Name several seed testing organizations. What is a Registered Seed Technologist?

General References

Association of Official Seed Analysts. 1991. Cultivar Purity Testing Handbook. (eds. M. B. McDonald and R. Payne). Contribution No. 33 to the Handbook on Seed Testing. Association of Official Seed Analysts. 78 pp.

Association of Official Seed Analysts. 1991. Rules for testing seeds. *Journal of Seed Technology* 12(3):1–109.

Douglas, J. E., ed. 1980. *Successful Seed Programs: A Planning and Management Guide*. Boulder, Colo.: Westview Press.

International Seed Testing Association. 1993. International rules for seed testing. *Seed Science and Technology* 21:1–288.

Justice, O. L. 1972. Essentials of seed testing. In: *Seed Biology*, vol. 3, pp. 301–370. New York: Academic Press.

McDonald, M. B., L. J. Elliot, and P. M. Sweeney. 1994. DNA extraction from dry seeds for RAPD analyses in varietal identification studies. *Seed Science and Technology* 22:171–176.

McDonald, M. B., R. Danielson, and T. Gutormson. 1992. *Seed Analyst Training Manual.* Association of Official Seed Analysts, Lincoln, NE.

Payne, R. C. 1986. Variety testing by official AOSA seed laboratories. *Journal of Seed Technology* 10:24–36.

Thompson, J. R. 1979. *An Introduction to Seed Technology.* New York: John Wiley and Sons.

U.S. Department of Agriculture. 1961. *Seeds: The Yearbook of Agriculture.* Washington, D.C.: U.S. Government Printing Office.

   1. Andersen, A. M., and C. M. Leach. Testing seeds for seedborne organisms, pp. 453–457.
   2. Carter, A. S. In testing, the sample is all-important, pp. 414–417.
   3. Colbry, V. L., T. F. Swofford, and R. P. Moore. Tests for germination in the laboratory, pp. 433–443.
   4. Davidson, W. A., and B. E. Clark. How we try to measure trueness to variety, pp. 448–452.
   5. Justice, O. L. The science of seed testing, pp. 406–413.
   6. Justice, O. L., and E. C. Houseman. Tolerances in the testing of seeds, pp. 457–462.
   7. Musil, A. F. Testing seeds for purity and origin, pp. 417–432.
   8. Zeleny, L. Ways to test seeds for moisture, pp. 443–447.

U.S. Department of Agriculture. Production and Marketing Administration in cooperation with Plant Industry, Soils, and Agricultural Engineering. 1952. *Testing Agricultural and Vegetable Seeds.* Washington, D.C.:U.S. Government Printing Office.

Wheeler, W. A., and D. D. Hill. 1957. *Grassland Seeds.* Princeton, N.J.:D. Van Nostrand Company.

Will, M. E., W. E. Kronsted, and D. M. TeKrony. 1967. A technique using lindane and cold treatment to facilitate somatic chromosome counts in *Lolium* species. *Proceedings of the AOSA* 57:118.

14

Seed Pathology and Pathological Testing

In many parts of the world, testing for seedborne diseases is an integral part of the routine inspection for seed quality. However, in North America, pathological testing has not been as important as purity and germination testing. This is partially because of uncertainty about whether the analytical results are significant enough to justify the expense of maintaining a pathological testing program. Additionally, large samples may have to be screened for results to be meaningful. Some pathogens can cause severe losses if as few as one seed in 10,000 to 50,000 seeds are infected. Problems in reliably detecting such small levels of infection and relating them to field losses are formidable, and these tests inevitably require much larger space and labor investments than do most other types of seed quality tests. Consequently, pathological testing of seed in North America has not developed to the level needed and is too often not a part of the routine seed laboratory analysis.

For pathological testing of seeds to be a reasonable requirement for sale of seeds, several conditions must be met. First, it must be established that seedborne infestation causes reduction in plant stands, leads to field diseases, or causes other problems. The mere existence of pathogens in seeds indicates little about the likelihood of the pathogen causing problems for the ensuing crop. Second, the level of acceptable infection must be established. For example, if 5% of the seeds in a given seed lot must be infected for significant economic loss to occur, it makes little sense to reject lots in which only 0.1% of the seeds are infected. Third, if the disease can develop explosively, seed pathogen testing may have to be accompanied by legal restrictions on the sale of untested or infested seed lots. Grower A may plant clean, healthy seeds, but may still suffer substantial losses if a neighbor plants infested seeds and an efficient vector is present (e.g., a tractor, an insect, or the farmer himself). However, if a disease is not present in an area, seeds infested with the pathogen should not be planted in that area, regardless of whether the pathogen can be demonstrated to infect plants from seedborne inoculum. Many diseases have no doubt been spread in this fashion and are still reaching new areas today. For example, pea seedborne mosaic virus reached the

United States this way, and snapdragon rust probably arrived in Australia via the same route.

Several developing forces promise to change the status of pathological testing in North America. One such force is the emergence of pathological testing in species for which seedborne diseases seriously endanger crop productivity. For example, seedborne bacterial diseases occur in several important crops in which pathological test results can be related to crop productivity and translated into profit and loss. Another factor causing increased attention to seed health testing is international trade in seeds. Almost all countries require phytosanitary certificates for imported seed to ensure that seedborne pathogens are not introduced from another country. Thus, domestic as well as international forces are contributing to the recognition of seed health testing as an important part of the seed testing process.

THE SEED—A MICROCOSM OF MICROBES

The seed has been described by Sinclair (1979) as a microcosm of microbes, with the potential for carrying a wide variety of fungi, bacteria, viruses, and (sometimes) nematodes, many of which can cause diseases in seedlings or plants. Some live on the seed surface and do not visibly affect the seeds' appearance. They may become harmful only when environmental conditions favor their growth and reproduction. Since such conditions normally exist in germination chambers, these may cause problems with seed rots and seedling blight and contribute to variability in germination results. Other microorganisms live in nonliving outer tissues of the seed such as bracts, pericarps, or seed coats and attack the germinating seedling when conditions become favorable. Still others are borne inside the seed, either on or inside the embryo or endosperm tissue. Although these do not usually kill the seed, they may delay germination and result in weak seedlings. Others survive in the embryo and resulting seedling to infect plants and crops when grown from such seed.

Seed microbes can be divided into fungi, bacteria, and viruses.

Fungi

Fungi have been present on this earth as long as any other living organism as evidenced by over 100,000 distinct species. Yet, with all of this diversity, they have several common characteristics that help us understand their important role as seed pathogens. First, they have no chlorophyll like most plants so they cannot make their own food. As a result, they live on substances provided by other plants or plant parts such as seeds. When these organisms used for energy are living, the fungi are parasitic. In other cases, the organisms or tissues they invade are dead in which case the fungi are saprophytes that cause rots or decay. Second, fungi produce rapidly growing mycelia which are microscopic branches that produce digestive enzymes and acids. These substances digest the nutritive material in which the mycelia are present and provide sustenance to the fungus for continued growth. Third, fungi produce spores which are the mycological reproductive equivalent of seeds from plants. These spores are produced in such astronomical numbers that fungal contamination can increase exponentially. Fourth, fungi require water, a favorable temperature, food, and oxygen for growth. Most fungi require a high water content. If water can be squeezed out of a tissue, then fungal growth is likely. They also grow best a fairly high temperatures, usually about 30°C for optimum growth. When temperatures are below 5–10°C or above 35°C, fungal growth is inhibited although some exceptions exist. Fungi that attack seeds can be divided into field and storage fungi based on their environmental requirements.

Field fungi invade seeds almost exclusively during development or after physiological maturity and usually have completed their damage prior to harvest. Common and serious examples of field fungi include *Alternaria, Cladosporium, Fusarium*, and *Helminthosporium*. For field fungi to cause seed damage, they require seed moisture contents in equilibrium with relative humidities greater than 90%. For cereal seeds high in starch such as barley, corn, oat, and wheat, this is usually at a seed moisture content of 20–25% or higher on a wet weight basis. These moisture contents are far above those encountered when seeds are stored. When the seeds dry down below 20% moisture and eventually go into storage, the continued growth of field fungi is reduced.

In contrast to field fungi, storage fungi actively invade seeds and cause damage under conditions that can be encountered during storage. Among the most prevalent and important storage fungi are those from the genera *Aspergillus* and *Penicillium*. They invade seeds in equilibrium with relative humidities between 65 and 90%. For starchy seeds, this is at moisture contents of 13–20%; soybeans at moisture contents of 12–19%; and for oil seeds at moisture contents of 5–12% (Table 14.1). Almost all storage fungi grow during conditions of seed storage that are favorable to their growth. The one exception is *Aspergillus flavus* which sometimes invades seeds of corn, peanut, and cotton in the field. Under normal conditions, the activity of storage fungi is not specific to seeds as a substrate. They will grow on almost anything, including chair stuffings, mattresses, pillows, and other materials when provided the appropriate environment.

Storage fungi are almost always classified as saprophytes (plants that live on dead tissue). They do not produce mycelia that release exocellular digestive enzymes and acids to digest invaded living tissue. So how do they create such widespread damage, including the decay of living tissue? The answer seems to be that these saprophytes initially colonize only those seed parts which are dead such as the seed coat. There they proliferate and through their normal metabolism synthesize toxins that are secreted to the outside of the fungus. The toxin kills the living tissue which provides additional substrate for digestion by the fungus. Perhaps the best known example of this process is the toxin produced by *Aspergillus flavus* called aflatoxin. This toxin not only is toxic to the seeds on which it grows but is also highly toxic to animals that digest the infected seed or grain. Because of the seriousness of this toxin, peanuts, corn, and cotton intended for feed are routinely inspected by the Food and Drug Administration for the presence of aflatoxin. Similar health problems exist with toxins produced by *Fusarium* species, *Aspergillus ochraceus*, and *Claviceps purpurea*.

Bacteria

A large number of bacteria exist in nature but only about 200 species are known to cause plant disease. In general, very few of these cause disease in seeds because

Table 14.1. Equilibrium Moisture Contents[a] of Common Grains, Seeds, and Feed Ingredients at Relative Humidities of 65–90% and Fungi Likely to Be Encountered

Relative Humidity (%)	Starchy Cereal Seeds, Defatted Soybean and Cottonseed Meal, Alfalfa Pellets, Most Feeds	Soybeans	Sunflower, Safflower Seeds, Peanuts, Copra	Fungi
65–70	13.0–14.0	12.0–13.0	5.0–6.0	*Aspergillus halophilicus*
70–75	14.0–15.0	13.0–14.0	6.0–7.0	*A. restrictus, A. glaucus, Wallemia sebi*
75–80	14.5–16.0	14.0–15.0	7.0–8.0	*A. candidus, A. ochraceus,* plus the above
80–85	16.0–18.0	15.0–17.0	8.0–10.0	*A. flavus, Penicillium,* plus the above
85–90	18.0–20.0	17.0–19.0	10.0–12.0	*Penicillium,* plus the above

From: Christensen and Sauer (1982).

[a]Percentage wet weight. The figures are approximations; in practice, variations up to ± 1.0% can be expected.

free water (high seed moisture content) and moderate to warm temperatures are required for bacterial growth and disease development. As a result, seeds typically are invaded by bacteria and used to assist in dispersal of the pathogen. They enter the seed through wounds created by insect vectors or they are translocated by the infected mother plant into the developing seed. As the seed matures and dries down, the bacteria become dormant. Then, during imbibition and subsequent germination, the bacteria rapidly resume growth, multiply enormously in the seedling and plant tissue. They cause necrosis of cells, abnormal growth of galls and tumors, breakdown of tissues (soft rots), and blockage of water-conducting vessels (wilts). Some of the worst bacteria that are seedborne and cause plant disease come from the genera *Agrobacterium* (crown gall), *Bacillus* (seed rot), *Corynebacterium* (wilt), *Erwinia* (soft rot), *Pseudomonas* (blight), and *Xanthomonas* (black rot/blight).

Viruses

Viruses are extremely small organisms usually composed of a nucleic acid strand encased in a proteinaceous shell. By definition, a virus is considered a transmissible parasite whose nucleic acid genome is less than 3×10^8 daltons and needs ribosomes and other components of the host cell for multiplication. Like bacteria, most viruses do not cause direct seed damage but use the seed for dispersal and subsequent plant infection. About 20% of the known plant viruses are seed transmitted. Plants can become infected during pollination and through wounds often caused by insects. The developing seed is then systemically infected by the mother plant. It is sometimes difficult to identify plant diseases caused by viruses because they often mimic nutritional deficiencies or plant response to other environmental stress. Typical conditions include chlorosis, stunting, wilting, mosaic patterns, and necrosis. The bleeding hilum found in soybean seeds is one visible example of a seedborne virus caused by the soybean mosaic virus.

CONTROL OF SEEDBORNE DISEASES

Preharvest Control

Preharvest control of seedborne diseases may be accomplished by one of three different methods: (1) selection of disease-free seed production areas, (2) cultural practices, and (3) point of origin inspection. In the first case, seed is produced under environmental conditions that restrict the occurrence of diseases. Thus, dry edible bean and snap bean seed produced in dry, irrigated areas of Idaho or California are more likely to be free of bacterial blight (*Pseudomonas vulgaris*) than seed produced in the Great Lakes states where the weather is more humid and favorable for disease development.

Regardless of seed production area, cultural practices are crucial in the prevention and control of seedborne diseases. These include the following:

   a. Planting disease-free seed.
   b. Treatment of seed with antibiotics.
   c. Spraying seed fields with fungicides, bacteriocides, and other antibiotics to prevent disease buildup.
   d. Hand roguing of diseased plants.
   e. Avoiding overhead irrigation which might otherwise create conditions favoring disease buildup.
   f. Use of resistant cultivars.
   g. Crop rotation.
   h. Isolation of seed fields from sources of potential infection.
   i. Chemical or biological control of insect vectors.

The third preharvest control of seedborne diseases is inspection of seed fields so that potential problems may be detected and eliminated prior to harvest. Diseased areas may be destroyed or diverted from seed use, or the entire field can be diverted. While these precautions may not completely prevent contamination by surfaceborne dusts, they do lower the probability of seed infection.

Postharvest Control

Postharvest control of seedborne diseases should be considered only as a last resort, since it is better to prevent the occurrence of seedborne diseases than to eradicate disease infection (or infestation) that is already present. However, several methods may help upgrade the phytosanitary quality of seed after harvest. These include (1) surface disinfectant by chemical seed treatment, (2) separation of diseased seed and foreign material, (3) hot-water treatment, and (4) organic solvent infusion of antibiotics.

Treating seed with antibiotics is usually effective only against surfaceborne pathogens, but in some cases systemic antibiotics (e.g., carboxin) can penetrate into the seed and eradicate internal infection. Sometimes penetration of antibiotics can be improved by organic solvent infusion. Separation measures are effective for eliminating seeds or foreign material in a seed lot that is disease infested. An example of this is ergot sclerotia in cereal seed, which can be eliminated by cleaning.

Hot-water treatment can be used to kill infection in the seed without destroying seed viability. Prior to the development of systemic fungicides (e.g., carboxin) this was the only effective method of controlling loose smut in wheat and barley seed.

FUNGI ASSOCIATED WITH SEED

Fungi cause the largest number of plant diseases and occur more commonly in or on seeds than bacteria or viruses. More than 8000 species of fungi have been identified as plant pathogens. Fungi associated with seeds consist of both saprophytic and pathogenic fungi. Saprophytic fungi are not specific to any particular host and may be found on seeds of various plants, whereas pathogenic fungi are usually confined to a limited host range. Both types may occur on the seed surface, in cracks, or inside the seed coat, but pathogenic fungi may also occur within the seed itself. While saprophytic fungi may cause problems in the seed testing laboratory by contaminating germination media, pathogenic fungi endanger crop productivity and are of great economic importance to agriculture because: (1) infected seeds may not germinate, (2) infected seeds provide inoculum for further spread of the disease, and (3) seed infection prior to harvest may cause reduction in both crop yield and seed quality.

Fungi are composed of a threadlike vegetative body called *mycelium*. They reproduce by means of spores that have a function similar to seeds in higher plants. Like seeds, fungal spores vary greatly in size, shape, and color; however, they are much smaller than seeds and microscopic in size. Some fungi do not produce sexual spores, but reproduce by means of vegetative structures such as sclerotia, which are hardened, compacted masses of mycelium. An example of a sclerotium is ergot, such as that formed by the fungus *Claviceps purpurea*.

METHODS OF DETECTION

Agar Testing

One of the oldest and most common pathogenic tests is the agar test for identification of seedborne fungi (see Figure 14.1b). Agar is a carbohydrate medium prepared from certain species of seaweed. It contains few nutrients for the growth of fungal pathogens; thus, it is usually supplemented with extracts from plants such as potato or various fruits and vegetables, which are in turn supplemented by sugars, salts, antibiotics, or other agents. The agar medium is prepared by mixing agar powder with an appropriate quantity of water and nutrient additives. This mixture is sterilized in an autoclave for 15 to 20 minutes and cooled to about 50°C, at which time an antibiotic may be added. This mixture is carefully poured into petri dishes by lifting the lid only enough to pour in the agar, thus avoiding contamination. It is then allowed to cool and solidify for 20 minutes after which it is ready for use.

Seed to be tested should be surface-disinfected by pretreating for one minute in a 1% available chlorine solution of sodium hypochlorite (NaOCl) prepared by diluting

Figure 14.1. *Agar plate test for seedborne fungi*

A. *Mycosphaerella* spp. on pea, agar plate test after 7 days: (1) Colonies of *Mycosphaerella* spp at 11 and 12 o'clock, large contaminant colony of *Alternaria tennusis* at 6 o'clock; (2 & 3) Close-up of colony at 12 o'clock, note peppery appearance of flask-shaped pycnida (fruiting structures) formed in agar surrounding the seed; (4) Close-up of colony at 12 o'clock showing pycnidia and pycnidial ooze containing spores (Courtesy of Jim Sheppard, Agriculture Canada).

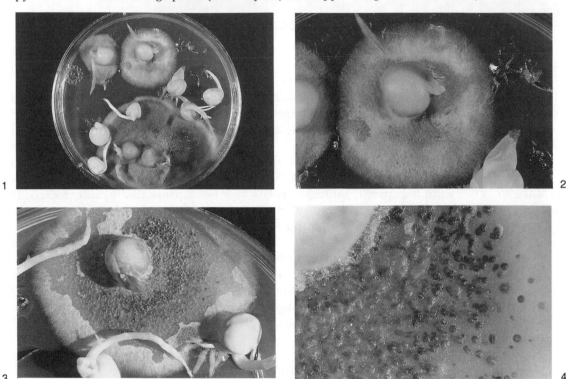

B. Same as in A, but *Ascochyta lentis* on lentil; note characteristic pycnidial ooze.

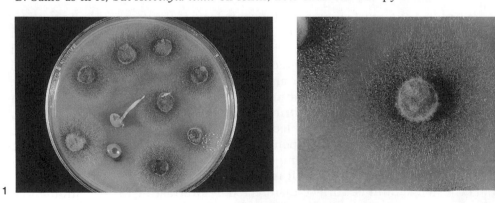

20 parts of laundry bleach (5.25% NaOCl) with 85 parts of water. For deep-seated infection, seeds may be treated with stronger solutions (e.g., 5.25% available chlorine). This eliminates contamination of the seed coat by saprophytes which tend to develop rapidly on the agar and may inhibit or completely obscure the slower-growing pathogens. Usually about 10 (depending on size) seeds to be tested are surface-sterilized and individually placed on the agar surface with a forceps. After placing each seed, the forceps tip is carefully disinfected by dipping into a 70% alcohol solution and passing through a flame.

Sometimes bacterial colonies develop on the agar or blotter (see below) and inhibit fungal growth, making identification difficult. This can be overcome by adding an antibiotic such as streptomycin sulfate to the autoclaved agar medium (after it cools to 50° to 55°C) or to the water used to moisten blotters in blotter tests.

After plating, the petri dishes are incubated at 20° to 25°C for about five to eight days, at which time seedborne pathogens (fungi) are identified on the basis of colony (vegetative) and spore characteristics.

Recently, disposable plastic petri dishes have almost completely replaced the glass petri dishes that had been used previously. These offer savings in labor, are convenient, and are more energy-efficient than glass dishes, which need to be emptied, cleaned, and sterilized before each use.

Blotter Tests

Blotter tests (see Figure 14.2) for pathogenic fungi are similar in technique to germination tests in that seeds are placed on moistened layers of blotter paper or filter paper and incubated under conditions that promote fungal growth. The procedure consists of saturating the blotters with sterile water, then allowing excess water to drain off briefly before the seeds are planted manually with a forceps, a vacuum counter, or planting board. Regardless of the placement method, the seeds should be evenly spaced to avoid contact with each other.

Profuse seedling growth in blotter tests may tend to make interpretations difficult.

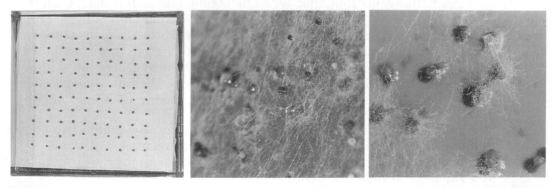

Figure 14.2. *Blotter test for Phoma lingam on* Brassica *seed (Rapeseed): 1. Arrangement of seed on blotter; 2 & 3. Pycnida growing on blotter after 10 days incubation. Note characteristic spore exudate (Courtesy of Jim Sheppard, Agriculture Canada).*

This difficulty can be overcome by adding enough 2,4-D sodium salt to provide a 0.2% moistening solution. Seedling growth can also be inhibited by a freezing technique that allows the seeds to imbibe for 24 hours (48 to 72 hours for corn) followed by a 24-hour exposure at $-15°C$ before normal incubation. (Normal incubation usually consists of temperatures of 20° to 25°C with a 12-hour day-night NUV light cycle for five to seven days.) Sporulation of many fungi is stimulated by alternating cycles of blue light and darkness. Following this period, the test is interpreted on the same basis as agar tests. However, killed seeds support luxuriant fungal growth and avoid the necessity of examining seeds through a tangle of seedlings. However, since freezing makes agar mushy, seeds must be placed on a paper substrate for freezing and afterwards transferred to blotters.

Virulence Tests

Virulence tests consist of making further isolations of suspected pathogens from blotter or agar tests, culturing them on agar for identification (spores, vegetative growth), inoculating the healthy plants (usually seedlings) for observation of pathogenecity symptoms, and subsequently reisolating and culturing the pathogen on a suitable substrate. This procedure follows Koch's postulates and is a method of positive identification of any pathogen. Inoculations may be by injection or other methods (e.g., spraying or dusting following mechanical abrasion).

Noncultural Tests

Several seedborne fungal pathogens can be detected by visual inspection of the seed sample, or special noncultural techniques. For example, ergot (*Claviceps purpurea*) contamination may be detected by the presence of large dark sclerotia. These sclerotia, when planted, produce fruiting bodies called apothecia. Apothecia further produce spores that provide inoculum in crops planted with ergot-infested seed.

Visual examination can also be effective for detecting smut balls of bunt, or stinking smut (*Tilletia foetida* or *caries*) of wheat. These consist of the pericarp of the caryopsis which has been completely replaced by black smut spores. Normally, many of the smut balls break open during threshing and spread spores to other seeds throughout the lot. These adhere to the caryopsis and serve as inoculum to the next generation. Although such spores are usually highly visible on the seed surfaces, the following procedure provides a systematic way of detecting and quantifying stinking smut infestation.

1. Wash the spores from the seed surface by shaking the seed sample in a known amount of water containing a small amount of detergent.
2. Centrifuge the washings from the sample and resuspend the resulting volume in a small, measured quantity of fluid.
3. Identify and quantify the spores under a high-powered microscope. Quantification of the spores can be made by counting the number of spores present in a given volume or over a given surface area of liquid.

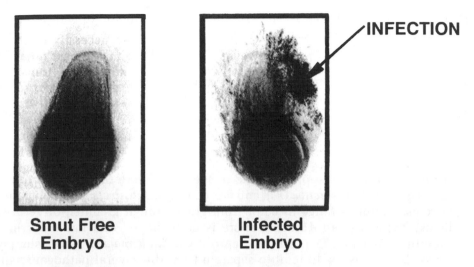

Figure 14.3. *The loose smut test of wheat showing noninfected versus infected embryo. (Courtesy of Gustafson Chemical Company.)*

Perhaps the most common noncultural test for seedborne fungal pathogens is for loose smut infection of wheat or barley (Figure 14.3). Infection is present as hyphal strands that infect the embryo internally. The presence of such infection and its incidence in a seed lot is determined by the following method:

1. Soften the seeds by soaking in sodium hydroxide overnight.
2. Isolate the embryo by washing in a stream of warm water and separate by a fine wire mesh sieve.
3. Repeat washing with a solution of lactophenol and water. The embryos will float while the chaff and endosperm tend to sink to the bottom and can be drawn off.
4. Place the separated embryos into thick-bottomed glass dishes and clear further by boiling in lactophenol for 10 to 20 minutes.
5. Arrange cleared embryos and examine under magnification for the presence of hypha.

IDENTIFICATION

Identification of fungal pathogens requires considerable study and expertise in mycology. Seed testing laboratories inspecting samples for fungal pathogens should have trained mycologists on their staff. Official laboratories in some countries do have such qualified staff and are well equipped for pathological testing. Unfortunately, few laboratories in North America have qualified staff available for routine detection of fungal pathogens, and are thus not equipped for identifying pathogens except for certain tests. However, tests for examining barley embryos for loose smut infection are routinely performed in several state and official laboratories.

If qualified staff are not available, technicians may be trained to make routine inspection of samples and identify at least commonly occurring fungal pathogens. In any case, a dissecting (i.e., stereoscopic) microscope capable of 10 to 90× magnification and high-intensity lighting are essential, along with basic laboratory supplies and equipment for preparation and incubation of agar and blotter tests.

The following descriptions were given by Kulik and Schoen (1977) for seven important genera of fungi that commonly invade seed of major crops and are relatively simple to detect and identify. Though separation of these genera into species can be made for a particular sample, such identification is usually not necessary when they are associated with the primary host crops listed below.

1. *Alternaria. Alternaria* spp. are found in seeds of crucifers (*Brassica* spp.), carrot (*Daucus carota* L. var. *sativa* DC), and flax (*Linum usisatissimum* L.), and many other species. Colonies range in color from gray through greenish-olive brown to black. Typically, an *Alternaria* spore is club-shaped and has transverse and longitudinal septa. The common saprophyte *A. alternata* (Fr.) Keissler (syn. *A. tenuis* C. G. Nees) is difficult to separate from the several pathogenic species, on the basis of colony characteristics. There is also some overlapping of spore features among the various species. However, the more important pathogenic species generally produce long-beaked spores that are solitary or in short chains and contrast with the longer chains of *A. alternata*. An important exception is *A. brassicicola* (Schw.) Wiltshire, a frequent isolated seedborne pathogen of crucifers. This fungus differs morphologically from *A. alternata* in generally having spores that are smooth and with fewer transverse septations.

2. *Ascochyta* (Perfect state: *Mycosphaerella*). *Ascochyta* spp. are often found associated with seeds of various legumes, including *Melilotus* spp., *Phaseolus* spp., *Pisum* spp., *Trifolium* spp., and *Vicia* spp. Seeds of the garden pea [*Pisum sativum* L.) may be invaded by three species of *Ascochyta:: A. pinodella* Jones (syn. *Phoma medicaginis* Malbr. & Roum. var. *pinodella* (Jones) Boerema] forms light to dark brown pycnidia from which spores are exuded in the light buff to pinkish brown mass. The spores average 3.5 × 8μ. *A. pinodes* Jones [perfect state is *Mycosphaerella pinodes* (Berk. & Blox.) Vestergr.] forms light to dark brown to black pycnidia with a light buff to pinkish spore extudate. The spores average 5 × 14μ. The pseudothecia formed by the perfect state are dark brown. To separate *A. pinodes* from *A. pinodella* it may be necessary to compare the size of the spores under a compound microscope. *A. pisi* Lib. forms brown pycnidia from which the spores are exuded in a carrot red mass.

3. *Diaporthe.* The most commonly encountered member of this genus is *Diaporthe phaseolorum* (Cooke & Ellis) Sacc., which is frequently seedborne in soybean. Colonies are wooly white, more compact and slower growing than those of *Fusarium* spp., which also are often present in soybean seeds. Mature colonies usually contain embedded, dark-colored perithecia or pycnidia (the latter produced by *Phomopsis sojae* Lehman, the imperfect state of *Diaporthe phaseolorum*) that exude milky or yellow droplets of spores. In contrast, these fruiting bodies may also occur scattered or more or less regularly spaced on or just below the surface of the seed coat, with scant development of mycelium or spore exuda-

tion. Although *D. phaseolorum* on soybean has been reported to consist of two varieties, i.e., *D. phaseolorum* var. *caulivora* Athow & Caldwell, and var. *sojae* (Lehman) Wehmeyer, it probably is not necessary in routine seed health testing to differentiate between them.

4. *Fusarium* (Perfect states include *Gibberella* and *Griphosphaeria*). *Fusarium* spp. are associated with seeds of many members of the Gramineae, including corn, rice (*Oryza sativa* L.), and wheat (*Triticum aestivum* L.): also with seeds of cotton (*Gossypium hirsutum* L.), and various legumes such as the soybean and *Phaseolus* spp. Colonies usually consist of fluffy white to pink or coral-colored mycelium, often reaching 25 mm in diameter after one week or incubation. Spores are transparent hyaline, canoe- or banana-shaped with pointed or rounded tips, usually subdivided into several cells with thin walls (septa), sometimes in spindle-shaped (elliptical) translucent to orange clusters, up to 0.5 mm in diameter. *F. moniliforme* Sheldon, a pathogen of corn, is an atypical species having spherical spores borne in long chains. Colonies of *Epicoccum* spp., found mainly on small grains, and of *Cephalosporium* spp., on corn, are similar in appearance to *Fusarium* spp., but *Epicoccum* spp. are easily separated by their dark spore clusters. Identification of *Cephalosporium* sp. is more difficult because the spore clusters may resemble those of *Fusarium* spp. although they are actually more globular. Individual spores of *Cephalosporium* sp. are capsule-shaped or sometimes ovoid. Its occurrence is relatively infrequent in most lots of seed corn.

5. *Helminthosporium* (Synonyms for the imperfect state include *Bipolaris* and *Drechslera, Cochliobolus, Ophiobolus*, and *Pyrenophora*, which are perfect states of this common imperfect genus.). *Helminthosporium* spp. invade seeds of many of the Gramineae, including cereals such as corn, rice, and wheat, and a variety of other plants. Colonies range in color from light to dark gray, but unless they are grown on certain agar media, their positive identification to genus based solely on colony characteristics is nearly impossible. This is due principally to the similar colony appearance of the ubiquitous and omnipresent saprophyte, *Alternaria alternata*. In both *Helminthosporium* and *Alternaria*, spores borne on short conidiophores often develop on the seed surface with little or no production of mycelium. In *Helminthosporium*, spores are usually dark-colored, cigar- or club-shaped, straight or slightly curved, and are subdivided into several cells having relative thick, transverse septa. *A. alternata* has dark, club-shaped, rough spores (sometimes ovoid) divided into several cells, usually with both transverse and longitudinal septa. The spores may occur separately, but usually form various-sized chains with the tapered end of each spore (the beak) giving rise to the next spore above it.

6. *Phoma* (Syn. *Phenodomus*; perfect states include *Leptosphaeria* and *Pleospora*.). *Phoma* spp. are associated with seeds of crucifers, beets (*Beta vulgaris* L.), and flax. Colonies growing from seeds of crucifers may be thin, or profuse, silvery-white, and after several days are penetrated from beneath by the relatively large (200μ) flask-shaped, dark-colored pycnidia that eventually exude pink to purple masses of spores. The single pathogenic species on *Brassica* spp., *P.*

lingam (Tode ex Fr.) Desm., can sometimes be separated from the saprophyte *P. herbarum* Westend, based on the smaller pycnidia and nearly colorless spore masses produced by the latter. An extremely low incidence of *P. lingam* in a seed lot may lead to extensive plant loss in the field. *Phoma betae* Frank, found in seeds of sugar beet, produces black pycnidia that may be almost spherical or flattened, up to 180μ in diamater, with colorless spores. Seeds of flax may be invaded by *Phoma exigua* Desm., which produces a colorless to gray mycelium, dark brown pycnidia, and colorless spores that may exude from the pycnidium in a cream-colored ribbon.

7. *Pyricularia.* *Pyricularia* spp. are found in seeds of rice and finger millet [*Eleusine coracana* (L.) Gaertn.] grown outside the United States. Colonies are very small, inconspicuous, grayish or icy green, consisting of groups of conidiophores bearing pear-shaped spores in terminal clusters. These clusters are usually restricted to the embryo end of the seed. Under low magnification, they are similar in appearance to the common seedborne saprophyte *Cladosporium* spp., but may be differentiated at higher magnification (80–90×) due to the olive-green, dry appearance of this saprophyte. Spores of *Pyricularia* spp. usually are pear-shaped and have two septa that divide them into three unequal-sized cells while *Cladosporium* spp. have smaller single-celled spores generally borne in short chains.

SEED-BORNE SAPROPHYTIC FUNGI

Saprophytic fungi are those that grow in dead tissue. Although they exist on all seed, they do not cause plant diseases as do parasitic fungi. Spores of saprophytic fungi are almost ubiquitous, occurring on seed as well as in the air. They are especially numerous on seed in storage, and will germinate and grow profusely any time the storage environment exceeds 75% relative humidity and 15°C.

According to Kulik and Schoen (1977), seed-borne or contaminating colonies of *Mucor, Rhizopus, Aspergillus,* and *Penicillium* spp. are often encountered when seeds are tested on agar or blotters for pathogenic fungi. Some *Aspergillus* spp. are pathogenic on stored seeds, and their presence may be indicative of improper (i.e., relatively moist) storage conditions. *Mucor* and *Rhizopus* are easily recognized by fast spreading cobwebby networks of aerial mycelium that produce large spore clusters borne on heavy stalks called conidiophores. *Aspergillus* and *Penicillium* produce slow-growing granular or velvety textured colonies. *Aspergillus* colonies may be white or creamy at first, then change rapidly to a wide variety of green or bluish green hues; this is accompanied by a profuse production of highly pigmented, powdery spores. *Aspergillus niger*, a common contaminant of cereal and cotton seeds, produces black spores (Figure 14.4).

Another common seed-borne saprophyte, *Chaetomium* spp., produces spore-bearing bodies called perithecia which are ornamented by various types of bristles or hairs. Although not as common as the genera cited above, *Chaetomium* might be mistaken for a pathogenic species.

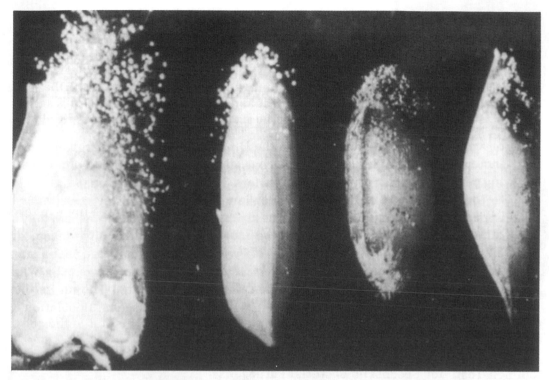

Figure 14.4. *Species of* Aspergillus *grown from maize, rice, wheat and barley. (Courtesy of C. M. Christensen.)*

FUNGAL ENDOPHYTE

Another interesting fungus that infects plants of interest to seed scientists is the fungal endophyte. Some cool-season grasses are infected by *Acremonium coenophiallum*, which causes forage grasses to be toxic to grazing animals but confers insect resistance to grasses. Tall fescue endophyte is the best known of these and is disseminated only by the seed. The fungal hyphae grow into the developing seed and are primarily located between the scutellum of the embryo and the aleurone layer of the mature seed (Bacon 1983). Biochemically, the fungus does not appear to alter the amino acid composition of the developing and mature seed (Bolesky et al. 1985). Developmentally, the fungus provides an advantage to the infected plants and seeds. Seeds from infected plants of *Lolium perenne* and *Festuca arundinacea* have higher rates of germination and produce more biomass and tillers than seeds from noninfected plants (Clay 1987; Rice et al. 1990). The fungal endophyte has been shown to enhance the production of indoleacetic acid in plant materials (DeBatista et al. 1990). Recently, because of the developmental advantages provided by the fungus, endophyte-enhanced seeds are being marketed for turf purposes.

Testing for Fungal Endophyte

Endophyte-free pastures may be established by planting endophyte-free seed. As a result, seed testing procedures have been developed to detect the presence of the fungal endophyte. These include taking the seed and allowing it to be digested in a 5% NaOH solution for 12 hours. After digestion, the glumes are removed from the seed, it is placed on a microscope slide, macerated, and stained with a one part 1% aniline blue solution in two parts 85% lactic acid. The aniline blue solution allows resolution of the fungal hyphae in the seed tissue under microscopic examination.

BACTERIAL PATHOGENS

Testing for seedborne bacterial pathogens is conducted in perhaps as many as 20 North American laboratories on a routine basis. While most procedures involve tests for detecting bacterial blight pathogens of field and garden beans, tests also exist for other crops, particularly for seeds of vegetables such as tomato and cabbage. Most of the tests used for the detection of seedborne bacterial pathogens can be considered modifications of one of four basic methods: (1) detection of infected seed by means of external characteristics visible on the seed coat, (2) diagnosis of plant symptoms resulting from inoculation (or injection) by material from suspect seed, (3) isolation of bacterial pathogen to identify the organism itself, and (4) a combination of the above methods.

Common Tests for Seedborne Bacterial Pathogens

Serological Techniques. Serological techniques are based on the reaction between antigens and antibodies. These techniques can be used for positive identification of both bacterial and virus organisms. Antiserum containing the specific antibodies is prepared by injecting an antigen (the bacteria or virus) into the bloodstream of live animals (usually rabbits). The blood immediately fights off the effect of the antigen by building up antibodies in the bloodstream (which can be recovered by bleeding the animal). The antiserum is used to test for the presence of the same bacteria in the homogenized seed by a precipitation test in agar. When antibodies and antigens are present, a precipitate occurs; if no precipitate appears, the causal agent is not present.

Different antisera are needed for serological testing for different pathogens. The serum may be prepared and kept for further use by freezing.

Although serological tests are used for detecting seedborne bacterial pathogens in several crops (e.g., bacterial blight of *Phaseolus vulgaris*), several problems exist. First, bacteria possess many antigenic materials, some of which are similar to those contained in other, nonpathogenic bacteria. This can cause confusion in identifying the precipitate designating the target pathogen. Second, the test is too detailed and time-consuming to be practical on an individual seed basis; thus, a homogenate of a seed sample is usually used. For testing bulk samples, techniques such as ELISA (enzyme-linked imunosorbent assay—page 347) offer promise of making serological tests quicker and more accurate, providing the problems of confusion over other antigenic materials cited above can be resolved.

Bacteriophage Technique. The bacteriophage technique is similar to the serological test in that it uses a specific reaction between two organisms to positively identify disease organisms. In this case, bacterial organisms can be identified by using special phage viruses which are added to homogenized seed material on the agar. If the bacterial organism is present, a clear plaque area appears on the agar due to attack on the bacteria by the virus.

The bacteriophage technique works well if contaminating bacteria are not present. Unfortunately, they usually are and will obscure the plaque area.

Plant Injection Tests. Plant injection tests can be useful for identifying certain seedborne organisms, particularly bacteria (see Figure 14.5) and viruses. The test is simple and easy to perform. Seeds to be tested are soaked in sterile water for a few hours, after which the leachate is injected into a young, healthy seedling. Sometimes individual seeds are homogenized in a liquid solution for injection into the test seedling. The injection may be placed in the vascular system (phloem) of the plant, which transports it throughout the plant. The entire plant is closely observed for development of disease symptoms.

Another method often used, especially for virus detection, is to rub a mixture of a homogenized seed (or a seed sample) and metal filings on the leaf surface. The metal filings puncture the plant tissue providing entry for any disease inoculum present, resulting in disease symptoms.

Young healthy plants of the species being tested are required for plant injection tests. Usually these are grown in a greenhouse under phytosanitary conditions and should be two to four weeks old when inoculated. About two weeks should be allowed for symptom development.

SEEDBORNE VIRUSES

According to Carroll (1979) about 200 plant viruses are known to cause diseases of plants, of which 100 are fairly well known. An additional 500, though nonpathogenic, are thought to be transmissible plant viruses. Of all these, only about 80 different viruses or viruslike organisms are considered to be seed-transmitted. A few of these, such as tobacco mosaic virus (TMV) on tomato, are carried on the surface of the seed or inside the seed outside the embryo. However, most, such as bean common mosaic virus (BCMV) and barley stripe mosaic virus, are carried inside the embryo.

Next to tests for seedborne bacterial pathogens, tests for seedborne viruses represent the most common tests for seedborne pathogens in North America. Two kinds of tests are commonly used—biological tests and serological tests.

Biological Tests

Biological tests include both grow-out tests and direct seed tests. Grow-out tests usually begin with a visual examination to detect abnormal-appearing seed. These, plus normal-appearing seed, are then planted in an appropriate sterile medium and placed in a favorable light and temperature environment for seedling germination and growth. Test seedlings are observed carefully for symptom development. This test is simple, inexpensive, and suitable for testing individual seeds. Biological tests may

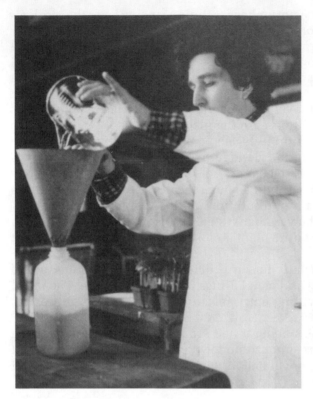

A.—Sterilizing beans.

C.—Pouring off solution to be injected into test plants.

B.—Soaking beans.

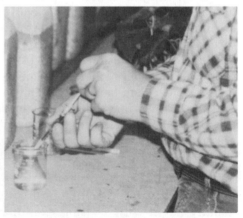

D.—Filling syringe.

Figure 14.5. *The seedling injection procedure used in Michigan for detection of seedborne bacterial blight in dry edible bean seed. (Courtesy of William Young, Michigan Department of Agriculture.)*

E.—Injecting bean test plants.   F.—Point of injection and lesion.

Figure 14.5. (Continued)

also be performed by grinding up seeds or seedlings from suspected seed lots and mechanically inoculating a sample of the homogenate into sensitive, healthy test seedlings and observing them for symptom development.

Direct seed tests are those in which both normal- and abnormal-appearing seed are soaked in an aqueous medium and ground into a slurry that is inoculated onto indicator or test plants. Seedborne viruses are indicated by the development of characteristic symptoms on the test seedlings. Single seeds or a composite sample from a seed lot may be tested. According to Carroll, this test is suitable for detecting a seedborne mosaic virus in pea seed, using purple pigweed (*Chenopodium amaranticolor*) as a test plant. Similar tests are used to screen lettuce seeds for the presence of lettuce mosaic virus; *Chenopodium quinosa* is the indicator plant in this case. It is illegal to plant seeds found to be infected by this or other assays in parts of California or Florida. All seeds planted in these areas must be tested.

Serological Tests

Serological tests for seedborne viruses are similar in principle to those described for detecting bacterial pathogens. They are based on the physiochemical reaction between the virus in the seed or homogenate from the sample to be tested and blood sera prepared from laboratory animals (usually rabbits). Five serological techniques were described by Carroll (1979); these differed slightly, depending on the method used for meeting of pathogens from virus-infected seed and antigens from the blood serum. These are described below:

Double Diffusion Test. The double diffusion technique has been used more extensively than any other serological method. It can assay a single seed or part of a seed. Usually the seed to be tested is soaked in tap water. Then the whole seed or its parts (e.g., embryo) is triturated and the resulting triturate is transferred to a well cut into the diffusion medium (commonly, agar gel). Thereafter, an antiserum specific for the suspect virus is placed in a separate well. In time the virus particles (antigen) and the antibody molecules diffuse toward one another. Since this diffusion is in two different directions, it is called double diffusion. When the two seroreactants reach a point in the gel at which the relative concentration of each is serologically equivalent, antigen and antibody molecules complex, precipitate, and become immobilized. The precipitate appears as a white band at some point between the two wells.

To detect viruses that have an elongated structure by the double diffusion technique, seed preparations must be treated with some kind of agent that breaks up or degrades the virus particles to enhance the diffusion of the virus particles in agar.

The double diffusion method is generally quite specific, reasonably sensitive, and can detect virus concentrations of 10 to 25 μg/ml. This test incorporates sodium dodecyl sulfate as a virus-degrading agent, and is used routinely at the Montana State Seed Testing Laboratory for serodiagnosis of barley stripe mosaic in barley embryos (Figure 14.6). Two hundred embryos from each seed lot are tested using filter-paper disks as seroreactant depots. Precipitate lines occur only between infected embryos on the peripheral disks and the antisera in the center disks. This test has been largely responsible for a significant reduction of barley stripe mosaic virus in commercial barley in Montana.

Radial Diffusion Test. The single or radial diffusion method is similar to the double diffusion technique except that only the virus antigen diffuses from a well out into the agar diffusion medium. The serum containing antibodies to a specific virus has been included in the agar in this test. Therefore, the procedure is to charge the wells with the seed or seedling preparations being tested. If virus molecules are present, they diffuse radially from the wells, and when they have reached a certain distance from the well surface they complex with the antibody molecules and precipitate in this region, forming a ring or halo around the charged well.

Normally a subunit antiserum (prepared by injecting rabbits with a degraded virus preparation) is used with the radial diffusion method, and therefore only degraded viruses can be detected. This method is sensitive, rapid, and can detect as little as 1 μg/ml of degraded virus. In a matter of minutes or a few hours, results can be obtained with preformed agar diffusion plates. This method has been used for the detection of seedborne barley stripe mosaic virus.

Latex Flocculation Test. Latex flocculation is another serological method in which 100 seeds are ground in a Wiley mill. Then about 0.1 g of the ground seed material is placed in 2 ml of buffer and further ground to a mortar with a pestle. About 20 μl of this seed extract are then placed in a 100-ml pipette to which 10 μl of tagged latex had been previously added. The tagged latex consists of a suspension of polystyrene latex spheres (about 0.81 μm in diameter). The latex spheres are sensitized, or covered with antibody molecules specific for a given virus. The pipette is oscillated for about 15 minutes and observed under a dissecting microscope. When virus is present in the sample, the latex suspension becomes flocculated or aggregated. (See Figure 14.7.)

The latex method is rapid, specific, and sensitive. Results can be obtained within

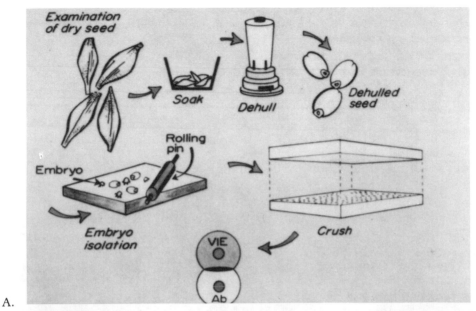

A.

B.

Figure 14.6. *The agar double diffusion test.(A) Schematic showing the preparation and serological analysis of barley embryo extracts. Antigen from the virus-infected embryo (VIE) formed a precipitate (lens-shaped area) with antibody (Ab) molecules. (B) Quadrant petri dish with white precipitates due to positive virus-antibody reactions. Precipitate lines occur only between infected embryos on the peripheral disks and the center antisera disks A, B, and C. (Reproduced by permission of the American Phytopathological Society from Carroll (1980).)*

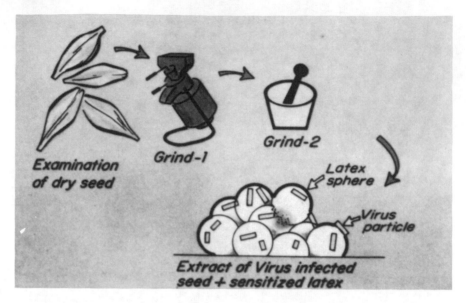

A.

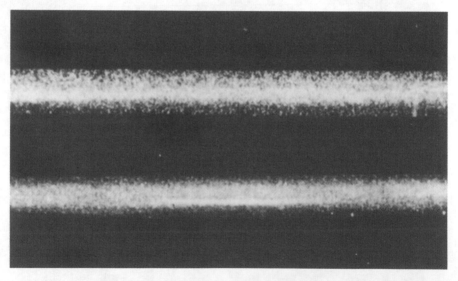

B.

Figure 14.7. *The latex flocculation test (A) Schematic of latex flocculation method, (B) Two 100-μl pipettes containing suspensions of antibody-sensitized latex. Virus presence in the upper pipette caused the suspension to become flocculated or aggregated. Absence of virus in the lower pipette was indicated by the dispersed appearance of the latex suspension. (Reproduced by permission of Zeitschr. Pflkrankh. Pflshutz. from Lundsgaard (1976).)*

15 minutes to 1 hour. The method can detect one infected seed per 100 seeds or 1 μg/ml of virus.

Enzyme-linked Immunosorbent Assay (ELISA). In this procedure, antiserum specific for a given virus is used to coat the polystyrene plate, and the antibody molecules become absorbed. Next, excess antibody molecules are washed with the plate, the seed sample is added, and the excess sample is washed away. This is followed by adding enzyme-labeled specific antibody and the enzyme alkaline phosphatase is conjugated to the antibody molecules specific for the virus for which the examination is made. The excess labeled antibody molecules are washed away, and finally, enzyme substrate is added. This enzyme hydrolyzes the substrate. The amount of substrate hydrolyzed is determined by measuring the extinction of 405-nm wavelength spectrophotometrically or by visual observation. For this assay, dry-milled or water-soaked seed is extracted with buffer and the seed preparations are then ground further with another buffer and a surfactant. Afterwards, the seed extracts are applied to polystyrene plates following the standard ELISA procedure. Virus-infected seed preparations should produce a yellow color.

High sensitivity is the main attribute of the ELISA method. Some workers claim to detect 0.1 μg/ml of virus. The method has been used experimentally to detect tobacco ring spot and soybean mosaic viruses in soybean seed. Low levels of infection (1 to 4%) could be detected in extracts from seed lots.

Serologically Specific Electro Microscopy (SSEM). This method and a few others comprise the category of methods known as immunoelectron microscopy. Essential steps in the SSEM procedure include: (1) the copper specimen support or grid is covered with a parlodion film, (2) the filmed grid is floated on antiserum diluted 1:100 to 1:5000 in tris buffer, (3) after 30 minutes the grid is washed repeatedly to remove unabsorbed serum proteins, (4) the grid is floated on an extract of virus-infected tissue from 10 minutes to 24 hours, (5) the grid is washed repeatedly to remove cellular debris and salts, (6) the grid is then stained with 5% uranyl acetate for 1.5 minutes or longer; (7) the grid is washed with distilled water or 95% ethanol, and finally (8) the grid is blotted dry and examined in the electron microscope. The procedure is shown in Figure 14.8. For seed diagnosis, dry seed is ground by mortar and pestle or Wiley mill, or soaked seed is homogenized. The seed preparation is applied to an electron microscope grid containing an antiserum specific for the virus in question.

Samples from infected seed lots reveal the presence of virus particles characteristic for the suspect virus. According to Carroll, when tobacco ring spot and soybean mosaic viruses were tested, each virus could be detected in a singly infected soybean seed. The SSEM method was sensitive enough to detect the two soybean-infecting viruses in a mixture of 1 infected seed per 1000 healthy seeds. The method is simple for trained personnel, and it is rapid, selective, and sensitive. However, it does require the use of a transmission electron microscope.

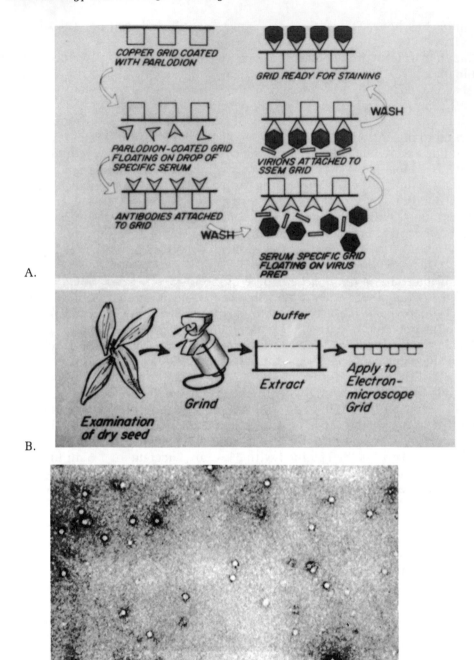

Figure 14.8. *The serologically specific electron microscopy (SSEM) method. (A) Principle of the SSEM method, (B) schematic of the SSEM method, (C) electron micrograph showing presence of the spherical particles of tobacco ringspot virus. (Reproduced by permission of the American Phytopathological Society from Brlansky and Derrick (1979).)*

General References

Agarwal, V. K., and J. B. Sinclair. 1987. *Principles of Seed Pathology*, Vol. 1 and 2. Boca Raton, Fla.: CRC Press.

Bacon, C. W. 1983. The fungal endophyte and tall fescue. In: *Proceedings of Tall Fescue Toxicosis Workshop*, Atlanta, Ga. pp. 34–42.

Baker, K. F. 1962a. Principles of heat treatment of soil and planting material. *Journal of Australian Institute of Agricultural Science* 28:118–126.

———. 1962b. Thermotherapy of planting material. *Phytopathology* 52:1244–1255.

———. 1969a. Seed pathology-concepts and control. *Journal of Seed Technology* 4(2):57–67.

———. 1969b. Aerated steam treatment of seed for disease control. *Horticultural Research* 9:59–73.

———. 1972. Seed pathology. In: *Seed Biology*, ed. T. Kozlowski, 2:317–416. New York: Academic Press.

Barnett, N. L., and B. B. Hunter. 1972. *Illustrated Genera of Imperfect Fungi*, 3rd ed. Minneapolis, MN: Burgess Publishing Company.

Bolesky, D. P., J. J. Evans, and S. R. Wilkinson. 1985. Amino acid composition of tall fescue seed produced from fungal endophyte (*Acremonium coenophealum*)-free and infected plants. *Agronomy Journal* 77:796–798.

Booth, C. 1971. The Genus *Fusarium*. Kew, Surrey, England: Commonwealth Mycological Institute.

Brlansky, R. H., and K. S. Derrick. 1979. Detection of seedborne plant viruses using serologically specific electron microscopy. *Phytopathology* 69:96–100.

Carpenter, P. L. 1977. *Microbiology*, 2d ed. Philadelphia: W. B. Saunders Company.

Carroll, T. W. 1979. Methods of detecting seedborne plant viruses. *Journal of Seed Technology* 4(2):82–95.

———. 1980. Barley stripe mosaic virus: Its economic importance and control in Montana. *Plant Disease* 64:138–140.

Carroll, T. W., P. L. Gossel, and D. L. Batchelor. 1979. Use of sodium dodecyl sulfate in serodiagnosis of barley stripe mosaic virus in embryos and leaves. *Phytopathology* 69:12–14.

Chidambaram, P., S. B. Mathur, and P. Neergaard. 1973. Identification of seedborne *Drechslera* species. *Friesia* 10:165–207.

Christensen, C. M., and R. A. Meronuck. 1986. *Quality Maintenance in Stored Grains and Seeds*. Minneapolis, Minn.: University of Minnesota Press.

Christensen, C. M., and D. B. Sauer. 1982. Microflora. In: *Storage of Cereal Grains and their Products*, 3rd ed., ed. C. M. Christensen, pp. 219–240. St. Paul, Minn.: American Association of Cereal Chemists.

Chualprasic, C., S. B. Mathur, and P. Neergaard. 1974. The light factor in seed health testing. *Seed Science and Technology* 2:457–475.

Clark, M. F., and A. N. Adams. 1977. Characteristics of the microplate method of enzyme-linked immunosorbent assay for the detection of plant viruses. *Journal of General Virology* 34:475–483.

Clay, K. 1987. Effects of fungal endophytes on the seed and seedling biology of *Lolium perenne* and *Festuca arundinacea*. *Oecologia* 73:358–362.

Coleno, A., A. Tregalet, and B. Digat. 1976. Detection des lots de semances cantamines par une bacterie phytopathogene. *Annales de Phytopathologie* 8:355–364.

A compendium of corn diseases. 1973. St. Paul, MN: American Phytopathological Society.

Compendium of soybean diseases. 1975. St. Paul, MN: American Phytopathological Society.

Cooper, V. C., and D. G. A. Waldey, 1978. Thermal inactivation of cherry leaf roll virus in tissue cultures of *Nicotiana rustica* raised from seeds and meristem tips. *Annals of Applied Biology* 88:273–278.

Cross, J. E. 1979. Importance of seed pathology in seed trade quality control programs. *Journal of Seed Technology* 4(2):99–102.

DeBattista, J. P., C. W. Bacon, R. Severson, R. D. Plattner, and J. H. Bouton. 1990. Indole acetic acid production by the fungal endophyte of tall fescue. *Agronomy Journal* 82:878–880.

Derrick, K. S., and R. H. Briansky. 1978. Assay for viruses and mycoplasma using serologically specific electron microscopy. *Phytopathology* 65:815–820.

Echandi, E., and M. Sun. 1973. Isolation and characterization of a bacteriophage for the identification of *Corynebacterium michiganense. Phytopathology* 63:1398–1401.

Ellis, M. B. 1971. Dematiaceous *Hyphomycetes*. Kew, Surrey, England: Commonwealth Mycological Institute.

Funder, S. 1961. *Practical Mycology*. 2d ed. New York: Hafner Publishing Company.

Guthrie, J. W. 1979. Routine methods for testing and enumerating seed-borne bacterial plant pathogens. *Journal of Seed Technology* 4(2):78–81.

Guthrie, J. W., D. M. Huber, and H. S. Fenwick. 1965. Serological detection of halo blight. *Plant Disease Reporter* 49:297–299.

Hamilton, R. I., and C. Nichols. 1978. Serological methods for detection of pea seed-borne mosaic virus in leaves and seeds of *Pisum sativum. Phytopathology* 68:539–543.

Hepperly, P. R., and J. B. Sinclair. 1978. Quality losses in *Phomopsis*-infected soybean seeds. *Phytopathology* 68:1684–1687.

Heydecker, W., J. Higgins, and Y. J. Turner. 1975. Invigoration of seeds? *Seed Science and Technology* 3:881–888.

Ilyas, M. B., O. D. Dhingra, M. A. Ellis, and J. B. Sinclair. 1975. Location of mycelium of *Diaporthe phaseolorum* var. *sojae* and *Cercospora kikuchii* in infected soybean seeds. *Plant Disease Reporter* 59:17–19.

Jeffs, K. A., ed. . 1978. *Seed Treatment*. Cambridge, U.K.: Heffers Printers Ltd.

Kulik, M. M., and J. F. Schoen. 1977. Procedures for the routine detection of seed-borne pathogen fungi in the seed testing laboratory. *Journal of Seed Technology* 2(1):23–39.

Letham, D. B. 1977. Seed treatment research. *New South Wales Nurserymen* 2(9):7.

Limonard, T. 1965. *Ecological Aspects of Seed Health Testing*. Wageningen: International Seed Testing Association.

Lister, R. M. 1978. Application of the enzyme-linked immunosorbent assay for detecting viruses in soybean seed and plants. *Phytopathology* 68:1393–1400.

Lister, R. M., S. E. Wright, and J. M. Kloots. 1978. Sensitive detection of barley stripe mosaic virus in barley seed and embryos by ELISA. *Phytopathology News* 12:198.

Lundsgaard, T. 1976. Routine seed health testing for barley stripe mosaic virus in barley seeds using the latex-test. *Zeitschrift Fur Zenkrankheiten Pflsanzen Pathologie* 83:278–283.

Malone, J. P., and A. E. Muskett. 1964. Seed-borne fungi. *Proceedings of the International Seed Testing Association* 29:179–384.

McGee, D. C. 1979. Epidemiological aspects of disease control. *Journal of Seed Technology* 4(2):96–98.

McGee, D. C. 1988. *Maize Diseases: A Reference Source for Seed Technologists*. St. Paul, Minn.: APS Press.

McGee, D. C. 1992. *Soybean Diseases: A Reference Source for Seed Technologists*. St. Paul, Minn.: APS Press.

Mink, G. I., and J. L. Parsons. 1978. Detection of pea seed-borne mosaic virus in pea seed by direct-seed assay. *Plant Disease Reporter* 62:249–253.

Nath, R., P. Neergaard, and S. B. Mathur. 1970. Identification of *Fusarium* species on seeds as they occur in the blotter test. *Proceedings of the International Seed Testing Association* 35:121–144.

Naumova, N. A. 1972. *Testing of Seeds for Fungus and Bacterial Infection.* 3rd ed. Washington, D.C.: U.S. Department of Agriculture and the National Science Foundation.

Neergaard, P. 1977. *Seed Pathology*, Vol. 1, New York: John Wiley and Sons.

Neergaard, P. 1977. *Seed Pathology*, Vol. 2. New York: John Wiley and Sons.

———. 1978. Detection of seed-borne pathogens by culture tests. *Seed Science and Technology* 1:217–254.

Noble, M. 1965. Introduction to series 3 of the handbook on seed health testing. *Proceedings of the International Seed Testing Association* 30:1045–1121.

Noble, M., and M. J. Richardson. 1968. *An Annotated List of Seed-borne Diseases.* 2d ed. Wageningen, Netherlands: International Seed Testing Association.

Parker, M. C., and L. L. Dean. 1968. Ultraviolet as a sampling aid for detection of bean seed infected with *Pseudomonas phaseolicola. Plant Disease Reporter* 52:534–548.

Phatak, H. C. 1974. Seed-borne plant viruses-identification and diagnosis in seed health testing. *Seed Science and Technology* 2:3–155.

Powell, C. C., Jr., and D. E. Schlegel. 1970. The histological localization of squash mosaic virus in cantaloupe seedlings. *Virology* 42:123–127.

Ralph, W. 1978. Enhancing the success of seed thermotherapy: Repair of thermal damage to cabbage seed using polyethylene glycol (PEG) treatment. *Plant Disease Reporter:* 62:406–407.

Rice, J. S., B. W. Pinkerton, W. C. Stringer, and D. J. Undersander. 1990. Seed production in tall fescue as affected by fungal endophyte. *Crop Science* 30:1303–1306.

Richardson, M. J. 1979. An annotated list of seedborne diseases, 3rd ed. *Commonwealth Mycological Institute Phytopathological Papers* 23:1–320.

Rodriguez-Marcano, A., and J. B. Sinclair, 1978. Fruiting structures of *Colletotrichum dematium* var. *truncata* and *Phomopsis sojae* in soybean seeds. *Plant Disease Reporter* 62:873–876.

Schaad, N. W. 1978. Use of direct and indirect immunofluorescence tests for identification of *Xanthomonas campestris. Phytopathology* 68:249–252.

Schaad, N. W., and W. C. White. 1974. A selective medium for soil isolation and enumeration of *Xanthomonas campestris. Phytopathology* 64:876–880.

Schneider, R. W., O. D. Dhingra, J. F. Nicholoson, and J. B. Sinclair. 1974. *Colletotrichum truncatum* borne within the seedcoat of soybeans. *Phytopathology* 64:154–155.

Schuster, M. L. 1972. Leaf freckles and wilt of corn incited by *Corynebacterium nebraskense* Schuster, Hoff. Mandel, Lazar. *Nebraska Research Bulletin* 270:1–40.

Shepard, J. F. 1972. Gel-diffusion methods for the serological detection of potato viruses X, S, and M. *Montana Agricultural Experiment Station Bulletin 662.*

Shepherd, R. J. 1972. Transmission of viruses through seed and pollen. *Principles and Techniques in Plant Virology*, ed. C. J. Kado and H. O. Agrawal, pp 267–292. New York: Van Nostrand-Reinhold Company.

Sheppard, J. W. 1979. Methods for routine detection of seedborne fungal pathogens. *Journal of Seed Technology* 4(2):74–77.

Sinclair, J. B. 1977. The microcosm of the soybean seed. *Illinois Research* 19(1):12–13.

———. 1979. The seed: A microcosm of microbes. *Journal of Seed Technology* 4(2):68–73.

Slack, S. A., and R. J. Shepherd. 1975. Serological detection of seedborne barley strip mosaic virus by a simplified radial-diffusion technique. *Phytopathology* 65:948–955.

Taylor, J. D. 1970. The quantitative estimation of the infection of bean seed with *Pseudomonas phaseolicola.* (Burkh.) Dowson. *Annals of Applied Biology* 66:29–36.

15
Seed Marketing

For centuries, farmers used their own seed. This approach offered the advantage of immediate availability of supply. However, it suffered from the disadvantage that farmers were oriented more toward production of grain or forage and their harvesting and storage facilities generally were not suitable for maintenance of optimum seed quality. Therefore, in some years, farmers experienced stand failures from planting inferior seed. Moreover, this process did not encourage the planting of seeds with improved genetic potential. The modern seed industry has filled this void with its emphasis on varietal development and increased yields which have economically benefited the farmer. It is for these reasons that most farmers now recognize that the purchase of seeds from a reputable seed company is in their best interest.

Seed marketing is the process through which seed moves from the farm where it is produced to the consumer who plants it. Depending on the type of seed and the proximity of its production to the site of its use, the marketing process may be simply a farmer exchange or it may be a complicated transaction involving several mid transactions and a highly organized seed industry. Figure 15.1 illustrates the seed marketing cycle in its most complicated form. Much of the seed produced in the world today, especially that which is produced outside its area of use, follows this kind of marketing scheme.

Successful seed marketing requires a demand for seed, a mechanism to supply the seed in a timely fashion, and an organizational structure to ensure that high-quality seed is produced.

DEMAND

One of the first questions a seed company must address when considering seed marketing is the size of the demand for seed. The determination of a *potential* market is based on cultivated acreage, seeding rates, and renewal rates. While the first two factors are relatively easy to determine, the renewal rate is more difficult to calculate because it is based on the return of the farmer to purchase seed on an annual or seasonal

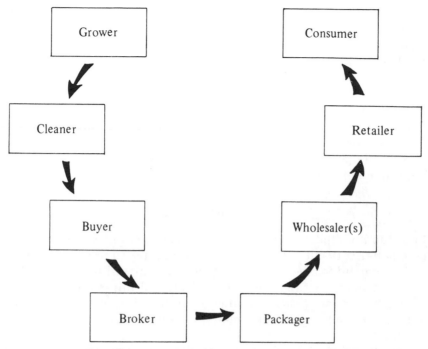

Figure 15.1. *A generalized seed marketing scheme.*

basis. When improved varieties are not available, farmers commonly set aside a portion of the harvested product for the next year's seed. Since the varietal purity and yield potential of high-quality seed of self-pollinating varieties in some instances may be satisfactorily maintained for several generations, replacement of seed may not be higher than 25–30% on an annual basis. This computation is particularly true in developing countries and is one of the reasons why the establishment of seed industries in lesser developed countries is slow to evolve. In contrast, the sale of hybrid crops is more stable since the renewal rate of hybrid seed is 100%.

Another important component of seed marketing is the creation of a demand for the improved seed product. Farmers tend to be conservative and must be convinced that new seed or any related cropping techniques (cultivation, fertilizer application, pest control, etc.) are superior to their present production practices, particularly when increased costs and extra effort are involved. Therefore, most seed companies provide local field demonstrations and yield trials to document the superiority of their newer, higher-yielding varieties. They also provide an opportunity for public relations and a chance to promote their public image to farmers.

SUPPLY

Successful seed marketing requires that seed be supplied at the right time in the correct quantity, at the place where it is used, and in appropriate packaging. Planting of seed generally occurs in an intense period of only a few weeks at the beginning of the growing season. If satisfactory quantities of seed are not available prior to planting, the seed will not be sold. To facilitate timely supply, farmers are encouraged to preorder their seed so that the seed company can anticipate the level of demand. Because many farmers tend to delay this important decision until varieties and costs of seed are compared, inventory management in a seed company is a difficult process. While seed can be delivered directly to the farmer, this often does not occur. Instead, most farmers obtain the product directly from a retailer. Therefore, the seed must be packaged in a manner which enables ease of transport and handling. Often, this means that seed is packaged in 50-lb paper, plastic, or burlap bags. More recently, many suppliers have provided seed in large bulk steel containers so that there is less handling of numerous bags and the time-consuming process of opening smaller, individual bags is avoided.

ORGANIZATION

Seed production often is centered in one area of a country. This is because certain locales have better soil and climatic regimes conducive to production of high-quality seed. Yet, the demand for the product may be national and international. As a result, the marketing of seed must require an organizational infrastructure where there is a system of distribution for a central production point to the customer. This often entails the use of brokers, wholesalers, and retailers to ensure satisfactory seed distribution (Figure 15.1).

Seed marketing is central to the success of any seed operation. The most efficient seed company will be organized so that all operations of the company flow through a central marketing office (Figure 15.2). This allows the marketing manager to forecast farmer demand, plan production dates and location of seed stock needs (inventory control of varieties and amounts), determine the level of sales promotion, and to communicate the status of seed supply with seed dealers. Such an organizational structure results in fewer management crises and permits smoother operation of the company.

Various organizations market seed. When no reliable and effective marketing system exists (e.g., in many developing countries), national governments often establish seed companies that are run by the government or its agency. This allows the government to encourage the development of improved varieties and to educate farmers about their benefits in producing higher yields. The rate at which a new variety is accepted by farmers depends on a number of factors. The first is agronomic superiority. The increased yield must cover the extra price for the seed and any other production inputs. In many cases, the government reduces the cost of the seed to the farmer to minimize the alternative use of feed grains which are often provided at below-market costs. Eventually, farmers understand that their profits are greater when they purchase improved seed rather than using their own seed from previous harvests. When demand increases, most governments encourage privatization of the seed industry. This can occur by groups of interested seed purchasers establishing cooperatives where the integrated marketing and distribution operations are shared by a central authority. More often, however, private seed companies recognize that profits are possible in a stabilized marketplace with a continuing demand for high-quality seed from a better-educated clientele. When this occurs, a private seed industry quickly develops.

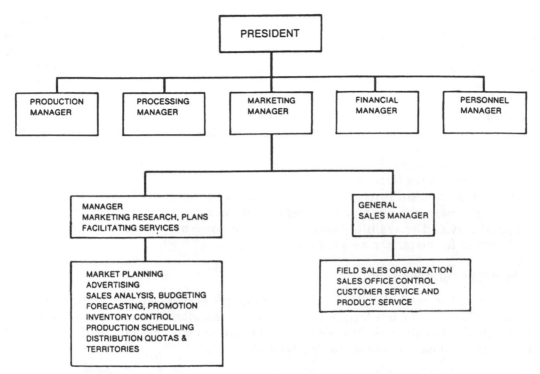

Figure 15.2. *Effective, market-oriented seed program organization which can focus all activities toward identifying, developing, and satisfying farmers' demand for improved seed. (From Law et al. (1971).)*

Seed marketing schemes in the United States and Canada for various kinds of seeds, including small-seeded grasses and legumes, hybrid corn and sorghum, cereal grains and soybean, and miscellaneous crops are described below.

Small-Seeded Grasses and Legumes

Since most of the forage and turf grasses and legumes is produced outside its major area of use, almost none is sold directly to the user; instead it is sold to local seed houses, and enters the marketing chain similar to that illustrated in Figure 15.1. The seed buyer may be an independent seed producer or part of a large seed company with a national or even international organization. The seed may be sold after it has been cleaned in the producer's farm cleaning plant or it may be taken "in the rough" directly from the field; in either instance, however, the grower is usually paid on a clean seed basis.

Although seed of the small-seeded grasses and legumes may be produced and sold on the open market, much of it is produced under contract with local seed dealers. These contracts are usually made before the seed field is established, and they specify the terms of the price to be paid, the quality expected, and responsibilities of both the grower and the contractor. Such contracts protect the seed producers by ensuring them a supply of seed at a reasonable, predetermined price. Contracts tend to stabilize seed prices, acting as a hedge against short supplies and rising prices. They provide growers with a reasonable margin of profit above their production costs and protect them against an unstable market. The contracts often specify the services of a field adviser, who advises the producer on cultural practices needed for maximum seed production.

If growers elect to produce seed without a contract, they may do so; however, the price they receive for their seed depends on the market at the time they are ready to sell. Sometimes this may work in their favor, although the odds may favor them if they choose to produce under contract. The success of contract production and the mutual satisfaction of both the seed producers and growers are reflected in the large amounts of the seed of small-seeded grasses and legumes produced under contract.

In Chapter 9 we learned that there is still some grass and legume seed produced within the area of utilization. Most of this is sold to local seed dealers on the open market or grown under contract similar to that discussed above; however, because of the proximity and accessibility to the consumers, some may be retailed directly by the seed grower to the farmer who uses it.

Hybrid Corn and Sorghum

Most of the seed of hybrid corn and sorghum is produced by selected growers under contract with a few large seed companies, who market the seed through farmer-dealers to their customers. The remainder is handled by numerous small seed companies who produce and market in localized areas.

Other Small Grains and Soybeans

Most seed of small grains and soybeans is marketed through local elevators or seed dealers. Seed of these crops is usually produced without contracts and is sold at the prevailing market price; however, contract production is becoming more common, especially by larger seed handlers, who wish to ensure themselves adequate supplies of seed for their customers at a reasonable cost.

Flower and Vegetable Seeds

Flower and vegetable seeds are produced almost entirely under contract and under rigid control of private seed companies, who usually determine the varieties to be grown, the quantity of seed needed, and the price of seed to the producer as well as the ultimate consumer. Rigid control is possible because the varieties are usually privately developed, have secret pedigrees, and are needed only in carefully regulated quantities.

Field Bean Seed

The marketing of field bean seed depends on its area of production. Perhaps one-half of the seed of field beans is produced in the western United States, where it is usually grown under contract to seed companies; however, much field bean seed is also produced in the commercial bean areas, particularly in Michigan, Ontario, and New York. In these areas, most seed is sold without contracts to local elevators at the prevailing market price. However, many of the elevators, commercial bean companies, and even some large seed growers arrange for contract production to hedge against large price increases and ensure adequate supplies.

In contrast to field beans, seed of garden beans is grown almost exclusively in the western United States—mostly under contract with local seed dealers who may represent independent businesses or parts of larger seed companies. Many of the independent seed producers also handle field bean seed as well.

Cotton Seed

In the southern United States, the great majority of cotton seed is controlled by a few large companies who contract with dependable farmers to produce the seed under certain restrictions and the state certification standards. In the West, most of the cotton seed is produced by grower-owned seed organizations who contract with farmers for the seed production.

MARKETING SEED OF PUBLIC VERSUS PRIVATE VARIETIES

Traditionally, access to seed of new crop varieties that have been developed and released from public experiment stations or university plant-breeding programs has been free and equally available to all growers and seed producers wishing to produce, promote, and market the seed. The initial seed of new varieties has traditionally been released through foundation seed organizations for increase and subsequent availability as certified seed. Thereafter, the seed enters the public domain and is made available to all growers and farmers desiring it. Consequently, no individual or seed company is able to monopolize the seed production and marketing of publicly developed varieties.

In contrast, the seed production and marketing of crop varieties developed by private seed companies is tightly controlled. This system has been relatively easy for varieties of vegetables, where the commercial product grown from the seed is also under company control. It has also worked well for corn and sorghum hybrids, for which seed stocks are under company control and loss of hybrid vigor incurred by replanting the progeny from seed comprises a type of biological varietal protection. Prior to passage of the Plant Variety Protection Act of 1970, no legal protection was available to owners or developers of varieties of sexually propagated crops. This Act, discussed more fully in Chapter 16, provides protection for sexually propagated crop varieties and provides legal protection to their owners in controlling the production and marketing of seed.

In recent years, considerable numbers of publicly developed varieties, particularly forage and turf varieties, have been released to private seed companies. Such exclusive releases are usually covered by formal memoranda of understanding setting forth the specific terms of the agreement. Normally, the exclusive release does not circumvent the usual role of foundation and certified seed organizations, but merely gives the company the right to control seed production and marketing of the variety. Some agreements stipulate that only certified seed may be sold under the variety name.

Exclusive releases seem to be justified for varieties for which: (1) the area of seed production is outside the area of commercial use, (2) seed production is a difficult or hazardous specialty (e.g., tobacco and sugar beet seed), or (3) the volume of seed needed of certain specialty crops is not large enough to justify participation of many different parties. In such cases, it is felt that no individual company or person can be expected to incur the expense of seed production, promotion, and marketing of unproved varieties without a guarantee that their competitors will not be able to capitalize on their efforts. Many superior, badly needed varieties have "died on the vine" because they were released openly and no one company would risk investing in seed production and promotion.

Since the passage of the Plant Varietal Protection Act, many public institutions are obtaining varietal protection on their new varieties. Such varietal protection obviously benefits the recipient company in the case of exclusive release by assuring them that another company cannot legally violate their assigned rights.

With the advent of variety protection, owners of protected varieties may sell ownership of their varieties or collect royalties from others whom they allow to produce and sell such seed.

THE SEED INDUSTRY

According to a survey of retail seed outlets in New York and New Jersey, 81% of the respondents indicated that the retail seed business was a sideline rather than a main enterprise. Aside from the farmer-dealer marketing of corn and sorghum hybrids and seed of certain other specialty crops, this is probably a fairly accurate reflection of the marketing of most field seeds throughout North America. Most crop seed is sold by local elevators or cooperatives as a sideline to the grain, feed, fertilizer, and agricultural chemicals industry. While much of this is purchased wholesale from local seed growers, a substantial portion, particularly of turf, forage, and vegetable seed, is purchased from wholesale seed dealers who trade almost exclusively in seeds.

Aside from the type of seed outlet described above, many companies exist that specialize in wholesaling seed. Much of their business is concerned with trading in seeds produced outside their marketing area, although they may handle locally produced seed as well. Most of the turf and forage seed produced in the western United States is eventually sold in the consuming states through these types of wholesale outlets. In turn, the wholesalers supply the seed to local elevators, cooperatives, and lawn and garden stores, who ultimately retail it to farmers, homeowners, sod growers, and hobbyists. Many seed wholesalers of this type are equipped with extensive conditioning facilities; therefore, they can purchase seed in bulk from larger seed companies or western seed outlets and reclean, blend, package, and prepare the seed according to the needs of their customers. Because of the competitive nature of this type of business, many of these companies have an aggressive sales program, which includes salespeople who regularly service and attend to the needs of their customers.

Another type of seed business is characterized by large seed companies that are organized on a regional, national, or even international basis. These may be active in the production as well as in the consuming areas; some marketing is done through elevators and cooperatives, and some seed is sold through their own retail outlets in the consuming area. The companies may be specialized in scope, such as seed of field and garden beans, vegetables, and flowers; others are highly diversified, with different divisions handling seeds of field crops, vegetables, turfgrass, flowers, and some specialty crops. These companies are active in the international seed trade, and have highly organized domestic seed production and marketing programs.

SEED TRADE ORGANIZATIONS

Like most professions, the seed trade is highly organized. Seed trade organizations provide their members an opportunity to meet and associate, on a professional and social level, with persons having mutual interests. More importantly, the organization gives the seed trade identity and power in influencing public opinion as well as state and national seed legislation.

United States

The seed trade within the United States is organized into state, national, regional, and specific interest associations. While most states have their own seed dealers associations, the national seed industry is organized into the American Seed Trade Association (ASTA). The ASTA was organized in 1883. Today it is divided into five regional associations representing the Pacific, Western, Northern, Atlantic, and Southern regions. It is divided by commodity interests into four divisions—the farm seed division, garden seed division, hybrid corn division, and lawn and turf division. Each division has its own staff of officers and committees. The ASTA has a board of directors, and officers are elected annually, but a permanent full-time executive vice president is employed to give continuity to its administration.

The activities of ASTA are many and varied and, of course, geared to the interests of the seed industry. The ASTA has been highly effective as a lobbying organization in influencing state and federal seed legislation. An example of its effectiveness was the passage of the Plant Variety Protection Act of 1970, for which the ASTA was largely responsible. In addition, ASTA sponsors educational meetings, such as its annual Farm Seed Conference and the Hybrid Corn and Sorghum, and Soybean Research Conferences. It also sponsors a national crops judging contest for agronomy students. Through a special organization, known as the American Seed Research Foundation, it sponsors seed research projects in public institutions. The results of this research are published and made available throughout the industry, and are doing much to broaden our knowledge of seeds.

Canada

The seed industry of Canada is organized into the Canadian Seed Trade Association (CSTA). It was organized in 1923, following a small gathering of seed trade representatives called by the Seed Commissioner of Canada to appoint trade representatives to the Advisory Board under the new Canadian Seeds Act, which was soon to become effective. Since that time, it has continued to be active as a lobbying agency in the interest of the seed trade of Canada; it has also fostered professional association within the Canadian seed industry and with that of the United States and other parts of the world.

CSTA maintains a full staff of officers—a president, first vice president, and full-time executive secretary. All officers assume responsibility for contacting members of the trade in their area and passing on information to their members in other parts of the country through the executive secretary's office.

Federation Internationale du Commerce des Semances (FIS)

The Federation International du Commerce des Semances is an international seed trade organization. Its purpose is to foster cooperation among the nations of the world in respect to facilitating the international commerce in seeds. It acts as a voice of the international trade in seeds to discourage national policies restricting the free movement of seeds across national boundaries. Although it has no policy-making powers, it is effective as an educational force in influencing national and international policies affecting the international movement of seed.

Questions

1. What are the advantages and disadvantages of growing seed under contract?
2. How is the system of marketing hybrid corn seed different from that of grass seed crops?
3. How much seed is retailed directly to the ultimate consumer by seed growers in your state?
4. Do you think it is good business for soybean seed producers in the Midwest to sell part of their production directly to consumers and to wholesale part of it through elevators and seed dealers? Why?
5. Do you think that exclusive release of publicly developed varieties is justified? If so, when and under what circumstances?
6. What class or species of seed generally follows the marketing cycle illustrated in Figure 15.1? How does the marketing of other kinds of seed deviate from this?
7. Do you feel the sale of seed in the local supermarket promotes high-quality seed?
8. If you wanted to establish a home lawn, where would you purchase your seed?

General References

Douglas, J. E., ed. 1980. *Successful Seed Programs: A Planning and Management Guide.* Boulder, Colo.: Westview Press.
Gregg, B. R. 1983. Seed marketing in the tropics. *Seed Science and Technology* 11:129–148.
Law, A. G., B. R. Gregg, P. B. Young, and P. R. Chetty. 1971. *Seed Marketing.* New Delhi, India: National Seeds Corporation.
Thompson, J. R. 1979. *An Introduction to Seed Technology.* New York: John Wiley and Sons.
U.S. Department of Agriculture. 1961. *Seeds: The Yearbook of Agriculture.* Washington, D.C., U.S. Government Printing Office.
    1. Carter, W. B., and E. P. Bugbee, Jr. How we get seeds of vegetables and flowers, pp. 493–499.
    2. Christensen, D. K., Earl Sieveking, and J. W. Neely, Handling seed of the field crops, pp. 499–506.

3. Heckendorn, W., and R. A. Edwards, Jr. The four types of seed trade associations, pp. 517–521.
4. Kuzelka, T. J., and W. H. Youngman. Statistics and trends, pp. 521–530.
5. McCorkle, C. O. Jr., and A. D. Reed. The economics of seed production, pp. 530–540.
6. Schiffman, J. F., and R. W. Schery. The responsibilities of the seedsman, pp. 514–517.
7. Schery, R. W. Grass seeds for lawns and turf, pp. 507–513.

Van Gastel, A. J. G. 1986. Seed marketing. In: *Seed Production Technology*, eds. J. P. Srivastava and L. T. Simarski, pp. 232–236. Aleppo, Syria: ICARDA.

Wagner, K. P., H. F. Creupelandt, and W. H. Verburgt. 1975. Seed marketing. In: *Cereal Technology*, ed. W. P. Feistritzer, pp. 108–129. Food and Agriculture Organization, Rome.

Wheeler, W. A., and D. D. Hill. 1957. *Grassland Seeds*. New York: D. Van Nostrand Company.

16

Seed Legislation and Law Enforcement

Thou shalt not sow thy fields with mingled seed—(Lev. 19:19)

Seed laws are designed to aid in the orderly marketing of seed. They establish regulations governing the sale of seed, thereby providing legal protection to both buyers and sellers. No country can expect to have a well-developed, effective seed industry without seed control regulations. In the United States, seed legislation exists at both the state and federal levels.

The basic purpose of both state and federal seed law is truth in labeling. Although some seed laws are designed to protect even the uninformed consumer, most require only that the seed be completely labeled for quality. It is assumed that buyers will read the label and select seed lots that meet their criteria. In many states it is legal to sell even poor-quality seed if it is properly labeled.

The truth-in-labeling concept was developed to avoid the type of *caveat emptor*, or "let the buyer beware," marketing philosophy prevalent in the early English markets. There, unscrupulous vendors sometimes used clever gimmicks to pass on virtual trash as agricultural seed to unsuspecting customers. There were even documented cases of "factories" established for the purpose of preparing so-called seed for sale. Often this was done by coloring small gravel and other inert material to substitute for, or mix with, the seed offered for sale. Another fraudulent scheme was to place sweepings, chaff, or other seed types in the center of seed containers to increase their bulk, a practice sometimes called "stove piping."

FEDERAL SEED LEGISLATION

History of Federal Legislation

Federal seed legislation dates back to 1905, when the Annual Appropriations Act was passed, giving the USDA authority to purchase seeds on the open market, test them for adulteration or mislabeling, and publish the test results together with the names of the persons offering the seed for sale. Under the provisions of this act, between 1912 and 1919 approximately 15,000 samples were tested, and 20% of the samples were found to be adulterated or mislabeled.

The second major federal seed legislation was the Seed Importation Act of 1912, restricting the importation of principal forage crops that were below minimum purity specifications and above maximum weed seed content. It was first amended in 1916 to require a minimum live seed requirement for imported seed, amended again in 1926 to require coloration of imported alfalfa and red clover, and further amended in 1926 to prohibit the shipment in interstate commerce of falsely or fraudulently labeled seed. The first amendment was added because much of the alfalfa and clover seed imported was unsuited for production in the United States, and coloration was intended to inform buyers of the possible lack of adaptation of crops grown from the seed.

The Federal Seed Act

Beginning in about 1936, discussions took place between several interested agencies with respect to further changes in the seed laws, which resulted in the enactment of the Federal Seed Act in 1939. This act is the single most important piece of seed legislation in United States history. It applies to all agricultural and vegetable seeds, to imported seeds, and to that sold in interstate commerce. A major improvement over the Seed Importation Act, it provided more detailed labeling requirements and did not require proof of intent to defraud in cases of mislabeling. The act was amended in 1956 to allow civil prosecution for complaints of violations and was amended again in 1960 to require labeling of pesticide-treated seed. Further amendments have been made as needed.

The Federal Seed Act is a truth-in-labeling law that governs the sale of seed in interstate commerce and seed imported into the United States. The aim of the act is to provide detailed regulations covering sale of seed on a national basis. Normally, it has no jurisdiction over seed marketed within state boundaries. The federal and state seed laws contain somewhat similar requirements. If seed is labeled to comply with the Federal Seed Act and is shipped in interstate commerce, it will normally comply with the labeling requirements of the state into which it is shipped. Thus, the Act helps maintain the integrity of each state seed law; however, no state may set standards for seed moving into the state from another below the minimum required by the Federal Seed Act. A state may have standards below those of the Federal Seed Act covering seed sold within the state's boundaries. However, it is a violation of the Federal Seed Act to move seed into a state in which it does not comply with the state's noxious weed seed restrictions; even though the seed may otherwise comply with the requirements of the federal law.

Though the Federal Seed Act is seemingly detailed and complex, most of its requirements are simple and should be well understood by everyone buying or selling seed. In general, the following information is required on the label (see Figure 16.1):

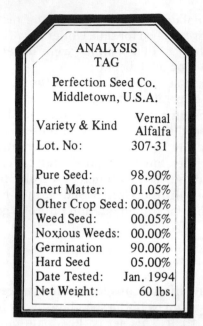

ANALYSIS TAG

Perfection Seed Co.
Middletown, U.S.A.

Variety & Kind	Vernal Alfalfa
Lot. No:	307-31
Pure Seed:	98.90%
Inert Matter:	01.05%
Other Crop Seed:	00.00%
Weed Seed:	00.05%
Noxious Weeds:	00.00%
Germination	90.00%
Hard Seed	05.00%
Date Tested:	Jan. 1994
Net Weight:	60 lbs.

Figure 16.1. A typical labeling tab

1. The name of the kind or kinds and variety for each agricultural seed component present in excess of 5% of the whole and the percentage by weight of each. If the crop is one that is regarded by the Secretary of Agriculture as generally labeled as to variety, the label shall bear either the name of the variety or the statement "Variety Not Stated."
2. Lot number or other identification.
3. Origin, if determined by the Secretary of Agriculture that the crop is one in which the origin is important from the standpoint of crop production. If the origin is unknown, that fact shall be stated.
4. Percentage by weight of weed seeds, including noxious types.
5. Kind and rate of occurrence (per lb. or oz.) of noxious weed seed, labeled in accordance with the law of the state into which the seed is shipped.
6. Percentage by weight of agricultural seed present other than those named in No. 1 above.
7. Percentage by weight of inert matter.
8. Germination percent for each agricultural seed in excess of 5% and percentage of hard seeds, and the calendar month and year the test was completed to determine such percentages.
9. Name and address of the seller, or the person to whom the seed is sold, together with a code designating the seller.
10. The year and month beyond which an inoculant, if shown on the labeling, is no longer claimed to be effective.

STATE SEED LAWS

The Federal Seed Act is important because it applies to all seed marketed on a national basis across state boundaries. However, it has no jurisdiction on seed sold within state boundaries. As a result, states have established their own seed laws.

Each of the 50 states has its own seed law that regulates the sale of seed within the state. Seed laws vary among states, since each is constructed to meet the unique needs of a particular state. All are primarily truth-in-labeling laws, although some are more protective and specify minimum quality standards that seed must meet to be eligible for sale.

In most states, seed laws are administered by the state department of agriculture. If a state does not have a department of agriculture, another state agency is responsible for seed law enforcement. For example, in Indiana seed law enforcement is the responsibility of Purdue University, which designates a faculty member as the Commissioner of Agriculture. Regardless of the specific arrangement, the responsible agency must maintain a staff of seed inspectors, who actually go to locations where seed is marketed and check for compliance with the seed law. Representative seed samples are routinely and randomly collected from seed dealers, elevators, railroad cars, lawn and garden stores (see Figure 16.2), and even supermarkets to be tested for compliance with the seed law. Violations are reported, and appropriate action is taken to correct the violation or to stop sale of the seed. A *stop sale* is the action by a state seed control agency that prohibits further sale of seed until violations are corrected. Usually the state seed law requires that a notice of all violations be periodically published and made available to the public.

The Need for Standardization

A major goal of the American seed industry has been to promote uniformity among state seed laws. Uniformity among laws is an advantage to the seed trade in that it simplifies interstate commerce in seed. Encouragement toward uniformity has also come from seed testing associations as well as federal seed control officials. The first suggested uniform state seed law was developed by the Association of Official Seed Analysts (AOSA) and the seed trade in 1907–1917. After the Federal Seed Act of 1939 was enacted, the USDA promoted the Recommended Uniform State Seed Law patterned after suggestions made by the AOSA and the American Seed Trade Association. The movement for uniformity among seed laws was aided in 1949 by the formation of the American Association of Seed Control Officials (AASCO), an organization of state and federal seed control officials. This organization has as its major goal the uniformity in seed laws, in rules and regulations, and in the uniform administration of laws pertaining to the sale and distribution of seeds.

In 1957 the AASCO published a "Recommended Uniform State Seed Law" (RUSSL) with the approval of its various member agencies. This law has no official jurisdiction, but is merely a model for states that desire to revise their seed laws. It provides both form and wording for revisions that can be made to promote uniformity among state seed laws. Although this law has been useful, it is recognized that each state has its

Figure 16.2. Sampling garden and flower seed at the point of sale.

own unique situations and its own interests to protect; consequently, no move toward complete uniformity is anticipated.

FEDERAL-STATE COOPERATION

Enforcement of the Federal Seed Act is the responsibility of the Seed Branch of the Livestock, Meat, Grain, and Seed Division, Agricultural Marketing Service of the USDA. To help carry out this responsibility, a federal seed laboratory is located at Beltsville, Maryland. This laboratory has historically assumed the additional responsibility for helping to standardize seed-testing activities and procedures throughout the United States. Federal-state cooperative agreements have been worked out through which apparent violations of the Federal Seed Act are brought to the attention of Seed Branch officials by various state officials. When an apparent labeling violation occurs, seed samples are exchanged and tested by both the federal and state laboratories. If a violation is established, the federal seed officials may make further investigations and send all information and evidence to the Seed Branch headquarters in Washington, D.C. Depending on the nature and seriousness of the violation, several alternatives can be followed. A warning letter may be sent to the violator about the apparent violation so that appropriate steps may be taken to prevent such action in the future. Another option is a *cease and desist* proceedings, which is similar to a court order to cease from such violations. In addition, the violator may be fined or even prosecuted under the criminal section of the act. Also, seed may be seized by the courts to prevent further sales. When seed is *seized* through court procedure, the court may permit the sale of seed after the violation has been corrected, or may permit its sale in another state where the seed would not be in violation of state laws. In extreme cases, the court may even order the seed destroyed. As with many state seed laws, the Federal Seed Act requires that a list of violations be maintained showing violators' names, the nature of the violations, and the penalty imposed.

PROVISIONS OF STATE AND FEDERAL SEED LAWS

Farmer Seed Exchange

Exchanges of seed between farmers are exempt from labeling laws; however, the laws carefully define such exemptions. If the seed is advertised in any way, the seed must be labeled.

Current Germination Tests

Since seed will deteriorate over a period of time, the month of the test on which the germination is based must appear on the label. Under the Federal Seed Act, no more than five months may elapse between the last day of the month in which the test was completed, and the date of transportation and delivery in interstate commerce. If the seed is in hermetically sealed containers, a period of 24 months is permitted. State seed laws vary in the length of time permitted since the last germination test.

Labeling Vegetable Seed Containers

Minimum germination standards are outlined for vegetable seeds in the Federal Seed Act. If seed being sold or shipped in interstate commerce has a germination as good or better than the minimum standard for that particular kind of seed, the germination of seed in packets of one pound or less does not have to be indicated on the label. If the germination is below the minimum standard, the statement "Below Standard" and the germination percentage must be given on the label.

Transport for Conditioning

If consigned to a seed cleaning or conditioning establishment, seed is exempt from labeling laws provided the statement, "Seed for Conditioning" is stated on the invoice and other shipping records and is clearly indicated on the seed container.

Disclaimers Not Allowed

Statements disclaiming responsibility for the information on the seed label are disallowed by all federal and state seed laws.

Collection of Damages

No damages may be collected through the seed laws even though a violation is established. Damage must be collected through separate action in a civil court.

Proof of Intent not Needed

Original seed laws were often ineffective because even when violations occurred the state had to prove intent to defraud, which was hard to do. After the 1956 amendment to the Federal Seed Act, violators could be penalized without proof of intent, or carelessness.

Coloration and Labeling of Treated Seed

A Federal Food and Drug rule requires that seed that has been treated with a chemical pesticide must be dyed a color that contrasts to the seed. Usually red or green dyes are used. In addition, under federal and state seed laws the seed container must bear a statement that the seed has been treated, along with the commonly accepted chemical or abbreviated chemical name of the substance; if the chemical is harmful to humans or animals in the amount applied, a statement such as "Do not use for food or feed or oil purposes" must also be included on the tag, along with an antidote to be used if the chemical is taken internally. When mercurials or similarly toxic materials are used, the word "Poison" in red letters on a contrasting background and the outline of a skull and crossbones are required on the label.

Noxious Weed Seeds

In interstate trade, it is a violation of the Federal Seed Act to ship seeds containing noxious weed seeds that are in excess of that allowed by the receiving state. Since each state seed law has established a list of prohibited and restricted noxious weeds (designated as primary and secondary noxious weeds in some states), the requirements vary among states. The sale of seed lots containing primary noxious weed seeds is prohibited. The sale of seed lots containing secondary noxious weed seeds is permitted, but the name and number (per ounce or pound) must be stated on the label. Some states limit the number of secondary noxious weed seeds that may be present per pound. Table 16.1 shows the classification of noxious weeds of Michigan.

Because of the importance of accurate identification of weed seeds, the Association of Official Seed Analysts (1993) has developed Handbook 25 on the Uniform Classification of Weed and Crop Seeds. This handbook was first published in 1952 and was subsequently revised in 1964, 1977, and 1993 because of continuing changes in scientific terminology. The purpose of the handbook is to (1) develop a better system of classifying contaminating species as other crop or weed seed, (2) provide a comprehensive listing and classification of over 2,000 species, (3) bring the scientific nomenclature of these species up to date, and (4) provide a comprehensive reference source to other scientific nomenclature systems. Handbook 25 is a valuable resource for standardizing the classification of noxious weed seeds found in a seed lot. Without this reference,

Table 16.1. Noxious Weeds of Michigan

Prohibited Noxious (Prohibited in all seed offered for sale in Michigan)	
1. Bindweed (*Convolvulus arvensis*)	5. Russion knapweed (*Centaurea picris*)
2. Canada thistle (*Cirsium arvense*)	6. Leafy spurge (*Euphorbia esula*)
3. Perennial sow thistle (*Sonchus arvensis*)	7. Quack grass (*Agropyron repens*)
4. Whitetop (*Lepidium draba*)	8. Horse nettle (*Solanum carolinense*)

Restricted Noxious (Restricted occurrence in all seed offered for sale in Michigan)	
1. Dodder (*Cuscuta* spp.)	6. Wild carrot (*Daucus carota*)
2. Fan weed (*Thlaspi arvense*)	7. Wild onion (*Allium* spp.)
3. Wild mustard (*Brassica kaber, juncea* and *nigra*)	8. Giant foxtail (*Setaria faberii*)
4. Hoary alyssum (*Berteroa incana*)	9. Yellow rocket or wintercress (*Barbarea vulgaris*)
5. Buckhorn plantain (*Plantago lanceolata*)	

incorrect evaluations could be made that would make perfectly acceptable seed lots unacceptable for sale because they contain weed seeds.

Imported Seeds

Special regulations of the Federal Seed Act apply to seeds imported into the United States. Screenings of any seed are not permitted to be imported except for screenings of wheat, oats, rye, barley, buckwheat, field corn, sorghum, and certain other edible crops that are not imported for seeding purposes and are declared for cleaning, conditioning, or manufacturing purposes. Screenings of other kinds of seeds may not be imported.

Imported seeds that contain 10% or more of seeds of alfalfa, red clover, or both must be stained to indicate that such seed is of foreign origin. When foreign-grown seed is mixed with domestically produced seed, each component must contain at least 10% stained seed.

If a seed lot contains any specified noxious weed seeds or more than 2% by weight of any weed seeds or less than 75% pure live seed (with certain exceptions), it is considered unfit and ineligible for import into the United States.

Keeping of Records

State and federal seed laws require that complete records for each lot of seed sold be maintained for varying lengths of time. Complete records refer to information relating to origin, treatment, germination, or purity of each lot of agricultural seed handled. The following information must also be maintained: declarations, labels, seed samples, records of purchases, sales, cleaning and bulking, handling, storage, analysis, and tests. Each person shipping seed in interstate commerce must retain a sample representing each lot of agricultural seed shipped. Federal regulations provide that any seed sample may be discarded one year after the entire lot represented by such sample has been sold.

Definition of Sale or Offer for Sale

Seed legislation is intended to cover regulations for seed sold or offered for sale. The term *offer for sale* includes representations of the salesperson, oral or written advertisements, including radio and television advertising, and price lists, newspapers, catalogs, and pamphlets that pertain to the seed for sale.

Definition of Seed Quality Terminology

The definition for seed quality terminology used in state and federal seed laws relies on the *Rules for Testing Seeds* of the Association of Official Seed Analysts (AOSA 1991). For example, the Federal Seed Act defines the term *germination* to mean the percentage of seeds capable of producing normal seedlings under favorable conditions, which is the definition in the *Rules for Testing Seeds* of the AOSA. Many other terms, such as *pure seed, other crop seeds, inert matter*, also were adopted from the AOSA rules.

Labeling

The Federal Seed Act and most state seed laws define the term *label* as the display or displays of written, printed, or graphic matter upon or attached to the container or seed. For example, any of the printing on the bag or on any cartons or on any tags attached to bags must be considered a part of the label. The term *label* is not confined to the information on the tag attached to the bag.

Amending Seed Laws

In establishing seed laws and regulations in the United States, state and federal officials give due consideration to the interest of the seed industry, for which many of the laws and regulations are intended. From the beginning of seed legislation in the United States, the seed industry and government officials have worked together in preparing legislation. Any regulations that are passed by state and federal seed laws, or any changes to these regulations, are adopted only after a public hearing at which any person in the state or United States is free to make comments in person or in writing. This makes certain that the views of interested parties—whether in industry, in seed law enforcement, or in the consumer field—are given due consideration. However, the final decision for the new law or changes in the law must be made by the duly constituted legislative bodies, relying largely on the advice of officials of the agency responsible for enforcing the act.

THE PLANT VARIETY PROTECTION ACT— LEGAL PROTECTION FOR CROP VARIETIES

The Plant Variety Protection Act, signed into law in 1970, provides legal protection in the production and sale of seed by owners or developers of new varieties of sexually propagated crops. Developers of varieties of asexually propagated (budding or grafting, etc.) crops have received protection since 1930 through the U.S. Patent Office. This protection has enabled them to control the propagation of their varieties and to collect a royalty for any use and propagation of their varieties by other parties. Prior to the Plant Variety Protection Act of 1970, such protection had not been available to originators of sexually propagated crop varieties.

The Plant Variety Protection Act is administered by the Plant Variety Protection Office within the U.S. Department of Agriculture. The office receives applications and evaluates each candidate on the basis of the following criteria: (1) *novelty*, (2) *uniformity*, and (3) *stability*. Novelty is described as distinctiveness (i.e., whether it can be distinguished from all known varieties on the basis of morphological, physiological, or cytological characteristics). Uniformity requires that all variations within the population must be describable, predictable, and commercially acceptable. Stability requires that its essential and distinctive characteristics remain unchanged throughout successive generations of seed increase.

The owner of a protected variety may assign his rights to others, or even sell them. Protection lasts for 17 years, at which time the variety becomes public property. In cases of emergency, or when needed to supply the country with sufficient food, fiber, or feed, the Secretary of Agriculture may declare a variety available to the public; however, in such rare instances the owner would be compensated for the public use of this variety.

The act provides two avenues of protection to the owner of a variety. First, it provides exclusive rights for the propagation and use of a protected variety. In cases of infringement, however, the owner is responsible for defending these rights in a civil court. Second, it gives the owner the right to stipulate in the application that the variety name be protected through seed certification. It is a violation of the Federal Seed Act to sell by variety name uncertified seed of varieties protected under this stipulation.

A special exemption for farmers grants them the right to produce seed of a protected variety for their own use. Plant breeders may also use protected varieties for breeding purposes. Use for breeding purposes does not constitute infringement of the Plant Variety Protection Act.

CANADIAN SEED LAWS

The Canadian Seed Control Act was enacted in 1905 to set minimum standards for pure seed, common and noxious weed content, and germination. It was amended in 1911 to establish more definite requirements. In 1923, a grading system was introduced for all seeds in commerce. The Seed Branch of Agriculture Canada (the Canadian Department of Agriculture) provides the seed inspection work for law enforcement.

The philosophy of Canadian seed legislation provides truth in labeling for seeds in commerce; however, the Canadian seed law is considerably more protective than the U.S. Federal Seed Act. Crop varieties in Canada are licensed to protect the seed user or consumer against losses that can occur from the purchase of unknown and inferior seed. Licensing is also intended to prevent deception from the sale of seed under modified or false variety names. The Canadian law provides a system whereby only varieties that have been tested and found agronomically and economically desirable for Canadian agriculture are allowed to be sold, advertised, or imported. The licensing of varieties is administered by the Plant Products Division of Agriculture

Canada and covers all agricultural and vegetable crop seeds. Seeds of root and vegetable crops, other than seed potatoes, are exempt.

EUROPEAN SEED LAWS

In contrast to the truth-in-labeling seed laws in the United States, most European seed laws are based on the assumption that the government knows what is best for the farmer. Many European seed laws strictly control what is to be sold within or imported into a country. This is done in various ways.

In France, seeds of varieties of most cereals, herbage crops, peas, vetches, horse beans, white clover, and the most important grasses cannot be sold unless the variety name appears on the official list. Only certified seed of these varieties can be sold. Field trials for approval of varieties are conducted by an official government agency. Five years of testing are required for approval. France has one of the most restrictive seed laws in effect, and its seed law is being used as a model by other countries.

England's varietal protection law has a varietal indexing system, official trials for testing all varieties submitted, and a list of acceptable varieties based on their performance. Although participation in the official tests is voluntary, seed cannot be sold by varietal name unless it has been entered into the testing program.

Most other European countries have rather restrictive seed laws similar to those in France and England. Until recently, Denmark had no seed law but relied almost entirely on an education approach to consumer protection. However, due to their membership in the European Common Market, they have recently drafted and adopted seed control legislation similar to that in other common market countries.

SEED LEGISLATION IN DEVELOPING COUNTRIES

Farmers in most developing countries generally save their own seed from year to year or exchange seed with their neighbors. This system of agriculture involves almost no commerce in seed and has little need for seed control legislation. However, as improved varieties become available and agricultural development occurs, trade in seed naturally follows, leading to the necessity for seed control legislation.

The first seed laws in developing countries are usually truth-in-labeling laws, requiring that seed be labeled as to quality, and that such labels truthfully represent the actual seed quality. For such laws to be effective, seed testing laboratories in which quality tests can be performed must be available, and farmers and seed dealers must be encouraged to use these services. Strong research, seed certification, and extension programs also promote effectiveness of seed control legislation.

As agriculture and the seed trade gain sophistication and become more complex, seed legislation tends to become more complex and restrictive. Most countries contemplating seed legislation seek advice from other countries that have long histories of successful seed legislation.

Questions

1. Can you list several noxious weeds in your state? What is meant by the term *caveat emptor*? How does this concept apply to seed labeling today?

2. What is meant by the concept of *truth in labeling*?
3. In State A, the minimum germination of agricultural seed allowed for sale is 60%. If a lot of alfalfa seed was shipped from State B into State A where it was offered for sale, sampled by state seed control officials, and found to germinate 45%, would the seed lot be in violation of the seed law in State A? The Federal Seed Act? Both? Why? Assume that a part of the same seed lot was shipped into State C, which has no minimum germination requirement. Would it still be a violation of the Federal Seed Act? Why?
4. What information is required on the seed label in your state?
5. What is the difference between a stop sale and a seizure?
6. What procedure must a seed regulation undergo to be amended?
7. Is it illegal to offer seed that has not been tested for germination and purity to a neighboring farm?
8. It is illegal to sell untested seeds in a retail outlet or store?
9. What is a disclaimer and what is its legality?
10. What is the law in your state pertaining to the labeling of treated seed?
11. What is the philosophical difference between seed laws in the United States compared to those in most European countries? Why do you think this difference exists?
12. Assume that a certified seed grower conditions his/her own seed, obtains a seed analysis, and labels the seed accordingly. The seed is then sold to a wholesaler, who in turn sells it to a local elevator where it is resampled and found to be mislabeled. However, the grower's name is still listed on the label with the labeling information. Who is responsible for the mislabeled information according to the seed law?
13. What is the basic purpose (or philosophy) of seed laws in the United States?
14. What kind of plants are covered by the Plant Variety Protection Act of 1970? Do you believe the law is fair in its limited coverage?

General References

Association of Official Seed Analysts. 1991. Rules for testing seeds. *Journal of Seed Technology* 12(3):1–109.

Douglas, J. E., ed. 1980. *Successful Seed Programs: A Planning and Management Guide*. Boulder, Colo.: Westview Press.

Larsen, A. L., J. H. Wiersema, and T. Handwerker. 1993. Uniform Classification of Weed and Crop Seeds. 137 pp. Contribution No. 25 to the Handbook on Seed Testing. Association of Official Seed Analysts, Lincoln, NE.

Thompson, J. R. 1979. *An Introduction to Seed Technology*. New York: John Wiley and Sons.

U.S. Department of Agriculture. 1961. *Seeds: The Yearbook of Agriculture*. Washington, D.C.: U.S. Government Printing Office.
    1. Clark, E. R. Sometimes there are frauds in seeds, pp. 478–482.
    2. Clark, E. R., and C. R. Porter. The seeds in your drill box, pp. 474–478.
    3. Crispin, W. R. Seed marketing services, pp. 470–474.
    4. Davidson, W. A. What labels tell and do not tell, pp. 462–469.
    5. Rollin, S. F. and F. A. Johnston. Our laws that pertain to seeds, pp. 482–492.

Glossary

AASCO. The initials of the American Association of Seed Control Officials representing state and federal seed law enforcement officials throughout the United States.

Abnormal seedling. A seedling (in a germination test) that does not have the essential structures indicative of the ability to produce a normal plant under favorable conditions.

Absorption. The uptake of moisture into the tissues of an organism (e.g., seed).

Achene. A small, dry, one-seeded fruit with a thin, dry wall that does not split open at maturity (e.g., sunflower seed).

Acorn. The fruit of an oak; see the definition of a nut.

ADP. Initials for adenosine-diphosphate, a complex sugar-phosphorus compound formed as the result of expenditure of energy and the loss of a phosphate group from the energy-rich ATP (adenosine-triphosphate) compounds.

Adsorption. The accumulation and adhesion of a thin layer of water (or gases) on the surface of another substance.

Adventitious embryony. A condition in seed in which the embryo arises from somatic (body) rather than reproductive tissue. This condition is common in certain grasses and often results in multiple embryos.

After-ripening. A term for the collective changes that occur in a dormant seed that make it capable of germination. It is usually considered to denote physiological changes.

Agamogony. A type off apomixis in which cells undergo abnormal meiosis during megasporo-genesis, resulting in a diploid embryo sac rather than the normal haploid embryo sac.

Agamospermy. A type of apomixis in which sporophytic tissue is formed, ultimately leading to seed development.

Aggregate fruit. Fruit development from several pistils in one flower, as in strawberry or blackberry.

Air screen cleaner. The basic piece of equipment for cleaning seed, utilizing air flow, and perforated screens (also called a fanning mill).

Albuminous seed. A seed having a well-developed endosperm or perisperm (nucellar origin).

Aleurone layer. The layer of high-protein cells surrounding the storage cells of the endosperm. Its function is to secrete hydrolytic enzymes for digesting food reserves in the endosperm.

Amino acid. Organic acid containing one or more amino groups ($-NH_2$), at least one carboxyl group ($-COOH$), and sometimes sulfur. Many amino acids are linked together by peptide bonds to form a protein molecule. Proteins are a fundamental constituent of living matter.

Amphitropous ovule. A type of ovule arrangement in which the ovule is slightly curved so the micropyle is near the funicular attachment.

Amylase. The enzyme responsible for catalyzing the breakdown of starch into sugars. It may be active in one of two forms: α-amylase and B-amylase.

Amylopectin. A type of starch molecule composed of long, branched chains of glucose units (a polysaccharide).

Amylose. A type of starch molecule made up of glucose units in long, unbranched chains (a polysaccharide).

Anatropous. A type of ovule arrangement in which the ovule is completely inverted, having a long funiculus with the micropyle adjacent to the base of the funiculus.

Androecium. Collectively, the stamens of a flower.

Angiosperm. A kind of plant that has seeds formed within an ovary.

Annual. The type of plant that normally starts from seed, produces its flowers, fruits, and seeds, and then dies within one growing season.

Annuoline. The fluorescent protein pigment exuded by the roots of ryegrass seedlings. The flurorescent nature of this material makes it useful in distinguishing annual and perennial ryegrass.

Anther. The saclike structure of the male part (stamen) of a flower in which the pollen is formed. Anthers normally have two lobes or cavities that dehisce at anthesis and allow the pollen to disperse.

Anthesis. The period of pollenization, specifically the time when the stigma is ready to receive the dispersed pollen.

Antipodal nuclei. Three of the eight nuclei that result from the megaspore by mitotic cell divisions within the developing megagametophyte (embryo sac). They are usually located at the base of the embryo sac and have no apparent function in most species.

AOSA. The initials of the Association of Official Seed Analysts, the organization of state and federal seed analysts of the United States and Canada.

AOSCA. The initials of the Association of Official Seed Certifying Agencies, the organization of certification agencies of the United States and Canada—formerly (prior to 1968) known as the International Crop Improvement Association (ICIA).

Apical placentation. A type of free-central placentation in fruit where the seeds are attached near the top of the central ovary axis.

Apogamy. A type of apomixis involving the suppression of gametophyte formation so that seeds are formed directly from somatic (body) cells of the parent tissue.

Apomixis. Seed development without the benefit of sexual fusion of the egg and the sperm cells.

Archesporial cell. The cell of the nucellus that differentiates and gives rise to cells ultimately destined to undergo meiosis and produce the megaspore mother cell.

Aril. A loose, papery appendage in some seeds (e.g., elm) originating as an extension (or proliferation) from the outer integument.

Aspirator. An air-blast separator. A seed conditioning (cleaning) machine that uses air to separate according to specific gravity (weight) and resistance to air flow.

Asexual reproduction. Reproduction by vegetative means without the fusion of two sexual cells.

ASTA. The initials of the American Seed Trade Association.

Astered embryo type. A type of embryo classification in which the terminal cell of the proembryo divides by a longitudinal wall and both the basal and terminal cells contribute to embryo development.

ATP. The initials for adenosine-triphosphate, an energy-rich complex, sugar phosphorus compound which provides energy for many metabolic reactions.

Auxins. A group of growth regulators that may stimulate cell growth, root development, and other growth processes including seed germination.

Awn. A slender appendage often associated with seeds, such as the "beards" of wheat or barley.

Axile placentation. The type of ovule attachment within a fruit in which the seeds are attached along the central axis at the junction of the septa.

Bacteriocide. A chemical compound that kills bacteria.

Bacteriophage. A virus that infects specific bacteria and usually kills them. Specific phages are used to identify certain bacterial plant pathogens.

Basal placentation. A type of free-central placentation in which the seeds are attached at the bottom of the central ovary axis.

Berry. A simple, fleshy, or pulpy and usually many-seeded fruit that has two or more compartments and does not burst open to release its seeds when ripe (e.g., blueberry).

Biennial. A kind of plant that produces only vegetative growth during its first growing season. After a period of storage or overwintering out-of-doors, flowers, fruits, and seeds are produced during the second year and the plant dies (i.e., a plant that requires two years to complete its life cycle).

Biological seed treatment. Use of fungi or bacteria to control soil and seed pathogens instead of synthetic chemical seed treatments.

Bitegmic testa. A testa (seedcoat) composed of two integumentary layers.

Blend. A term applied to seed mixtures of different crop varieties (or species) that have been mixed together to fulfill a specific agronomic purpose.

Brand. A legal trademark registered by a particular company or distributor for its exclusive use in marketing a product such as seeds or plants.

Breeders' rights. Varietal protection—the legal rights of a breeder, owner, or developer in controlling seed production and marketing of crop varieties.

Brick grit test. A type of seedling emergence (vigor) test utilizing uniformly crushed brick gravel through which seedlings must emerge to be considered vigorous.

Buckhorn machine. A machine that mixes water and sawdust with seed lots containing seeds of buckhorn plantain. The mucilagenous seedcoat of the buckhorn seed attracts the water and sawdust and changes its size and specific gravity, allowing it to be separated by the air screen machine or gravity separator.

Bulb. An enlarged, fleshy, thick, underground part of a stem surrounded by thickened, leafy scales and shortened leaves. Roots develop at the base of a bulb (e.g., wild onion).

Bulbil. A small bulb or bulblet produced above the ground, as in wild garlic.

Bumper mill. A machine designed to clean timothy seed by a continuous bumping action on an inclined plane. The uncleaned seed is metered onto the plane, which is continuously bumped by sets of knockers. The cylindrical timothy seeds are rolled into separate grooves while noncylindrical contaminants are jarred off the end of the inclined plane and separated.

Callus. A hard or thickened layer at the base of certain grass florets.

Calyx. A collective term for all the sepals surrounding a flower; it forms part of the covering of some seeds.

Campylotropous ovule. A type of ovule arrangement in which the ovule is slightly curved and the micropylar end is pointed slightly downward so the funiculus and micropyle are close together on the mature seed on opposite sides of the hilum, as in legumes.

Capsule. A dehiscent fruit with a dry pericarp usually containing many seeds.

Carpel. Female reproductive organ of flowering plants. One or more carpels may be united to form the pistil.

Caruncle. A fragile appendage or outgrowth of the outer integument of the seed of some species (e.g., leafy spurge).

Caryopsis. A dry, indehiscent one-seeded fruit (as in grasses) in which the pericarp and integuments are tightly fused.

Catalase. An enzyme that catalyzes the degradation of hydrogen peroxide to water and the oxidation by hydrogen peroxide of alcohols to aldehydes during seed germination.

Catalyst. A substance that can induce or accelerate a chemical reaction without undergoing any change itself.

Catkin. A spike influorescence with a single unisexual flower arising from the peduncle, as in *Alnus rubra* (red alder).

Cellular endosperm. A type of endosperm in which the early development is characterized by cell wall formation accompanying each nuclear division.

Cellulose. A long-chain complex carbohydrate compound (polysaccharide) with the general formula $(C_6H_{12}O_6)_n$. It is the chief substance forming cell walls and the woody parts of plants.

Certified seed. Seed produced under an officially designated system of maintaining the genetic identity of, and including provisions for, seed multiplication and distribution of crop varieties. It also refers to the class of certified seed which is the progeny of registered or foundation seed. It is identified by a blue tag; thus, it is sometimes called "blue-tag" seed.

Chaffy grass divider. A subsampling device used to divide a sample of chaffy grass seed into a working sample.

Chalaza. The part of an ovule where the integuments originate. In orthotropous ovules the chalaza is directly underneath the funicular attachment. In other types of ovule arrangement it can sometimes be distinguished on the outside of the seed near the hilum (e.g., campylotropous (legumes)).

Chromosome. A rodlike-bearer of hereditary material (genes) inside the nucleus of all cells.

Chenopodiad embryo type. A type of embryo classification in which the terminal cell of the pro-embryo divides by a transverse wall and both the basal and terminal cells contribute to embryo development.

Circadian rhythm. See endogenous rhythms. A type of rhythmic plant or animal growth response which appears to be independent of external stimuli.

Circinotropous ovule. A type of ovule arrangement in which the funiculus is very long and completely encircles the ovule, which otherwise has an orthotropous (straight) arrangement.

Circumscissle capsule. A capsule which at maturity splits open at the middle so that the top comes off like a lid (e.g., plantain).

Clone. A group of individuals (plants) of common ancestry that have been propagated vegetatively, usually by cuttings or by multiplication of bulbs or tubers.

Cold test. A type of stress (vigor) test that tests the performance of seeds in cool, moist soil in the presence of various soil microorganisms. The test is conducted by planting the seeds in moist, unsterilized field soil, exposing them to cool (5–10°C) temperatures for about a week, then allowing them to germinate in the same soil at warmer temperatures.

Coleoptile. A transitory membrane covering the shoot apex of certain species that protects the plumule as it emerges through the soil. The coleoptile is photosensitive and stops growth when exposed to light, allowing the plumule to break through and continue growth.

Coleorhiza. A transitory membrane covering the emerging radicle (root apex) in some species. It serves the same function for the root as the coleoptile does for the plumule.

Color separator. A machine that separates seeds on the basis of their color differences.

Coma. A tuft of hairs attached to a seed (e.g., "brush" on wheat).

Complete flower. A flower that has pistils, stamens, petals and sepals.

Complete hybrid. A legal designation for a seed lot indicating that at least 95% of the seed represents hybrid seed.

Compound cyme. A determinate inflorescence where there is secondary branching, and each ultimate unit becomes a simple cyme (e.g., *Sapanoria officinales*).

Conditioning. The term used to describe the process of cleaning seed and preparing it for market. Formerly called processing.

Conductivity test (of seed leachates). An electrical conductivity test that associates the concentration of leachates from seeds, after soaking in water, to their quality.

Constancy of destination. The theory of embryo development that suggests that each part of the mature embryo inevitably arises from predetermined cells of the proembryo.

Corymb. An indeterminant inflorescence in which the lower pedicels arising from the penduncle are successively longer than the upper ones, giving a rounded or flat-topped appearance (e.g., *Prunus emarginata*).

Cotyledon. Seed leaves of the embryo. In most dicotyledon seeds they are thickened and are storage sites of reserve food for use by the germinating seedling.

Coumarin. A chemical growth inhibitor that has germination-inhibiting capability.

Caveat emptor. A Latin expression meaning "let the buyer beware," often applied to seed marketing prior to the consumer protection era.

Crassinucellate. The nuclear condition in which the embryo sac originates and develops deep (several cell layers) under the nucellar epidermis.

Cristae. The inner folds of the mitochondrial membranes where many enzymes catalyzing metabolic processes are located.

CSTA. The initials of the Canadian Seed Trade Association.

Crucifer embryo type. A type of embryo classification in which the terminal cell of the proembryo divides by a longitudinal wall and the basal cell plays only a minor part (or none) in subsequent embryo development.

Cultivar. A variety of a cultivated crop.

Cutin. A complex fatty or waxy substance found on the surface of certain seeds or leaves, often making them impermeable to water.

CVPO. Certification of seed on the basis of varietal purity only, referring to a type of seed certification that certifies seed as to genetic purity without specific criteria for purity, germination, and other aspects of mechanical seed quality.

Cyme. A type of inflorescence in which the main axis ends in a flower. Further growth is by lateral branches, which may also terminate in a flower.

Cytoplasm. The contents of a cell between the nucleus and the cell wall.

Damping-off. A condition in which young seedlings are attached (parasitized) and killed by soil-borne fungi immediately following germination. This condition is sometimes mistaken for poor seed quality.

Day-neutral plants. Plants that have no daylength requirements for floral initiation.

Desiccant. A chemical applied to crops that prematurely kills their vegetative growth; often used for legume seed crops so the seed can be harvested prior to normal plant senescence.

Dehiscence. The splitting open at maturity by pods or capsules along definite lines or sutures.

Detasselling. Artificially removing (cutting or pulling) the tassel of the female parent to prevent selfing during hybrid seed corn production.

Determinant flower. A floral axis which terminates as a flower rather than a bud.

Dicot. An abbreviated name for dicotyledon which refers to plants having two seed leaves.

Dioecious (diecious). Refers to plants having stamens and pistils on different unisexual plants. Therefore, both sexes must be grown near each other before seed can be produced (e.g., American holly).

Diploid (2n). Refers to two sets of chromosomes. Germ cells have one set and are haploid; somatic cells have two sets and are diploid (except for polyploid plants).

Diploid apogamety. A type of diplospory (apomixis) in which seed develops from some cell other than the egg but does not require the stimulus of pollination.

Diploid parthenogenesis. A type of diplospory (apomixis) in which seed develops from the unfertilized egg of the embryo sac without need for pollination.

Diploid pseudogamy. A type of diplospory (apomixis) in which the stimulus of pollination is needed for seed development.

Diplospory. A type of agamospermy (apomixis) in which a diploid embryo sac is formed from archesporial origin.

Disclaimer. A statement (legally invalid) on a seed container or label disavowing responsibility for the information contained on the label.

Disinfectant. A chemical treatment used to disinfect seed for planting. It is especially useful for surface-borne pathogens.

Diverticulae. The tendrillike forks projecting from the ends of the haustoria (adsorptive arms) of the developing embryo or endosperm.

DNA. Deoxyribonucleic acid, a component of the nucleus (chromosomes) and the basic building blocks of genes. It carries the hereditary information of a cell.

Dormancy. A physical or physiological condition of a viable seed that prevents germination even in the presence of otherwise favorable germination conditions.

Double cross hybrid. The first generation progeny of a cross between two single-cross hybrids.

Drupe. One-seeded, stone fruit (e.g., cherry, peach, plum).

Ecotype. A strain within a given species adapted to a particular environment.

Electrophoresis test. A method for separating and mapping protein bands from homogenized plant (or seed) preparations. The separations are made within a jel preparation across an electrical field. The test may have potential use in variety identification and tests for varietal purity.

Electrostatic separator. A machine that separates seed on the basis of their ability to accept and retain an electrical charge.

Elite seed. The class of Canadian pedigreed seed corresponding to the foundation seed class in the United States.

Embryo. The generative part of a seed that develops from the union of the egg cell and sperm cell and during germination becomes the young plant.

Embryo sac. The female sexual spore of the ovule; also known as the mature female gametophyte or megagametophyte.

Embryogeny. Embryo growth and development.

Endogenous rhythm. A type of rhythmic plant or animal response or growth capacity that is not affected by external stimuli (see circadian rhythm).

Endocarp. Inner layer of the fruit wall (pericarp).

Endosperm. The tissue of seeds that develops from sexual fusion of the polar nuclei of the ovule and the second male sperm cell. It provides nutrition for the developing, growing embryo.

Enzyme. A catalyst produced in living matter. Enzymes are specialized proteins capable of promoting chemical reactions without themselves entering into the reaction; consequently, they are not changed or destroyed.

Epicotyl. The portion of the embryo or seedling above the cotyledons.

Epidermis. The outer layer of cells in plants that protects them against drying and mechanical injury.

Epigeal germination. A type of germination in which the cotyledons are raised above the ground by elongation of the hypocotyl.

Epistase. The development of a well-defined nucellar on integumentary tissue (or plug) in the micropylar or chalazal area of an ovule.

Ergot. Dark spur-shaped sclerotium that develops in place of a healthy seed in a diseased (fungus-infected) inflorescence. Ergot sclerotia are toxic to both man and livestock and were an original source of the hallucinatory drug, LSD.

Exalbuminous seeds. Seeds with only small amounts of endosperm.

Excised embryo test. A quick test for evaluating the growth potential of a root-shoot axis that has been detached from the remainder of the seed.

Exhaustion test. A type of vigor test that measures the ability of seeds to grow rapidly under rigidly controlled conditions of high temperature, relative humidity, and moisture content in continuous darkness.

Exocarp. Outermost layer of the fruit wall (pericarp).

F_1 hybrid. Denotes the first generation offspring from the mating of two parents.

F_2 seed. The second generation progeny from the mating of two parents.

Fanning mill. The air-screen machine that utilizes air flow and sieving action in separating and cleaning seeds.

Far-red light. The radiant energy in the long wavelength range of the visible spectrum between 700 and 760 nanometers.

Fat. An ester of three fatty acids and glycerol (or another alcohol) found in plants or animals. When they exist in liquid form, they are frequently called oils.

Fatty acid. Organic compound of carbon, hydrogen, and oxygen that combines with glycerol to make a fat.

Fermentation. Chemical change in sugar induced by the activity of the enzyme systems or microorganisms under anerobic conditions. For example, in the brewing industry, yeast enzymes produce carbon dioxide and alcohol from sugar by fermentation.

Field burning. The practice of burning the stubble and plant residue of seed fields after threshing. It is a cultural practice for aiding insect and disease control as well as stimulating tiller production and delaying sodbinding.

Filament. The stalk that supports the anther in the stamen (male part) of a flower.

Film coating. Application of seed additives in a "sticky" polymer directly to the seed in one to multi-layers that increase the seed's weight 1 t 10% to improve seed performance.

Firm seeds. A term sometimes applied to grass caryopses that are dormant due to seedcoats that are impervious to water or gases.

First generation hybrid. An F_1 hybrid.

FIS. Federation International du Semances—an international federation of seed trade.

Floral induction. The physiological changes in response to external stimuli (light quality, daylength, etc.) that occur in vegetative meristems and subsequently allow them to become reproductive meristems and undergo floral initiation.

Floral initiation. The morphological changes in the development of a reproductive meristem from a vegetative meristem.

Floret. The smallest unit of a flower. In grasses it consists of the lemma, palea, stamens, and pistil.

Florigen. The universal hormone that supposedly causes plants to change from the vegetative to the reproductive state.

Fluid drilling. Pregermination of seeds in gels such as hydroxyethyl cellulose followed by furrow planting to enhance germination, stand establishment, and seedling growth.

Follicle. A fruit with a simple (single) pistil that at maturity splits open along one suture (e.g., milkweed, larkspur).

Foundation seed. The progeny of breeder seed; used as planting stock for registered and certified seed.

Free central placentation. The type of ovule attachment within a fruit that bears seeds along a free central axis with no separations (septa).

Fruit. A mature ovary and any associated parts.

Fungicide (seed). A chemical that disinfects the seed and/or protects it from soil-borne fungi during germination.

Fungiculus. The stalk that connects an ovule (seed) to the placenta of the ovary wall.

Gamete. A sex cell that unites with another sex cell to form a zygote.

Gamet precision divider. A type of mechanical halving device for subdividing a large seed sample to obtain a smaller working sample for germination or purity analysis. It has an electrically operated rotating cup into which the seed is funneled to be spun out and into one of two spouts.

Gametophyte. The part of the flower that produces gametes or sex cells.

Gene(s). Units of inheritance located in linear order on chromosomes.

General seed blower. A precision seed blower used to aid in separating light seed and inert matter from heavy seed.

Genetic drift. A gradual (or sometimes abrupt) change in the germplasm balance of a cross pollinated variety causing a change in its characteristics. Usually applies to grass or legume varieties when seed is produced outside their area of adaptation. The shift may be caused by selective differences in plant mortality or flowering habit under the different environment.

Genetic purity. Trueness to type or variety, usually referring to seed.

Genotype. The hereditary makeup of a plant (or variety) which determines its inheritance.

Germ. A term for the embryo of some seeds, especially the cereal grains.

Germ plasm. An expression used in a broad sense to denote the hereditary properties of an individual plant or plant population that are transmitted from one generation to another.

Germ tube. The tube that grows out from the pollen grain, usually into the stigma, down the style and into the ovary to permit sexual fusion.

Germination. The resumption of active growth by the embryo culminating in the development of a young plant from the seed.

Gibberellic acids. A group of growth promoting substances first discovered in the *Gibberella spp*. They regulate many growth responses and appear to be a universal component of seeds as well as other plant parts.

Glomerule. A very compact cyme (e.g., *Saxifraga integrifolia*).

Glumes. The pair of chaffy bracts that occur at the base of a grass spikelet, often completely closing it.

Gravity separator. A machine utilizing a vibrating porous deck and air flow to separate seeds on the basis of their different specific gravity.

Growth regulator. A synthetic compound produced in the laboratory which controls growth responses in plants and seeds.

Gymnosperm. A kind of plant that produces seeds but no fruits. The seeds are not borne within an ovary and are said to be naked (hence the name).

Gynoecium. The female part of a flower or pistil formed by one or more carpels and composed of the stigma, style, and ovary.

Hard seed. A seed that is dormant due to the nature of its seedcoat, which is impervious to either water or oxygen.

Haploid (1N). A term indicating one-half the normal diploid complement of chromosomes.

Haustoria. In seeds, a type of armlike absorptive organ sometimes projecting from the developing endosperm or embryo into other seed parts to gather nutritive support.

Head. An inflorescence in which the floral units on the peduncle are tightly clustered, surrounded by a group of flowerlike bracts called an involucre (e.g., sunflower).

Helobial endosperm. Intermediate between the nuclear and cellular endosperm types in which development is characterized by free nuclear division as well as cell wall formation in some areas.

Hemianatropous ovule. A type of ovule arrangement in which the straight ovule axis orientation is perpendicular to that of the funiculus.

Hemicellulose. Complex cell wall constituent that is similar in appearance to cellulose but more easily broken down to simple sugars. Common forms include xylan, mannans, and galactans.

Hesperidia. Berrylike fruit with papery internal separations (septa) and a leathery, separable rind (e.g., citrus fruits).

Hilum. The scar remaining on the seed (ovule) at the place of its detachment from the seedstalk (funiculus).

Hormone. A chemical substance that is produced in one part of a plant and used in minute quantities to induce a growth response in another part.

Hybrid vigor. The increase in vigor of hybrids over their parental inbred types; also known as heterosis.

Hydrogen peroxide (H_2O_2) test. Quick test to determine seed viability. In response to a H_2O_2 soak, viable seeds elongate their roots through a cut in the seedcoat; a commonly used quick test for conifer seeds.

Hygroscopic. In seeds, the high tendency to take up moisture, even as water vapor.

Hypha. A thread of a fungus mycelium. Plural-hyphae.

Hypocotyl. The part of the embryo axis between the cotyledons and the primary root which gives rise to the stalk of the young plant.

Hypogeal germination. A type of germination in which the cotyledons remain below the ground while the epicotyl grows and emerges above the ground.

ICIA. The initials of the International Crop Improvement Association which in 1968 was renamed the Association of Official Seed Certifying Agencies (AOSCA).

Imbibition. The initial step in seed germination involving the uptake of moisture by absorption from the germination media and hydration of the seed tissue.

Imperfect flower. Unisexual flowers; flowers lacking either male or female parts.

Inbred. A plant from successive self-fertilizations of parents throughout several generations.

Inclined draper. A device for separating seeds using an inclined endless belt onto which seeds are metered; seeds are separated on the basis of their different tendencies to roll down the plane or to catch and be carried up and into a separate discharge spout.

Incomplete flower. A flower that lacks any of the four basic parts (pistils, stamens, sepals, petals).

Increase (seed). To multiply a quantity of seed by planting it, thereby producing a larger quantity of seeds which may also be called an increase.

Indehiscent. Not splitting open at maturity.

Indent cylinder separator. A seed separator utilizing a rotating indented cylinder through which seeds are passed for cleaning. It lifts shorter seeds from longer seeds, thus separating them.

Indent disk separator. A seed separator utilizing multiple rotating disks inside a cylinder through which seeds are moved. It lifts shorter seeds from longer seeded types, thus separating them.

Indeterminate flower. A flower which terminates in a bud which continues to be meristematic throughout the growing season, resulting in flowers of different maturity within the same inflorescence.

Inert matter. One of the four components of a purity test; it includes nonseed material and seed material that is classified as inert according to the Rules for Testing Seeds.

Inhibitor. A chemical substance that retards or prevents a growth process such as germination.

Indexing. The process used to test vegetatively reproduced planting stock for freedom from virus diseases. Disease evaluations are based on greenhouse or field plot growout tests.

Inflorescence. The flowering structure of a plant (e.g., umbel, spike or panicle).

Inoculant. A preparation containing a specific nitrogen-fixing bacteria that is added to legume seed prior to planting to assure that the resulting crop will have nitrogen fixation ability.

Inoculum. Any material such as spores, bacteria or fungus bodies that serve as a means of propagating or spreading a pathogenic disease. In legume seed inoculation, the inoculum is the bacterial inoculant (see above).

Integumentary tapetum. The layer of nucellar cells with radial elongation and two nuclei that immediately surround and provide nutritive support to the developing embryo sac of some species. Later it becomes hardened and provides a protective layer to the mature seed.

Integuments. The tissues covering or surrounding the ovule, usually consisting of an inner and outer layer which comprises the seedcoat of the mature ovule.

Interagency certification. The certification of seed through the cooperation of two different agencies. Usually the term indicates that the field inspection is performed by one agency and the seed inspection is made by a second agency which completes the certification process and issues tags.

Intermediate day plants. Plants initiate flowers best under intermediate day-lengths.

Involucre. A close collection of bracts surrounding an inflorescence or flower.

ISTA. The initials of the International Seed Testing Association.

Isolation requirement. The spatial separation required between a seed field and other sources of mechanical and genetic contamination, especially between cross pollinated varieties.

Isotherm (absorption). A graph showing the curve of the relationship between seed moisture content and its equilibrium relative humidity; also called moisture equilibrium curve.

Keel (flower). The two fused anterior petals of a legume flower.

Legume. A member of the pea family characterized by having dry, multiseeded pods that dehisce along two sutures at maturity.

Lemma. One of two bracts of the grass floret; it is located on the side nearest the embryo and opposite the rachilla.

Lignin. The extremely complex strengthening or deposition material in plants that tend to make them hard and woody. Chemically, lignin shows both phenolic and alcoholic characteristics.

Locule. The cavity of an ovary.

Loculicidal capsule. A type of capsule that at maturity splits open through the midrib of the carpel into the locules (e.g., iris, tulip).

Long day plants. Plants that initiate flowers best under long day (short night) regimes.

Longevity. Length of life or viability of organisms; often used in terms of seed longevity.

Macrogametophyte. A name sometimes applied to the megagametophyte (see megagametogenesis, below).

Magnetic separator. A machine that separates seeds on the basis of their magnetic characteristics (whether they are attracted or repelled by a magnet).

Male sterile. Producing no functional pollen.

Malpighian layer. A protective layer of cells in the coats of many seeds characteristically comprised of close-packed, radially positioned, heavy-walled, columnar cells without intercellular spaces. They are usually heavily cutinized or lignified and relatively impervious to moisture and gases.

Matriconditioning. Hydrating seeds in low water potential solid carriers such as clays or vermiculite followed by subsequent drying to enhance germination, stand establishment, and seedling growth (also known as solid matrix priming).

Megagametogenesis. The development of the female gametophyte (megagametophyte) from a functional megaspore.

Megaspore. One of the four cells (of archesporial cell origin) formed in the ovule of higher plants as a result of meiosis, or sexual cell reduction division. One of these later undergoes mitosis to give rise to the female gamete (megagametophyte, or embryo sac).

Megasporogenesis. The development of the megaspore from the archesporial cell.

Meiosis. Cell division during which homologous chromosomes pair; one member of each pair separates and passes to daughter cells, each having one-half the original chromosome number. Also called reduction division.

Mericarp. One-half of a two-sectioned fruit known as a schizocarp. Characteristic of the carrot family.

Meristem. Undifferentiated tissue located at the tips or growing points of vegetative or reproductive organs capable of undergoing cell division and elongation.

Meristematic cells. Undifferentiated cells in plant meristems which are capable of undergoing cell division.

Mesocarp. Middle layer of the fruit wall (pericarp).

Messenger RNA. The form of ribonucleic acid that conveys the genetic messages coded in the deoxyribonucleic acid of the nucleus to the ribosomes of the cytoplasm where proteins are formed.

Metabolism. The chemical changes within a living cell.

Microgametogenesis. The development of the microgametophyte (pollen grain) from a microspore.

Microgametophyte. A mature pollen grain, or male gamete.

Microspore. The male spore in the anther from which the male gametes develop.

Microspore mother cell. One of many cells in the microsporangium (anther) which undergoes microsporogenesis to yield four microspores.

Microsporogenesis. The development of microspores from the microspore mother cell.

Micropyle. The integumentary opening of the ovule through which the pollen tube enters prior to fertilization.

Middlings. The seed from the middle portion of the gravity separator. This portion usually contains a mixture of light and heavy seed and is recycled for further conditioning.

Mitochondria. Microscopic rod-shaped or spherical organelles present in living cells. They contain the enzyme systems active in the respiration process.

Mitosis. Normal cell division in which each daughter cell has exactly the same chromosome number as the mother cell.

Monocot. An abbreviated name for monocotyledon, referring to plants having single seed leaves (cotyledons). Examples are bamboo and corn.

Monoecious (monecious). Having stamens and pistils in different unisexual flowers on the same plant (e.g., corn).

Monogerm seed (sugar beet). A sugar beet "seed" (botanically a fruit) containing only one ovule in contrast to a multigerm "seed" which represents aggregate fruit containing several ovule units.

Mother cells. Special cells in the anther and ovule that give rise to pollen or egg cells.

Multigerm seed (sugar beet). An aggregate fruit containing several ovules.

Mucilages. The gummy (sticky when wet) complex carbohydrate substances which cover the seeds, bark or stems of some plant species (e.g., buckhorn plantain seeds).

Multiline. A composite (blended) population of several genetically related lines of a self-pollinated crop.

Multiple fruit. Developed from a cluster of flowers on a common base (e.g., fig).

National Foundation Seed Project. A project administered by the USDA Agricultural Research Service to contract for and provide foundation seed at certain forage varieties, especially where seed of a variety is produced outside its area of adaptation.

National Seed Storage Laboratory. The laboratory at Ft. Collins, Colorado, that is operated by the USDA as a permanent repository for storing germ plasm of all kinds of plants under controlled temperature and relative humidity conditions.

Nick. In hybrid seed production, the condition existing when two inbred plants flower and are ready for sexual crossing at the same time (i.e., the majority of the pollen is ready when the flowers of the other parent are ready to be fertilized).

Nonrecurrent apomixis. A type of apomixis in which a haploid embryo sac is produced as a result of apomeiosis (abnormal cell division during megasporogenesis). It results in seed with a haploid embryo and thus haploid plants that are usually sterile.

Noxious weed. A weed species that is defined by law as being noxious; usually highly objectionable when found in crop seed lots.

Nucelius. The tissue of the ovary wall in which the archesporial cell arises and where megasporogenesis, megagametogenesis, and ovule development occurs.

Nuclear endosperm. The type of endosperm in which the early development is characterized by rapid cell enlargement accompanied by nuclear division without cell wall formation.

Nucleic acid. A highly complex organic molecule found in the nucleus of cells; believed to be the substance that determines heredity and governs the behavior of all cells.

Nucleus. The part of the cell bearing the chromosomes.

Nut. A dry, indehiscent, one-seeded fruit with a hard, woody shell.

Nutlet. A small, dry, indehiscent fruit composed of one-half a carpel enclosing a single seed; developed by folding and splitting the carpel into a compound pistil.

OECD. The initials of the Organization for Economic Cooperation and Development, an international agency which, among other things, has developed specifications, procedures, and standards for an international seed certification scheme.

Open pollinated variety. A heterogeneous variety of a cross pollinated crop that is allowed to interpollinate freely during seed production. In contrast to hybrids, representing controlled crosses, open pollinated varieties are not common in modern crop production.

Operculum (seed). A type of epistase (integumentary proliferation) that is deposited inside the ovule, forming a tight-fitting micropylar or chalazal plug in the mature seed; contributes to water impermeability and hard seedcoat dormancy.

Orthotropus ovule. The simplest type of ovule arrangement in which the ovule is erect, with the micropyle at one end and the funiculus at the other.

Osmoconditioning. Soaking seeds in aerated, low water potential osmotica such as polyethylene glycol or salts followed by subsequent drying to enhance germination, stand establishment, and seedling growth (also known as priming).

Other crop seed (percentage). One of the four components of a purity test; the total percentage (by weight) of seed of all crop species each comprising less than 5% of the seed lot.

Ottawa seed blower. A type of seed blower developed by C.W. leggitt of the Canadian Department of Agriculture that has a slender metal blowing tube used for small seeded crops.

Ovary. The part of the pistil containing the ovule.

Ovoid. Egg-shaped.

Ovule. The structure within the ovary of the flower that becomes the seed following fertilization and development.

Ovum. An egg cell.

Palea. One of the thin bracts of a grass floret enclosing the caryopsis that is located on the side opposite the embryo.

Palisade layer. In seeds, this term is used interchangeably with malpighian layer.

Panicle. An inflorescence in which the lateral branches arising from the peduncle produce flower-bearing branches instead of single flowers (e.g., *Avena sativa*).

Paper-piercing test. A stress test for seedling vigor utilizing sand covered by filter paper through which the seedlings must emerge to be considered vigorous.

Pappus. A tuft of delicate fibers or bristles such as the feathery appendage on a ripe dandelion seed representing a modified calyx.

Parietal cell. The sister cell of the megaspore mother cell originating from the division of the archesporial cell. It is nonfunctional and usually degenerates.

Parietal placentation. A type of placentation in which the seeds are attached in the ovary near the outer ovary wall; usually associated with vestiges of septa rather than along the ovary axis as in other types of placentation.

Parthenocarpy. Production of fruit without seeds as in bananas and some grapes.

Partial hybrid. A legal designation for seed lots representing at least 75 but less than 95% hybrid seed.

Pathogen. Any organism capable of causing disease by obtaining its nutrition either partially or wholly from its diseased host.

Pedicel. The stalk of a floret.

Pelleted seeds. Seed that are commercially prepared for precision planting by pelleting them inside a special preparation to make them more uniform in size. Sometimes special nutrient or growth-promoting substances are placed in the pellets to aid in seed germination and growth.

Pepo. A fruit with a hard rind without internal separations or septa (e.g., cantaloupe, watermelon, cucumber).

Perennial. A plant which survives and produces vegetative growth and flowers year after year without being replanted.

Perfect flower. A flower having both staminate (male) and pistillate (female) organs.

Perianth. A collective term for all the petals of a flower.

Pericarp. The ovary wall. It may be thin and fused with the seedcoat as in corn, fleshy as in berries, or hard and dry as in pods of legumes.

Perisperm. A type of endospermlike storage tissue in a mature seed that develops from the nucellus of the parent plant—thus it has the 2n chromosome number. Examples of species with well-developed perisperm tissue include beet and pigweed.

Petal. A unit of the inner perianth whorl or corolla.

Petiole. The stem of a leaf.

Phage-plaque. A clear area caused by a bacteriophage in a bacterial colony, caused by dissolving of specific bacterial cells.

Physiological dormancy. Seed dormancy caused by internal physiological conditions that prevent germination. Often referred to as epicotyl or embryo dormancy.

Physiological maturity. The maturity of a seed when it reaches its maximum dry weight. This usually occurs prior to the normal harvest date.

Phytase. An enzyme that catalyzes the breakdown of phytin, the source of inorganic phosphorus in seed metabolism.

Phytochrome. The bluish photoreversible protein pigment responsible for the photoperiodic control of flowering and seed germination. It exists in two forms in plants; the biologically active P_{F-R} (receptive to far-red light) and the biologically inactive P_R (receptive to near-ed light).

Phytosanitary certificate. A certificate issued by a legally constituted authority of federal or state government stating that a seed lot has been inspected and found to be free of disease infestation. These certificates are frequently used in international seed trade agreements to prevent the spread of seed-borne diseases among countries.

Photoperiodism. The response (e.g., flowering, germination) of organisms to the relative length of daily periods of light and darkness.

Pipered embryo type. A type of embryo classification in which the second wall of the zygote (fertilized egg) is longitudinal, or nearly so.

Pistil. The female, or seed-bearing organ of the flower. It is composed of the ovary, style, and stigma.

Placentation. The method of attachment of the seeds within the ovary.

Plastid. Small cytoplasmic organelles containing pigments (e.g., chloroplasts, which give the green color to plant leaves).

Plenum chamber. The chamber, or air chest, associated with (usually underneath) a crop (seed) dryer into which air is moved and allowed to distribute immediately prior to its entry into the drying bed.

Plumule. The major leaf bud of the seed or seedling. That part of the embryonic plant axis above the cotyledons. Also known as epicotyl.

Pneumatic conveyor. A method of conveying seed within a conditioning plant utilizing air as the driving force.

Pod. A fruit that is dry and nonfleshy when ripe, and splits open at maturity to release its seeds.

Polar nuclei. Two nuclei of the female gametophyte (sex cell) that unite with one of the sperm cells to form the endosperm of a developing seed.

Pollen. The small, almost microscopic, yellow bodies that are borne within the anthers of flowers and contain the male generative (sex) cells. The mature microgametophyte.

Pollen tube. A microscopic tube that grows down the stigma from the pollen grain through which the sperm cells are deposited into the embryo sac.

Pollination. The process by which pollen is transferred from the anther where it is produced to the stigma of a flower.

Polyembryony. The condition in which an ovule has more than one embryo. This condition is common to certain grasses.

Polymer. A large molecule formed by joining together many small identical molecular units (e.g., starch formed by long chains of glucose).

Polyploid. An individual (plant) that carries two or more complete sets of homologous (pairs) chromosomes.

Pome. A fruit in which the floral cup forms a thick outer fleshy layer and that has a papery inner pericarp layer (endocarp) forming a multi-seeded core (e.g., apple, pear).

Poricidal capsule. A capsule that at maturity splits open at pores near the top, releasing mature seed (e.g., poppy).

Prechill. The practice of exposing imbibed seeds to cool (5–10°C) temperature conditions for a few days prior to germination at warmer conditions. See definition of stratification.

Prehydration. Soaking seeds in water or gels prior to planting to enhance germination, stand establishment, and seedling growth.

Priming. Soaking seeds in aerated, low water potential osmotica such as polyethylene glycol or salts followed by subsequent drying to enhance germination, stand establishment, and seedling growth (also known as osmoconditioning).

Primorida. Organs in their earliest stage of development as a leaf primordia or meristem.

Processing. The term formerly used to describe the technology of cleaning seed and preparing it for the market. See conditioning.

Proembryo. The young embryo in its early stages of development.

Progeny. Offspring.

Propagule. Any type of plant part used to reproduce another individual plant asexually.

Protein. An essential constituent of all living cells. Proteins occur naturally and are complex combinations of amino acids linked by peptide bonds.

Protoplasm. The essential complex living substance of cells on which all vital functions of nutrition, secretion, growth, and reproduction depend.

Pseudocarpic fruit. A fruit consisting of one or more ripened ovules attached or fused to modified bracts or other nonfloral structures (e.g., sandbur).

Pseudogamous apogamety. A type of diplospory (apomixis) in which the seed develops from some cell of the diploid embryo sac other than the egg, but one in which the stimulus of pollination is required before development will begin.

Pseudogamy. A type of apomixis in which the diploid egg cell develops into the embryo without fertilization of the egg cell, although only after fertilization of the polar nuclei with one of the sperm cell from the male gamete to form a normal triploid (3n) endosperm.

Pubescence. A covering of short, soft hairs.

Pure line variety. A variety (cultivar) of a self-pollinated species derived from a single plant.

Pure live seed (PLS). The percentage of pure seeds in a seed lot that have the ability to germinate. The percentage of PLS is determined by multiplying percent germination by percent pure seed and dividing by 100.

Pure seed content (percentage). The percentage of each crop species that comprise five percent or more (by weight) of a seed lot.

Quick test (seed testing). A type of test for evaluating seed quality, usually germination, more rapidly than standard laboratory tests.

Quicker method (of purity testing). The method of purity testing that distinguishes between seed and inert matter purely on the basis of physical characteristics. See stronger method.

Quiescence. The absence of growth, usually inferring the absence of environmental conditions favoring growth; although dormant seeds are quiescent, quiescence is distinguished from dormancy, which implies the inability to germinate even in the presence of environmental conditions favoring growth.

Racerne. A type of inflorescence in which the single-flowered pedicels are arranged along the sides of a flower shoot axis.

Rachilla. The central axis of a grass floret.

Rachis. The main axis of a flower (or leaf).

Radicle. The rudimentary root of the seed or seedling that forms the primary root of the young plant.

Raphe. A ridge (seam), sometimes visible on the seed surface, which is the axis along which the ovule stalk (funiculus) joins the ovule.

Receptacle. The basal structure to which the flower parts are attached, sometimes forming part of the mature fruit, as in apple.

Registered seed. A class of certified (generic sense) seed which is produced from foundation seed and planted to produce certified (blue-tag) seed. It is identified by a purple tag.

Release. A crop variety (or germ plasm) that is released and designated to be reproduced, marketed, and made available as seed for public use.

Renovation (seed production). Usually refers to the mechanical removal of plants from a very dense, unproductive, or sodbound stand for the purpose of revitalizing its productivity.

Respiration. The metabolic process by which a plant (or animal) oxidizes its food and provides energy for assimilation of breakdown products.

Restorer line. An inbred line that permits restoration of fertility to the progeny of male sterile lines to which it is crossed.

Rhizome. A nonfleshy, more or less horizontal, underground stem.

RNA. The initials for ribonucleic acid, a component of the nucleus, and to some extent, the cytoplasm, which relays the genetic messages coded in the DNA of the nucleus to the ribosomes of the cytoplasm, which in turn form amino acids into proteins.

Rogue. A noun referring to an offtype plant. When used as a verb it refers to the act of removing such plants.

RST. The initials for Registered Seed Technologist, a designation for a private (commercial) seed analyst who has passed tests and met other professional and academic requirements to merit a seal and the designation "Registered Seed Technologist."

Rudimentary. Incompletely developed.

RUSSL. The initials of the Recommended Uniform State Seed Law, a model for state seed laws prepared by the American Association of Seed Control Officials (AASCO).

Samara. An indehiscent, winged fruit in which the seedcoat is loose inside the pericarp (e.g., maple, ash).

Sampling. The method by which a representative sample is taken from a seed lot to be sent to a laboratory for analysis. It is most commonly accomplished using triers, or seed probes, although hand methods and mechanical sampling methods are also used.

Scalper. A screening or air blast machine used to separate the very large, very small, or very light contamination from a seed lot—a rough cleaner.

Scopoletin. The growth promoting substance reported to stimulate seed germination of some plant species.

Scarification. The process of mechanically abrading a seedcoat to make it more permeable to water. This process may also be accomplished by brief exposure to strong acids (e.g., sulfuric acid).

Schizocarp. A dry, two-seeded fruit of the carrot family that separates at maturity along a midline into two mericarps. Each mericarp has a dry, indehiscent pericarp enclosing a loose fitting ovule.

Sclerotium (pl. sclerotia). Compact mass of fungus hyphae usually with black outer surface and white inner surface. Capable of remaining dormant for long periods and eventually giving rise to fruiting bodies.

Scorpioid cyme. A determinate inflorescence in which the lateral buds on one side are suppressed during growth, resulting in a curved or coiled arrangement.

SCST. The initials of the Society of Commercial Seed Technologists, an organization of commercial and private registered seed technologists and seed analysts of the United States and Canada.

Scutellum. A shield-shaped organ of the embryo of grass. It is often viewed as a highly modified cotyledon in monocotyledons.

Secondary dormancy. A type of dormancy imposed by certain adverse environmental conditions in previously nondormant seeds, or seeds in which primary dormancy has been broken.

Seed. A mature ovule consisting of an embryonic plant together with a store of food, all surrounded by a protective coat.

Seed-borne. Carried on or in seeds.

Seedcoat. The protective covering of a seed usually composed of the inner and outer integuments. Also called the testa.

Seed coating. Application of substances such as fungicides, insecticides, safeners, micronutrients, etc. directly to the seed that do not obscure its shape.

Seedling. A young plant grown from seed.

Seed pellet. Obscuring the shape of the seed with an amalgam of fillers and cementing additives (sometimes containing other substances such as plant growth regulators, innoculants, fungicides, etc.) to a specific size to enhance mechanical planting and seed performance.

Seedstocks. Seed used as a source of germ plasm for maintaining and increasing seed of crop varieties.

Seizure. Legal court action that takes seed off the market and makes it subject to court-ordered disposition by law enforcement officials. It usually results from serious seed law violations.

Seminal organs. Pertaining to the seed or germ, or those already developed in the embryo within the seed (i.e., seminal roots are roots arising from the embryo compared to adventitious roots that arise later).

Sepals. A floral part—the outer whorl, referred to collectively as the calyx.

Septicidal capsule. A capsule that at maturity splits along the septa and releases the seeds.

Septum. A partition, as between the locules of a fruit.

Short day plants. Plants that initiate flowers best under short day (long night) regimes.

Sillique (sillicle). A fruit, characteristic of the mustard family, which has two valves that at maturity split away from a persistent central partition. If it is several times longer than wide it is termed a sillique. If it is broad and short it is called a sillicle.

Simple cyme. The simplest branched determinant inflorescence where the lateral flowers develop later than the terminal flower (e.g., mouse-eared chickweed).

Simple fruits. Developed from a single pistil or ovary that may be simple or compound (e.g., berry, as in blueberry).

Single cross hybrid. A hybrid between two inbred lines.

Sodbinding. A condition of older grass seed fields where high plant population (interplant competition), thatch build-up, and certain unknown factors create conditions in which seed production decreases. This condition can often be relieved by fertilization or mechanical renovation.

Solanad embryo type. A type of embryo classification in which the terminal cell of the proembryo divides by a transverse wall and in which the basal cell plays a minor part (or no part) in subsequent embryo development.

Solid matrix priming. Hydrating seeds in low water potential solid carriers such as clays or vermiculite followed by subsequent drying to enhance germination, stand establishment, and seedling growth (also known as matriconditioning).

Solitary flower. The simplest expression of a determinant inflorescence.

Somatic cells. Pertaining to cells of the plant body other than reproduction tissue.

South Dakota blower. A popular type of seed blower used in purity testing of seeds. Air is passed through plastic tubes to help in the separation of seeds according to their specific gravity and resistance to air flow.

Sperm cell. The male generative cell that fertilizes the egg cell and unites with the polar nuclei.

Spadix. A special type of spike with a fleshy inflorescence axis.

Spathe. A large bract surrounding an inflorescence, especially a spadix.

Spike. A basic type of inflorescence in which the flowers arising along the rachis are essentially sessile (stalkless).

Spikelet. The unit of the grass flower that includes the two basal glumes subtending one to several florets.

Spiral separator. A type of seed separator with no moving parts. The seeds enter at the top and slide or roll down an inclined spiral runway. The speed of seed movement and centrifugal force allows separation of the heavier, round, fast-moving seeds from those that move slower. It is sometimes called a vetch separator.

Spore. In seed plants, the spore is the first cell of the gametophyte generation. The two kinds, microspore and megaspore, produce male and female gametes, respectively.

Stamen. The part of the flower bearing the male reproductive cells composed of the anthers on the filament (stalk).

Stigma. The upper part of the pistil that receives the pollen.

Stock seeds. See seedstocks.

Stoner. A modification of the gravity separator especially constructed to separate stones from crop seeds such as beans.

Stop-sale. An official administrative action by a state seed control agency which prohibits further sale of seed. It is issued when evidence of mislabeling or a seed law violation is found.

Stratification. The practice of exposing imbibed seeds to cool (5–10°C) temperature conditions for a few days prior to germination in order to break dormancy. This is a standard practice in germination testing of many grass and woody species.

Strain. A term sometimes used to designate an improved selection within a variety.

Stronger method (purity testing). The method of purity testing previously used by European seed analysts which attempted to determine if a crop seed was capable of germinating before it was classified as pure seed or as inert matter. See quicker method.

Strophiole. A rather rare appendage arising from the seedcoat of some species near the hilum area. It may be variable in shape and has no apparent function.

Style. The stalk of the pistil between the stigma and the ovary.

Subsampling. The procedure (usually by halving methods) by which a smaller representative working sample is obtained from the larger sample submitted for seed analysis.

Suspensor. The group or chain of cells produced from the zygote that pushes the developing proembryo toward the center of the ovule in contact with the nutrient supply.

Swath. A windrow. A row of cut or pulled crop usually waiting for drying or curling before further harvesting.

Synergid nuclei. Two of the eight cells of the embryo sac—usually remaining nonfunctional.

Syn-1 Syn-2. The first and second generation progenies, respectively, of a synthetic variety after the individual lines are composited.

Syngamy. Sexual fusion of the sperm and egg cells.

Synthetic seeds. Seeds (often a somatic embryo surrounded by a synthetic encapsulation) produced from vegetative tissue (usually by tissue culture) that are clones possessing identical genotypes.

Synthetic variety. A variety composed of an interbreeding population of several cross-pollinated plant lines.

Tailings. Partly threshed material that has passed through the coarse shakers or "straw walkers" and is eliminated at the rear of a threshing machine.

Tenuinucellate. The nucellar condition in which the embryo sac originates and develops only one cell layer beneath the nucellar epidermis.

Testa. The outer covering of the seed; the seedcoat.

Tetrad. A group (quartet) of four spores formed by division of the same mother cell, as in a tetrad of microspores.

Tetrazolium (TZ). Indicates a class of chemicals that have the ability to accept hydrogen atoms (and undergo reduction) from dehydrogenase enzymes during the respiration process in viable seeds. This is the basis of the tetrazolium test during which the tetrazolium chemical undergoes a color change, usually from colorless to red.

TZ test. Quick test to determine seed viability (and sometimes vigor).

Thermoperiod. A period of the proper temperature to elicit a specific growth or developmental response—as in flowering or seed germination.

Three-way hybrid. A hybrid between an inbred line and single cross hybrid.

Tiller. A branch arising from the base of a monocot plant, especially in the grass family.

Tolerance. The amount by which a second test may differ from a first test without being attributed to an actual difference in seed quality. Tolerances are usually based on normal random variation, or sampling error.

Trier. A hand manipulated probe for sampling seeds.

Umbel. A type of inflorescence in which the minute flower units are arranged into flat or umbrella-shaped heads as in carrots and celery.

Unitegmic testa. A testa (seedcoat) made up of only one integument.

Unsaturated fatty acid. A fatty acid that has a double bond between the carbon atoms at one or more places in the carbon chain; hydrogen can be added at the site of the double bond.

Utricle. A small, thin walled, one-seeded fruit in which the seed is only loosely attached to the pericarp.

Varietal protection. Legal protection (breeders rights) to developers or owners of crop varieties giving them exclusive right to control seed production and marketing.

Vegetative. Referring to asexual (stem, leaf, root) development in plants in contrast to sexual (flower, seed) development.

Vermiculite. A porous form of mica, a mineral, which makes good rooting media for seed germination because of its capacity to retain moisture and permit aeration.

Vernalization. Bringing into a spring condition. In reference to flowering, it is the process by which floral induction is promoted. It is sometimes (perhaps erroneously) applied to seeds to indicate stratification in order to break dormancy, enabling them to germinate.

Viable (viability). Alive. Seed viability indicates that a seed contains structures and substances including enzyme systems that give it the capacity to germinate under favorable conditions in the absence of dormancy.

Vibrator separator. A machine utilizing a vibrating deck for separating seeds on the basis of their shape and differing surface textures.

Vigor. The AOSA has defined vigor as "those seed properties which determine the potential for rapid uniform emergence and development of normal seedlings under a wide range of field conditions."

Vitascope. A commercial device for helping speed up the tetrazolium test utilizing a mechanical seed slicing mechanism, a timer, and a controlled temperature bath.

Viviporous. In seeds, it denotes the condition in which they are able to germinate while still attached to the mother plant.

Volunteer plants. Unwanted plants growing from residual seeds of previous crops.

Wavelength. The distance between two corresponding points on any two consecutive waves. For light it is very small and is measured in nanometers (nm), which equal about 0.04 millionths of an inch.

Weed. Any plant in a place where it is a nuisance might be considered a weed. The term is usually used to denote noncultivated plants growing in fields, lawns, gardens, or other areas used by man.

Weed seed (percent). The total percentage (by weight) of a seed lot which is composed of seed of plants considered to be weeds. One of the four components of a purity test.

Windrow. A loose, continuous row of cut or uprooted plants usually allowed to dry in place until harvested. A swath.

Wing. A membrance, or thin, dry expansion or appendage of a seed or fruit.

Xenia. The direct, visible effects of the pollen on endosperm and related tissues in the formation of a seed (e.g., seed color). It results in hybrid characteristics of form and color.

Zygote. A fertilized egg.

Index